AF848895

PHYSICS RESEARCH AND TECHNOLOGY

ADVANCED QUANTUM MECHANICS UPON THEOREMS

PHYSICS RESEARCH AND TECHNOLOGY

Additional books in this series can be found on Nova's website
under the Series tab.

Additional e-books in this series can be found on Nova's website
under the e-book tab.

PHYSICS RESEARCH AND TECHNOLOGY

ADVANCED QUANTUM MECHANICS UPON THEOREMS

FU-SUI LIU

nova publishers
New York

Copyright © 2014 by Nova Science Publishers, Inc.

All rights reserved. No part of this book may be reproduced, stored in a retrieval system or transmitted in any form or by any means: electronic, electrostatic, magnetic, tape, mechanical photocopying, recording or otherwise without the written permission of the Publisher.

For permission to use material from this book please contact us:
Telephone 631-231-7269; Fax 631-231-8175
Web Site: http://www.novapublishers.com

NOTICE TO THE READER

The Publisher has taken reasonable care in the preparation of this book, but makes no expressed or implied warranty of any kind and assumes no responsibility for any errors or omissions. No liability is assumed for incidental or consequential damages in connection with or arising out of information contained in this book. The Publisher shall not be liable for any special, consequential, or exemplary damages resulting, in whole or in part, from the readers' use of, or reliance upon, this material. Any parts of this book based on government reports are so indicated and copyright is claimed for those parts to the extent applicable to compilations of such works.

Independent verification should be sought for any data, advice or recommendations contained in this book. In addition, no responsibility is assumed by the publisher for any injury and/or damage to persons or property arising from any methods, products, instructions, ideas or otherwise contained in this publication.

This publication is designed to provide accurate and authoritative information with regard to the subject matter covered herein. It is sold with the clear understanding that the Publisher is not engaged in rendering legal or any other professional services. If legal or any other expert assistance is required, the services of a competent person should be sought. FROM A DECLARATION OF PARTICIPANTS JOINTLY ADOPTED BY A COMMITTEE OF THE AMERICAN BAR ASSOCIATION AND A COMMITTEE OF PUBLISHERS.

Additional color graphics may be available in the e-book version of this book.

Library of Congress Cataloging-in-Publication Data

ISBN: 978-1-62948-586-7

Published by Nova Science Publishers, Inc. † New York

Contents

Preface

The main purpose of this textbook is to establish the advanced quantum mechanics and the bases of quantum field theory upon theorems, and thus make them to be free from postulates, axioms, guesses, principles, assumptions, and paradoxes. In chapter 1, the 52 theorems, 18 corollaries, one criterion, and one law in "Quantum Mechanics upon Theorems" are listed. For advanced quantum mechanics and bases of quantum field theory, this textbook adds 20 new theorems and 7 new corollaries. For example:

(1). The commutation relations between canonical field coordinates and canonical conjugate field momentums are no longer assumptions, and they are given by exactly proved theorem;

(2). This textbook gives standard six steps to quantize any field;

(3). This textbook concludes that wave-particle duality≡field-particle duality in essence. The object's motion (object = large or small massy particle, and massless photon and phonon) obeys the guide of wavefunction (in the language of quantum mechanics) or quantized field (in the language of quantum field theory) according to the spontaneous equiprobability symmetry breaking rule.

This textbook has the following four prominent features:

(1). Easy to understand. Due to exact demonstrations, any concepts and pictures are clear, and is not drift from one place to another place. For example, after reading chapter 9, every reader will feel that the quantum field theory is a natural and simple extension of quantum mechanics.

(2). Strictness in mathematics. Let me take the scattering theory as an example to explain the necessity of the strictness. This textbook proves exactly that in scattering theory of all other textbooks there are a seriously wrong limit $t \to \infty$, and this mathematical mistake leads to the following differences.

In all other textbooks conclude *a priori*:

(i). Scattering processes in first order perturbation keep energy conservation.

(ii). Scattering processes in second perturbation do not keep energy conservation, but call these processes the virtual processes.

(iii). Scattering processes can be treated in principle by time-independent Schrödinger equation, and, correspondingly, can use Green's function method.

This textbook proves:

(i). Scattering processes are generally energy-nonconservative.

(ii). One cannot call the second order perturbation processes the virtual processes due to the energy nonconservation. The producing process of, for example, meson should connect with a real process rather than the so-called "virtual" process. One cannot *a priori* believe energy conservation, because the system in scattering process is open rather than isolated.

(iii). The time-independent Schrödinger equation and Green function equation always keep energy conservation, and thus they are, in principle, wrong for scattering processes.

(3). Novel four applications.

(i). The microscopic superconductivity can be applied at room temperature.

(ii). In some quantum transition processes the energy might be strongly nonconservation.

(iii). The quantum relaxation theory can explain both the KWW non-exponential relaxation, and the origin of the pink noise. (The KWW relaxation and origin of the pink noise are open problems lasting more than two centuries.)

(iv). Add new five theorems to the fractional quantum mechanics.

(4). To shorten the distance between study and research. Even you are a student or a beginner, you already can write high quality papers, and can bring forth ideas in new technologies. It might be that no one reader believes this at this stage. Let me give some examples:

(i). After studying chapter 10, readers can prove more than 50 new conclusions, including theorems, corollaries, and criteria, for the fractional quantum mechanics.

(ii). After studying chapter 4, readers can find some wrong conclusions accepted now by many references.

(iii). After studying chapters 6 and 7, readers can have some new ideas to get energy purely fictitiously, and can design some new electronic devices.

Author predicts nearly with confidence that the publications of "Quantum Mechanics upon Theorems" and "Advanced Quantum Mechanics upon Theorems" will lead the following three developments:

(1). "Energy nonconservation" enters into the everyday life of people. Energy source problem is no longer a problem.

(2). 70% of semiconductor devices and 80% materials applied in strong electricity will be substituted by the materials being of microscopic superconductivity.

(3). There will be a grand unified theory, being nearly with confidence.

Author of this textbook likes to give readers bigger imagination space and initiative space for the more possible influences of the our two textbooks.

Chapter 1

Quantum Mechanics upon Theorems

1.1 Introduction

This textbook has a viewpoint that the advanced quantum mechanics and the bases of quantum field theory are natural and simple extensions of quantum me4chanics. Therefore, this chapter list the 52 theorems, 16 corollaries, 1 criterion, and 1 law in the published "Quantum Mechanics upon Theorems", and make a little interpretations, if it is very necessary. These theorems make the quantum mechanics a subject without any hypotheses, postulates, axioms, assumptions, principles, paradoxes, and with clear pictures. However, now even the newest textbooks such as Refs. [1, 2] still establish quantum mechanics on postulates, and thus have many misunderstandings for the quantum mechanics, such as: connection of quantum Zeno effect with the so-called measurement, connection of decoherence with environment, and so on [2].

In this textbook an "object" represents both massy particle (such as big universe, small electron and quark) and massless particle (such as photon and phonon), is also called wavepacket.

The labels of the following equations and theorems are the same as that in the published "Quantum Mechanics upon Theorems".

Theorem 0: Watershed theorem: There is only one watershed between classical and quantum mechanics. If one describes any object by using both coordinates and momenta at the same time, then this is classical mechanics. If one describes any object by using either coordinates or momenta, and not by both at the same time, then this is quantum mechanics.

1.2 Bases of Quantum Mechanics

Theorem I: Existence of universal constant: The dimension of product of canonical coordinate and canonical conjugate momentum is always (energy$\times$ time). In nature there is a universal constant, which's dimension is (energy$\times$ time). For memory, we call this constant Planck constant, and is expressed by $\hbar = h/2\pi$. (Note that the dimensions of angular momentum and action are (energy$\times$ time) as well.)

Corollary I: Momentum and energy: Momentum $p = \hbar k$. Energy $E = \hbar\omega$. The dimensions of k and ω are $1/length$ and $1/time$, respectively.

Proof:
From theorem I, we have:
(canonical coordinate×canoniucal conjugate momentum/$\hbar$)
is a quantity without dimension, and the dimension of
(canonical conjugate momentum/$\hbar$) is the inverse of the dimension of (canonical coordinate). Therefore, we have the following conclusion:

If q_i represents the canonical coordinate, for example, the coordinate x, then the corresponding canonical conjugate momentum is $p_i = p = \hbar k$, where k is called wavenumber, and its dimension is $1/length$. At this stage, we cannot say more about the physical meaning of k.

If q_0 represents the canonical coordinate, for example, the time t, then the corresponding canonical conjugate momentum is $p_0 = -E \equiv -\omega\hbar$, where ω is called angular frequency, and its dimension is $1/time$. At this stage, we cannot say more about the ω. QED.

Note that from theorem I we still do not know how to determine the value of the constant. At this stage we just know that, for example, if $p = \hbar k$, $p = mv$ is momentum of x direction, then the dimension of k is $1/length$, and if energy $E = \hbar\omega$, then the dimension of ω is $1/time$. After establishing the wavepacket law in subsection 3.4.7, we prove a corollary, according to which we know that k represents the average wavenumber of a particle (particle=wavepacket). The average wavenumber of a particle can be determined by experiments. Of course, the momentum of a particle can be determined. Thus, from $p = \hbar k$ the $\hbar$ is determined. Similarly, we can determine $\hbar$ from $E = \hbar\omega$. The present most precise value is $\hbar = h/(2\pi) = 1.05457266 \times 10^{-27}$ erg·second. One often calls $\hbar$ and h Planck constant for memory. Some books also call $\hbar$ the reduced Planck constant.

Theorem II: Operator representation of observable: The two variables in Fourier pair variables construct two theirs own spaces. Any variable in its own space is a number, but in space of partner variable is an operator.

Theorem III: Non-commutation property: The product of Fourier pair variables, AB, is without dimension, and $[A, B] = i$.
For Fourier pair variable compositions $[A, B] = iC$, where C is a Hermitian operator.

Theorem IV: Preparation for uncertainty relations: If W is a real number, and $\Delta\alpha$, $\Delta\beta$ are two quantities, which satisfy

$$|\Delta\alpha|^2 - \lambda W + \lambda^2 |\Delta\beta|^2 \geq 0 \qquad (3.3.1.5.3)$$

for all real number λ, then the product of these two quantities satisfies

$$|\Delta\alpha|^2 |\Delta\beta|^2 \geq \frac{1}{4} W^2 \qquad (3.3.1.5.4)$$

Theorem V: Uncertainty relations: Suppose that A and B represent the Fourier pair variable compositions or the Fourier pair variables (or considering $p = \hbar k$, $E = \hbar\omega$, and so on, the above statement becomes: Suppose that A and B represent the canonical conjugate pair variables or the canonical conjugate pair variable compositions) and W is given by

$$\begin{aligned} W &= i \int_{-\infty}^{+\infty} dx dy dz \Psi^*(x, y, z, t)[A, B]\Psi(x, y, z, t) \\ &\equiv i < \Psi|[A, B]|\Psi >, (3.3.2.1.1) \end{aligned}$$

then

$$(\Delta A)^2(\Delta B)^2 \geq \frac{|W|^2}{4}. \tag{3.3.2.1.2}$$

The lowest bound in Eq. (3.3.2.1.2) is independent of time t, although the left hand side of Eq. (3.3.2.1.2) can be dependent on t through Ψ. The often used form of Eq. (3.3.2.1.2) is

$$(\Delta A)(\Delta B) = \sqrt{(\Delta A)^2}\sqrt{(\Delta B)^2} \geq \frac{|W|}{2}. \tag{3.3.2.1.3}$$

(In history, Heisenberg first proposed the uncertainty relation between coordinate and momentum. This and the following corollaries give a general proof for the uncertainty relations. For memory, we call Eq. (3.3.2.1.3) either uncertainty or Heisenberg uncertainty relation.)

Corollary II: Variables of function: Any function $\in \mathcal{M}(\mathbb{R})$ cannot depend on the two variables of Fourier pair variables or the two compositions of the Fourier pair variable compositions commonly at the same time, or, equivalently, any wavepacket-function $\in \mathcal{M}(\mathbb{R})$ cannot depend on the two variables of canonical conjugate pair variables or the two compositions of canonical conjugate pair variable compositions commonly at the same time.

Corollary III-1: Object volume: If any position space function $\in \mathcal{M}(\mathbb{R})$ represents the distributions of mass or energy or charge or momentum of any object, including both massy and massless particle, then this object has to be three dimensional in position space.

Corollary III-2: Zero point energy: Any object has zero point energy, which is caused by Heisenberg uncertainty relation. This zero point energy is independent of temperature, and is not equal to zero even at 0 K.

Theorem VI: A free wavepacket compositions: A wavepacket in free motion consists of product with three factors: guide-factor; front-factor; and body-factor.

Theorem VII: Free wavepacket non-spreading: Free wavepackets do not spread with time.

Theorem VIII: Average wavenumber and frequency: The average wavenumber of wavepacket is $< k_x >= k_{x\circ}$. The average frequency of wavepacket is $< \omega >= \omega(k_{x\circ})$.

Corollary IV: Average momentum and energy: Average momentum of wavepacket is $< p_x >=< \hbar k_x >= \hbar k_{x\circ}$. Average energy of wavepacket is $< E >=< \hbar\omega >= \hbar\omega(k_{x\circ})$.

Corollary V: Possible wave property: A free object (universe, electron, photon, phonon, and so on) has wave property in principle. The wavenumber is determined by (*average momentum*)$/\hbar$, and frequency is determined by (*average energy*)$/\hbar$. (In history, de Bloglie first proposed the wave property of massy particle as a hypothesis in 1925. This corollary proves that the possible wave property of massy and massless particle.)

Theorem IX-1: Wavepacket compositions: Even we use the plane waves with wavenumbers of discrete values to expand wavepacket-function, a free wavepacket in three dimensional position space still consists of product with three factors: guide-factor; front-factor; and body-factor.

Theorem IX-2: Wavepacket compositions: Even we use the spherical coordinate to expand wavepacket-function, a free wavepacket in three dimensional position space still consists of product with three factors: guide-factor; front-factor; and body-factor.

Theorem IX-3: Wavepacket compositions: For a system consisting of many different particles the wavepacket composition theorem VI is still correct.

Theorem X: Schrödinger equation: The guide-factor (=wavefunction) of wavepacket in potential $V(\mathbf{r},t)$ satisfies generally the three dimensional time-dependent differential equation

$$i\hbar\frac{\partial}{\partial t}\Psi(\mathbf{r},t) = \left[-\frac{\hbar^2}{2m}\nabla^2 + V(\mathbf{r},t)\right]\Psi(\mathbf{r},t). \qquad (3.4.5.2.1)$$

Eq. (3.4.5.2.1) is called Schrödinger equation, because Schrödinger proposed this equation in 1926.

Proof:

The proof in Ref. [3] is not clear. Here we make the proof once again. We use two methods. The first method is like the method in Ref. [3]. To avoid misunderstanding, for any operator o we write it as $\hat{o}$. The labels of equations are same as that in Ref. [3].

(1). First method.

The expression of a three dimensional free wavepacket is given by Eq. (3.4.4.2.27), which is

$$\begin{aligned}
\Psi(x,y,z,t) &\overset{V=0}{=} \mathrm{e}^{i\mathbf{k}_\circ\cdot\mathbf{r}-i\omega(\mathbf{k}_\circ)t}\times\mathrm{e}^{-i\mathbf{k}_\circ\cdot\mathbf{r}'}\times\Psi(\mathbf{r}',0) \\
&\overset{V=0}{\equiv} I\times II\times III \\
&= guide\ factor\times front\ factor\times body\ factor. \qquad (3.4.5.2.2)
\end{aligned}$$

It is easy to verify that in case of $V=0$, guide-factor I satisfies the following three dimensional differential equation in representation of coordinate space and time:

$$\begin{aligned}
i\hbar\frac{\partial}{\partial t}\mathrm{e}^{i\hat{\mathbf{k}}_\circ\cdot\mathbf{r}-i\omega(\hat{\mathbf{k}}_\circ)t} &= \hbar\omega(\hat{\mathbf{k}}_\circ)\mathrm{e}^{i\hat{\mathbf{k}}_\circ\cdot\mathbf{r}-i\omega(\hat{\mathbf{k}}_\circ)t} \\
&= \frac{\hat{p}^2}{2m}\mathrm{e}^{i\hat{\mathbf{k}}_\circ\cdot\mathbf{r}-i\omega(\hat{\mathbf{k}}_\circ)t} \\
&= -\frac{\hbar^2}{2m}\nabla^2\,\mathrm{e}^{i\hat{\mathbf{k}}_\circ\cdot\mathbf{r}-i\omega(\hat{\mathbf{k}}_\circ)t}. \qquad (3.4.5.2.3)
\end{aligned}$$

Eq. (3.4.5.2.3) is the three dimensional time-dependent guide-factor equation of free wavepacket, and is called three dimensional time-dependent Schrödinger equation of free particle.

Let us seek the equation satisfied by guide-factor $\Psi(x,t)$ in case of $V(\mathbf{r},t)\neq 0$.

Theorem II in subsection 3.3.1.4 has proved that in spacetime representation

$$\mathbf{p}\equiv\hbar\mathbf{k}\to -i\hbar\nabla, \quad E\to i\hbar\frac{\partial}{\partial t}, \qquad (3.4.5.2.4)$$

where E represents the total energy of the system.

ThusEq. (3.4.5.2.3) can formally be written as

$$E_{total}\Psi = \frac{p^2}{2m}\Psi = E_{kinetic}\Psi. \qquad (3.4.5.2.5)$$

If $V\neq 0$ and using spacetime representation, then Eq. (3.4.5.2.5) is transformed naturally into

$$E_{total}\Psi = (E_{kinetic}+V)\Psi = E_{total}\Psi. \qquad (3.4.5.2.6)$$

Considering the operator representations in Eq. (3.4.5.2.3) yields Eq. (3.4.5.2.1). QED.

Eq. (3.4.5.2.1) is called three dimensional time-dependent Schrödinger equation of the guide-factor of a non-free wavepacket. (guide-factor≡wavefunction)

(2). Second method.

If an object is in a general potential field $V(x,t)$, then we can expand the wavepacket-function in representation of momentum and time exactly as follows:

$$\begin{aligned}
\Psi(\hat{x},t) &= \frac{1}{\sqrt{2\pi}}\int_{-\infty}^{\infty} dk\overline{\psi}(k)e^{ik\hat{x}-i[\omega(k)+V(\hat{x},t)/\hbar]t} \\
&= e^{-iV(\hat{x},t)t/\hbar}\frac{1}{\sqrt{2\pi}}\int_{-\infty}^{\infty} dk\overline{\psi}(k,t)e^{ik\hat{x}-i\omega(k)t} \\
&= e^{-iV(\hat{x},t)t/\hbar}e^{ip_\circ\hat{x}/\hbar-iE(p_\circ)t/\hbar}\times e^{-ip_\circ\hat{x'}/\hbar}\times\psi(\hat{x'},0) \\
&= e^{ip_\circ\hat{x}/\hbar-i[E(p_\circ)+V(\hat{x},t)]t/\hbar}\times e^{-ip_\circ\hat{x'}/\hbar}\times\psi(\hat{x'},0) \\
&\equiv \mathcal{W}_p(\hat{x},t,p=p_\circ)\times II\times III \\
&\equiv guide\ factor\times front\ factor\times body\ factor. \qquad (3.4.5.2.2)'
\end{aligned}$$

Next, we use the representation of coordinate and time, and make $\mathcal{W}_p(\hat{x},t,p=p_\circ)\to\Psi(x,t)$. Note that the notation Ψ in different cases can represent different quantity.

The guide-factor satisfies the following equation:

$$\begin{aligned}
i\hbar\frac{\partial}{\partial t}\Psi(x,t) &= i\hbar\frac{\partial}{\partial t}e^{i\hat{p}_\circ x/\hbar-i[E(\hat{p}_\circ)+V(x,t)]t/\hbar} \\
&= E(\hat{p}_\circ)+V(x,t)\Psi(x,t) \\
&= \left[\frac{\hat{p}_\circ^2}{2m}+V(x,t)\right]\Psi(x,t) \\
&= H\Psi(x,t). \qquad (3.4.5.2.1)'
\end{aligned}$$

QED.

Corollary VI: Wavepacket non-spreading: Any wavepacket never spreads in weak external field.

Corollary VII: General wavepacket compositions: The wavepacket (= object = particle = wavepacket-function) in potential field consists still of the three factors as the case of $V(\mathbf{r},t)=0$. The sole difference is that the guide-factor in cases of $V\neq 0$ obeys Schrödinger equation of $V\neq 0$.

Wavepacket motion law:

Any objects, including both massy and massless particles, are wavepackets.

Any objects have both wave and particle (corpuscle) properties because theirs wavepackets consist of guide-factor (which is also called wavefunction. In general cases, this wavefunction has to contain all informations, such as spin, polarization, spinor components, parity, and so on) and body-factor (which is substance of the object).

The motion of the body-factor accepts guide of the guide-factor. This guide is of equiprobability symmetry.

The actual motion of the body-factor is determined by spontaneous equiprobability symmetry breaking.

This spontaneous equiprobability symmetry breaking process is a still more microscopic process. In essence, this process is stochastic or random. In this process the momentum conservation, the energy conservation, the general and special theory of relativity and so on, determined by theories belonging to determinism, might be violated.

Corollary VIII: Matter wave: Massy particle (rest mass$\neq 0$) has wave property. The wavenumber of the matter wave is $k = p/\hbar$, where k and p represent the average wavenumber and momentum of the marry particle, respectively. The angular frequency of the matter wave is $\omega = E/\hbar$, where ω and E represent the average angular frequency and momentum of the marry particle, respectively.

Corollary IX: Particle property of wave: Wave, such as light and acoustic wave, has particle property. The energy and momentum of wave quantum are $E = \hbar\omega$ and $p = \hbar k$, respectively. The wavenumber of the wave's particle is k, which is the average wavenumber of the wavepacket of the wave's particle. The angular frequency of the wave's particle is ω, which is the average angular frequency of the wavepacket of wave's particle.

Theorem XI: State superposition: Linear composition of possible states (= wavefunctions = solutions of Schrödinger equation = guide-factors≠wavepacket-function) is also a possible state. In mathematics, if $\Psi_i(x,t)$, $(i = 1, 2, \cdots, N)$ satisfy Schrödinger equation Eq. (3.4.5.2.1)

$$\left[i\hbar\frac{\partial}{\partial t} + \frac{\hbar^2}{2m}\frac{\partial^2}{dx^2} - V(x,t)\right]\Psi_i(x,t) = 0, \quad (i = 1, 2, \cdots, N) \qquad (3.5.3.1)(3.4.5.2.1)$$

then the wavefunction

$$\Psi(x,t) = \sum_{i=1}^{n\leq N} c_i\Psi_i(x,t) \qquad (3.5.3.2)$$

also satisfies Schrödinger equation Eq. (3.5.3.1).

Corollary X: Interference between component states: There are interference between component states of superposition state, or say, the state superposition is a coherent superposition.

Criterion: Criterion of quantum or classical mechanics: Suppose that the system is one dimensional.

$$\left\{\begin{array}{c} \lambda_x > L_x \\ or,\ equivalently, \\ h > mv_{xg}L_x \end{array}\right\} \quad quantum\ effect\ prevails\ over\ classical\ effect, \qquad (3.6.1.1)$$

$$\left\{\begin{array}{c} \lambda_x << L_x \\ or,\ equivalently, \\ h << mv_{xg}L_x \end{array}\right\} \quad classical\ effect\ prevails\ over\ quantum\ effect, \qquad (3.6.1.2)$$

$$\left\{\begin{array}{c} \lambda_x < L_x < 10\lambda_x \\ or,\ equivalently, \\ mv_{xg} < h < 10mv_{xg}L_x \end{array}\right\} \quad quantum\ effect\ \ still\ \ manifests\ a\ little, \qquad (3.6.1.3)$$

where λ_x, L_x, m, v_{xg}, and h are the average wavelength, size, group velocity in x direction of wavepacket, and Planck constant, respectively.

Corollary XI: If $\hbar \to 0$: If $\hbar \to 0$, then even a microscopic particle such as electron obeys classical mechanics. On the contrary, so long as $\hbar$ is big enough, then even our Earth has to obey quantum mechanics.

Corollary XII: Velocity and mass effects: From $h < mv_{xg}L_x$ we see that the larger the v_{xg} or m are, the weaker the quantum mechanical effects are, i. e., the high velocity and large mass of particle does not benefit quantummechanical effect.

If a macroscopic particle with mass of one gram and with $v_{xg} = 1$ cm per second, then the condition for that the quantummechanical effect prevails over the classical mechanical effect in x direction is the particle's length $L_x < 6.6 \times 10^{-27}$ cm, which is too short. The actual size of this particle is much much larger than this value of L_x. Therefore, the properties of the so-called macroscopic particle always corresponds to the average values of quantum mechanics, and thus there is no quantummechanical wave effect, i. e., the classical effect prevails over the quantum effect.

If $\lambda_x << L_x$, then we call this object macroscopic object. If $L_x < 10\lambda_x$, then we call it microscopic object.

Here, we would like to mention our viewpoint coming from this criterion and the zero theorem.

Classical physics includes: Newton classical mechanics, special and general theory of relativity, and so on. In these theories, any objects are described by both momenta and coordinates at the same time, the variations of objects are deterministic, there are no coherence between two macroscopic objects, and so on.

Quantum physics includes: quantum mechanics, quantum field theory, and so on. In these theories, any objects cannot be described by both momenta and coordinates at the same time, the variations of objects are stochastic, there are coherence between two microscopic objects (except macroscopic quantum phenomena), and so on.

Theorem XII: Ehrenfest average value theorem: Newton equations, written in the forms

$$\frac{d\mathbf{r}}{dt} = \frac{\mathbf{p}}{m} \tag{3.6.2.1}$$

and

$$\frac{d\mathbf{p}}{dt} = -\bigtriangledown V, \tag{3.6.2.2}$$

are exactly satisfied by the average values of corresponding operators ($\mathbf{p}$, $\mathbf{r}$, V) on wave-function (guide-factor). Here, V represents real potential energy. (Ehrenferst proved this theorem in 1927.)

1.3 Stationary States of a One Dimension System

Theorem XIII-1: If $\Psi(x)$ is a solution of Eq. (4.1.3), and eigenvalue is E, then $\Psi^*(x)$ is also a solution, and its eigenvalue is also E.

Corollary XIII: If for some E the solution of Eq. (4.1.3) is not degenerate (i. e. just has one independent solution), then the solution can be taken as real except a constant real or complex factor. QED.

Proof:

If $\Psi(x)$ is a solution with energy E and the energy level is not degenerate, then $\Psi^*(x)$ and $\Psi(x)$ describe the same quantum state. Thus, $\Psi^*(x) = C\Psi(x)$. Its complex conjugate is $\Psi(x) = C^*\Psi^*(x) = |C|^2\Psi(x)$. Therefore, $|C| = 1$ and $C = e^{i\alpha}$. α is real number. For convenience, we take $\alpha = 0$, then $\Psi^*(x) = \Psi(x)$ which means that $\Psi(x)$ is a real function.

Theorem XIII-2: Suppose that $V(x)$ is invariant under space reflection, $V(-x) = V(x)$. If $\Psi(x)$ is a solution of Eq. (4.1.3) and its energy is E, then $\Psi(-x)$ is also a solution with energy E.

Corollary XIV: If $V(x)$ is invariant under space reflection, and $\Psi(x)$ is not degenerate, then $\Psi(x)$ has definite parity.

Theorem XIII-3: If $V(-x) = V(x)$, then for any energy eigenvalue E, one can always find a group of complete solutions of Schrödinger equation. Every solution has certain parity.

Theorem XIII-4: For some energy eigenvalue E, one can always find a group of complete real solutions, i. e., any solution with energy E can be expressed as a linear superposition of these real solutions in the group.

Theorem XIII-5: If $V(x)$ is step square potential, i. e.,

$$V(x) = \begin{cases} V_1, & x < a \\ V_2, & x > a \end{cases} \tag{4.1.12}$$

and $V_2 - V_1$ is finite, then energy eigenfunction $\Psi(x)$ and its derivative $\Psi'(x)$ are continuous certainly.

Theorem XIII-6: If for one dimensional particle $\Psi_1(x)$ and $\Psi_2(x)$ are the solutions of Eq. (4.1.3), and belong to the same energy, then

$$\Psi_1\Psi_2' - \Psi_2\Psi_1' = constant. \tag{4.1.16}$$

Theorem XIV: Quantization of bound state: For bound system, its ground state has no node. Its excited state is an oscillation function in space, has thus nodes. The n-th excited state has n nodes. The energy spectrum of excited states is discrete, i. e., energy quantization. (Note that the label of ground state is 0.)

1.4 General Formalism

Theorem XV: Eigenvalue of Hermitian operator: The eigenvalue of Hermitian operator are real.

Theorem XVI: Orthogonality of eigenfunction: The eigenfunctions of Hermitian operator are orthogonal to each other if the corresponding eigenvalues are unequal.

Theorem XVII-1: Real average value: The average value of Hermitian operator on any state is real.

Theorem XVII-2: Hermitian operator: If for any state Ψ the average value of an operator is real, then the operator is Hermitian.

Corollary XV: Average value of O^2: If O is Hermitian operator, then for any state

$$< O^2 >= (\Psi, O^2\Psi) \geq 0. \tag{5.2.28}$$

Theorem XVIII: Commutation: If two linear and Hermitian operators, O and P, have the same eigenfunctions, then they commute, that is, $[O, P] = 0$.

Theorem XIX: Common eigenfunction set: If operators O and P commute, then a common set of eigenfunctions can be chosen for them.

Theorem XX: Inverse theorem: If Ψ is common eigenfunction of O and P, then $[O, P] = 0$

Theorem XXI-1: Normalization of eigenfunction of continuous spectrum: The continuum spectrum eigenfunction is orthogonal and is normalized to Dirac δ function.

Theorem XXI-2: Normalization of eigenfunction of continuous spectrum: The discrete spectrum eigenfunction is orthogonal and is normalized to Kroniker δ notation.

Theorem XXII: Periodic boundary condition: To ensure that the momentum operator $p_x = -i\hbar\partial/\partial x$ in domain $[-L/2 \leqq x \leqq L/2]$ is Hermitian operator, $\Psi(x)$ has to satisfy periodicity boundary condition $\Psi(-L/2) = \Psi(L/2)$. (Refer to Ref. [4] for proof.)

Theorem XXIII: Motion equation of operator: An operator O, corresponding to an observable, satisfies

$$\frac{dO}{dt} = \frac{\partial O}{\partial t} + \frac{1}{i\hbar}[O, H].$$

Theorem XXIV: Average energy conservation: The uniformity of time leads to conservation of average value of Hamiltonian, i. e., average energy conservation.

Let us make the following four discussions on the energy conservation and so on:

(1). The invariance of the operator H under certain transformation, defined by the operator F, means that the action of the operator F on the wavefunction $H\Psi$ is equivalent to the action of H on the wavefunction $F\Psi$, i. e.

$$FH\Psi = HF\Psi. \tag{5.6.2.2}$$

In other wards, the invariance of H with respect to the transformation realized by the operator F reduces to the condition that F commutes with the Hamiltonian:

$$FH = HF. \tag{5.6.2.3}$$

Let us introduce a time-shift operator T_τ which shifts the time by an amount τ. By definition $T_\tau = t + \tau$, and we have from Eq. (5.6.1.6)

$$T_\tau\Psi(t) = \Psi(t - \tau). \tag{5.6.2.4}$$

The quantity τ is a parameter of the operator T_τ. The fact that the time is uniform for the system considered by us can mathematically be expressed by the commutation relation (i. e., commutator)

$$[T_\tau, H] = 0. \tag{5.6.2.5}$$

Instead of the time-shift operator it is convenient to use a generating function for the transformation or the infinitesimal time-shift operator $I(t)$, which is defined as

$$I(t) = \frac{\partial}{\partial \tau} T_\tau \bigg|_{\tau=0}. \tag{5.6.2.6}$$

We can easily find that the explicit form of the operator $I(t)$, if we bear in mind that

$$I(t)\Psi(t) = \frac{\partial}{\partial \tau} T_\tau \Psi(t)\bigg|_{\tau=0} = \frac{\partial}{\partial \tau}\Psi(t-\tau)\bigg|_{t=0} = -\frac{\partial}{\partial t}\Psi. \tag{5.6.2.7}$$

We have thus

$$I(t) = -\frac{\partial}{\partial t}. \tag{5.6.2.8}$$

The energy conservation law is connected with the fact that the operator H commutes with the infinitesimal time-shift operator $I(t)$. The operator

$$-i\hbar I(t) = i\hbar\frac{\partial}{\partial t}, \tag{5.6.2.9}$$

which has the dimension of energy is in this connection sometimes called the energy operator.

(2). Many authors such as Ref. [5] at this stage will point out that: "One should, however, bear in mind the conditions under which one can use this terminology, i. e., the above energy operator. The energy in quantummechanical system in a stationary state is determined by the eigenvalues of Hamiltonian operator. Hamiltonian, i. e., a function of the operators of coordinates and momenta, is thus the operator of energy of system. In contrast to the spatial coordinates, the time coordinate is not an operator, and is only a parameter."

To this viewpoint, we have different viewpoint. We have proved in theorem II that in time t representation the energy has to be an operator; on the other hand, in energy representation the time has to be an operator as well.

(3). The exact meaning of the energy in the quantummechanical energy conservation is average value of Hamiltonian. An average value of Hamiltonian comes from contributions of infinitely still more microscopic processes. Therefore, the energies in some still more microscopic processes might violate this classical energy conservation law.

(4). The necessary condition of theorem XXIV is uniformity of time, which requires that the system is closed or isolated. In quantum transition theory under external perturbation. The existence of external perturbation will destroy the time uniformity, and thus the average energy nonconservation can occur in transition processes in principle.

Theorem XXV: Average momentum conservation: The spatial uniformitye leads to conservation of average value of momentum operator, i. e., average momentum conservation.

Our two comments for theorem XXV.

(1). The exact meaning of the momentum in the above quantummechanical momentum conservation is average value of momentum operator. An average value of momentum operator comes from contributions of infinitely still more microscopic processes. Therefore, the momentum in some still more microscopic processes might violate the classical momentum conservation law.

(2). The necessary condition of theorem XXV is uniformity of position space. If an external perturbation destroys the position space uniformity, and then the average momentum nonconservation can occur in principle.

Theorem XXVI: Average angular momentum conservation: The spatial isotropy leads to conservation of average value of angular momentum operator, i. e., average angular momentum conservation.

Our two comments for theorem XXVI.

(1). The exact meaning of the angular momentum in the above quantummechanical angular momentum conservation is average value of angular momentum operator. An average value of angular momentum operator comes from contributions of infinitely still more microscopic processes. Therefore, the angular momentum in some still more microscopic processes might violate the classical angular momentum conservation law.

(2). The necessary condition of theorem XXVI is isotropy of position space. If an external perturbation destroys the position space isotropy, and then the average angular momentum nonconservation can occur in principle.

Theorem XXVII: Parity conservation: The invariance of spatial inversion leads to parity conservation.

Comment for theorem XXVII.

The exact meaning of the parity in the above quantummechanical parity conservation is the average value of the parity operator. In chapter 8, we will demonstrate that the solutions of Dirac equation are spinor functions. Therefore, we should use spinor wavefunction to describe electrons, and the parity operator is of matrix form.

Up to 1956, it was assumed that the parity conservation was a universal law of nature. It was established in 1956 by Yang and Lee and by Wu and co-workers that in the β-decay of atomic nuclei and in the decay of muon, pion, and hyperon the parity violations ware observed. These violations showed that weak interactions, which determine these decay processes, do not have spatial inversion symmetry.

Theorem XXVIII: Unitary transformation: If U is a unitary transformation operator, then $U^+ = U^{-1}$.

Theorem XXIX: Three pictures of quantum mechanics: If H does not depend on time explicitly, then define

$$U(t, t_\circ) = e^{-iH(t-t_\circ)/\hbar}. \tag{5.7.1.9}$$

If H depends on time explicitly, then

$$U(t, t_\circ) = Te^{-i\int_{t_\circ}^{t} dt' H(t')/\hbar}, \tag{5.7.1.9$'$}$$

where T is time-ordering operator.

If $H = H_\circ + V(x,t)$ and $H_\circ$ does not depend on time t explicitly, then we define

$$U_0(t,t_1) = e^{-iH_\circ(t-t_1)/\hbar}, \tag{5.7.1.9$''$}$$

If $H = H_\circ(t) + V(x,t)$ and $H_\circ(t)$ depends on time t explicitly, then we define

$$U_0(t,t_1) = Te^{-i\int_{t_1}^{t} dt' H_\circ(t')/\hbar}, \tag{5.7.1.9$'''$}$$

Schrödinger picture:

$$|\Psi_S(t) >= U(t,t_\circ)|\Psi_S(t_\circ) > . \tag{5.7.1.10}$$

O_S is independent of time.

$$i\hbar\frac{\partial|\Psi_S(t)>}{\partial t} = H|\Psi_S(t)> . \tag{5.7.1.11}$$

$$\frac{dO_S}{dt} = 0. \tag{5.7.1.12}$$

Haisenberg picture:

$$|\Psi_H >\equiv |\Psi_S(t_\circ)> . \tag{5.7.1.13}$$

$$O_H(t) = U^+(t,t_\circ)O_S U(t,t_\circ). \tag{5.7.1.14}$$

$$i\hbar\frac{\partial|\Psi_H>}{\partial t} = 0. \tag{5.7.1.15}$$

$$i\hbar\frac{dO_H(t)}{dt} = [O_H(t),H]. \tag{5.7.1.16}$$

Interaction picture:

Assume that $H = H_\circ + V(t)$, where $H_\circ$ represents kinetic energy operator, and is independent of time.

$$|\Psi_I(t) >= U_0^+(t,t_1)|\Psi_S(t) >= U_I(t,t_\circ)|\Psi_I(t_\circ)> . \tag{5.7.1.17}$$

$$O_I(t) = U_0^+(t,t_1)O_S U_0(t,t_1). \tag{5.7.1.18}$$

$$U_I(t,t_\circ) = U_0^+(t,t_1)U(t,t_\circ)U_0(t_\circ,t_1) = T\exp\left(-\frac{i}{\hbar}\int_{t_\circ}^{t} dt' V_I(t)\right). \tag{5.7.1.19}$$

$$i\hbar\frac{\partial|\Psi_I(t)>}{\partial t} = V_I|\Psi_I(t)> . \tag{5.7.1.20}$$

$$V_I(t) = U_0^+(t,t_1)V(t)U_0(t,t_1). \tag{5.7.1.21}$$

$$i\hbar\frac{dO_I(t}{dt} = [O_I(t),H_\circ]. \tag{5.7.1.22}$$

Theorem XXX: Three formulations of quantum mechanics: Just from the proved commutation relation $[x,p] = i\hbar$, all the three formulations of quantum mechanics can be derived exactly and without any assumptions, postulates, hypotheses, and axioms. The three formulations are:

(1). Heisenberg operator (matrix) formulation,

$$\frac{dO}{dt} = -\frac{i}{\hbar}[O,H] + \frac{\partial O}{\partial t}, \tag{5.8.1}$$

(2). Schrödinger wave equation formulation,

$$i\hbar\frac{\partial}{\partial t}\psi(x,t) = \left[\frac{-\hbar^2}{2m}\frac{\partial^2}{\partial x^2} + V(x)\right]\psi(x,t), \tag{5.8.2}$$

(3). Feynman path integral formulation, (3-1). Propagator $K(x_f, x_0; t_f, t_0)$ in phase space (=Feynman Hamiltonian path integral):

$$K(x_f, x_0; t_f, t_0) = \int_{x(t_0)=x_0}^{x(t_f)=x_f} \mathcal{D}p\mathcal{D}x \exp\left[\frac{i}{\hbar}\int_{t_0}^{t_f} dt[p\dot{x} - H(p,x)]\right], \tag{5.8.3}$$

where $p = p(t)$, $x = x(t)$, $\dot{x} = dx(t)/dt$,

$$\mathcal{D}p\mathcal{D}x \equiv \lim_{N\to\infty}\left[\prod_{k=1}^{N}\frac{dx_k dp_k}{2\pi\hbar}\right]\frac{dp_0}{2\pi\hbar}, \tag{5.8.4}$$

$$\int_{x(t_0)=x_0}^{x(t_f)=x_f} \mathcal{D}p\mathcal{D}x = \lim_{N\to\infty}\left[\prod_{k=1}^{N}\int_{-\infty}^{\infty}\int_{-\infty}^{\infty}\frac{dx_k dp_k}{2\pi\hbar}\right]\int_{-\infty}^{\infty}\frac{dp_0}{2\pi\hbar}, \tag{5.8.5}$$

(3-2). Propagator in configuration space (=Feynman Lagrangian path integral)

$$K(x_f, x_0; t_f, t_0) = \int_{x(t_0)=x_0}^{x(t_f)=x_f} \mathcal{D}x \exp\left(\frac{iS}{\hbar}\right), \tag{5.8.6}$$

$$S = \lim_{\substack{N\to\infty \\ \varepsilon\to 0}} \sum_{k=0}^{N}\left(m\frac{(x_{k+1}-x_k)^2}{2\varepsilon^2} - V(x_k)\right)\varepsilon = \int_{t_0}^{t_f}\left(\frac{m\dot{x}^2}{2} - V(x)\right)dt, \tag{5.8.7}$$

$$\mathcal{D}x = \lim_{\substack{N\to\infty \\ \varepsilon\to 0}} \frac{1}{\sqrt{2\pi i\hbar m^{-1}\varepsilon}}\prod_{k=0}^{N}\frac{dx_k}{\sqrt{2\pi i\hbar m^{-1}\varepsilon}}. \tag{5.8.8}$$

Theorem XXXI: Eigenfunction of L_z: To ensure that L_z is a hermitian operator, any eigenfunction of L_z must satisfy periodic boundary condition (or, single value condition).

1.5 Hellmann-Feynman Theorems

Theorem XXXII: HF theorem of time-independent wavefunctions:

$$\frac{\partial E_{n\lambda}}{\partial\lambda} = \int \Psi_{n\lambda}\frac{\partial H_\lambda}{\partial\lambda}\Psi_{n\lambda}d\mathbf{r}, \tag{6.3.1.1}$$

where H_λ is Hamiltonian operator depending upon a continuous parameter λ, $\Psi_{n\lambda}$ is a wavefunction (eigenfunction) of the Hamiltonian, depending implicitly upon λ, $E_{n\lambda}$ is the energy (eigenvalue) of the wavefunction, $d\mathbf{r}$ implies an integration over the domain of the wavefunction.

Theorem XXXIII: HF theorem for time-dependent wavefunctions: For a general time-dependent wavefunction satisfying the time-dependent Schrödinger equation, the Hellmann-Feynman theorem is not valid. However, the following identity holds:

$$\left\langle\Psi_{n\lambda}(t)\left|\frac{\partial H_\lambda}{\partial\lambda}\right|\Psi_{n\lambda}(t)\right\rangle = i\hbar\frac{\partial}{\partial t}\left\langle\Psi_{n\lambda}(t)\left|\frac{\partial\Psi_{n\lambda}(t)}{\partial\lambda}\right.\right\rangle. \tag{6.3.2.1}$$

1.6 Identical Particles

Theorem XXXIV: Pauli theorem: The criterion that distinguishes between the two kinds of particles is their spins: Fermions have half-integer spin, and obey Fermi-Dirac statistics; Bosons have integer and zero spin, and obey Bose-Einstein statistics.[6, 7]

Theorem XXXV: Identity of identical multiparticle system: The multiparticle wavefunction (which can also depend on time) of identical particle systems are separated into two kinds: Symmetric if the eigenvalue of permutation operator is +1; Antisymmetric if the eigenvalue of permutation operator is -1. The particles in identical multiparticle system, described by an antisymmetric wavefunction, are of half-integer spin, are called fermions, and obey Fermi-Dirac statistics; The particles in identical particle system, described by a symmetric wavefunction, are of integer or zero spin, are called bosons, and obey Boson-Einstein statistics. The exchange symmetry of multiparticle wavefunction of the identical multiparticle system does not change with time, and the state transition between different symmetry states is impossible.

In some references our identity theorem is called "indistinguishability principle of identical multiparticle system". This textbook thinks:

(1). The so called indistinguishability is not a necessary concept or even is a wrong concept. Because the particles in a identical multiparticle system are distinguishable. The identity defined in this textbook is enough to deduce all the properties of identical multipleparticle system;

(2). The unique characteristics of the identical multipleparticle system can be proved. Therefore, we use "identity theorem" other than "indistinguishable principle".

Theorem XXXVI: Pauli exclusion: It is not possible to have a multiparticle fermion state in which there are two fermions in the same single particle states.

Theorem XXXVII: Nonlocality of quantummechanical multibody states: The quantummechanical multibody states such as entangled states have nonlocality (nonlocalty=distance-correlativity).

1.7 Time-Dependent Tr nsition Theory

Theorem XXXVIII: Boundary transition: For the boundary absorption, if $|H'_{k'k}|^2\rho(k') = A\omega_{k'}^{\beta}$ $(\beta \gtrless 0)$ in Eq. (10.3.1.3)$'$, then

(i). The transition probability (density) $P_{k'k}(t)$ in Eq. (10.3.1.3)$'$ has maximum value at $\omega_{k'k} - \omega = \omega_{k'} \gtrless 0$, which means that quantum transition of boundary absorption mainly occurs in the process bing of energy nonconservation.

(ii). The time-dependence of transition probability $P_{all\leftarrow k}(t)$ is $t^{1-\beta}$.

Theorem XXXIX: Energy in transition processes: If the energy-dependence of the integrand in the formula of $P_{k'k}(t \geq t_\circ)$ is

$$\begin{aligned} P_{k'k}(t \geq t_\circ) &\propto E_{k'}^{\beta} \frac{\sin^2 \frac{(E_{k'k}-E)t_\circ}{2\hbar}}{(E_{k'k}-E)^2} \\ &\equiv E_{k'}^{\beta} II, \qquad (10.3.2.4) \end{aligned}$$

then the maximum values of $P_{k'k}(t \geq t_\circ)$, which correspond to the probable transitions, satisfy the following relations

$$\begin{array}{lll} E_{k'} > E_k + E & \text{if} & \beta > 0, \\ E_{k'} = E_k + E & \text{if} & \beta = 0, \\ E_{k'} < E_k + E & \text{if} & -1 < \beta < 0, \end{array}$$

which means that the system energy in the most probable transition increases (i. e., nonconservation), maintains (i. e., conservation), and reduces (i. e., nonconservation) if $\beta > 0$, $\beta = 0$, and $-1 < \beta < 0$, respectively.

Corollary XVI: Condition of approximate energy conservation in transition processes: If the perturbation energy is vary large, and the k'-dependence of the vacuum support ability is weak enough.

Theorem XXXX: Time-dependence of transition rate: If $|H'_{k'k}|^2\rho(k') \propto \omega_{k'}^{\beta}$ ($\beta \neq 0$) in Eq. (10.3.2.2), then the transition rates, $w_{all\leftarrow k} \equiv \sum_{k' \neq k} P_{k'k}/t$ is time-dependent, i. e., the Fermi golden rule does not hold water. Fermi golden rule holds just in the case of $\beta = 0$.

Comments for theorem XXXX:

(1). If we define $w_{all\leftarrow k} \equiv d\sum_{k'\neq k} P_{k'k}/dt$, then $w_{all\leftarrow k}$ is time-dependent for $\beta \gtreqless 0$, and Fermi golden rule does not hold always.

(2). In some cases the time dependences of $|H'_{k'k}|^2\rho(k')$ are much more complex than $|H'_{k'k}|^2\rho(k') \propto \omega_{k'}^{\beta}$. Let us illustrate with an example. The photoeffect in subsection 10.4.5 demonstrates that

$$|H'_{k'k}|^2\rho(k') \propto \frac{\omega_{k'}^{1.5}}{(1+k'^2a^2)^6}, \quad k' = \sqrt{\frac{2\mu\omega_{k'}}{\hbar}}, \quad a = 0.53 \times 10^{-8} cm, \quad \mu = electron\ mass. \tag{10.3.2.6}$$

Substituting Eq. (10.3.2.6) into third equality of Eq. (10.3.2.5) and making numerical integration, we obtain a curve of $w_{all\leftarrow k}$ versus t, which is shown in Fig. 10.1.

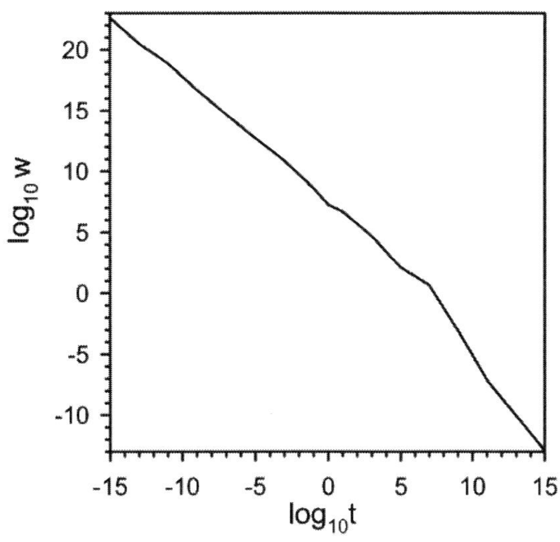

Figure 1.1: Theoretical curve of $\log_{10} W \equiv \log_{10} w_{all\leftarrow k}$ versus $\log_{10} t$ (t/second).

If Fermi golden rule holds water, then the curve in Fig. 10.3 should be a horizontal line.

(3). This textbook prove Fermi golden rule just under one condition, i. e., $|H_{k'k}|^2\rho(k')$ is independent of k', and all other steps in our derivations are exact. However, all until now references and textbooks prove Fermi golden rule under two conditions: both the $|H_{k'k}|^2\rho(k')$ is independent of k' and the time limit condition $t_\circ \to \infty$. They use the following mathematical formula

$$\lim_{t_\circ \to +\infty} \frac{\sin^2 \alpha t_\circ}{\pi t_\circ \alpha^2} = \delta(\alpha). \qquad (10.3.2.7)$$

Substituting Eq. (10.3.2.7) into Eq. (10.3.2.5), the integrand has a delta function $\delta(E_{k'} - E_k - E)$, which is an energy conservation factor, and after integration the expression is proportional to a factor $t_\circ$. From these, they think that both energy conservation and Fermi golden rule in transition processes are proved.

As is well known, the less the conditions in a theory are, the better the theory is. Therefore, the derivations for the Fermi golden rule in this textbook are certainly better than the method of until now all references and textbooks. Next we will point out the purely mathematical mistakes in all until now references and textbooks.

(4). The physical meaning of $t_\circ$ is duration time of a time-dependent perturbation, the magnitude order of which should be the observed relaxation time. The magnitude orders of relaxation time are generally from 10^{-6} *to* 10^{-15} second. Therefore, the condition $t_\circ \to \infty$ is not reasonable in physics, and is dangerous in mathematics (Every body knows the famous words on arbitrarily coming from taking infinite limit.).

(5). Based on the two conditions, all until now references and textbooks conclude that in the integrand of the total transition probability formula there is a delta function which can ensure energy conservation of transition processes. Some references think that energy conservation of transition processes is a matter of course, and in the past 80 years some references do not have a wee bit of doubts for this energy conservation in quantum transition processes. Even some authors found the possible energy nonconservation in transition processes, but still looked for some "reasons" to argue the energy conservation. For example, Schiff in Ref. [8] thought that the energy nonconservation comes from Heisenberg uncertainty relation and/or the Fourier expansion of a time-dependent perturbation function. However, in our above derivations the integrand in formula of total transition probability the energy conservation's delta function never appears. On the contrary, our proof for Fermi golden rule needs, exactly the inverse, the contributions of processes coming from even very strong energy nonconservation.

(6). We estimate that in the past 80 years (from 1930 to now) some references made probably 10^3 conclusions by using Fermi golden rule and the mathematical mistake taking infinite time limit, but 500 in the 1 000 are of clear mistakes. If readers check many calculations drawn from transition processes, including scattering processes, once again, then readers might modify the conclusions, and thus might have important new discoveries.

Theorem XXXXI: Energy nonconservation in transition processes: In quantum transition processes the energy is more or less not conservative except some physical conditions are imposed.

Comment 1: Although in some cases the magnitudes of the energy nonconservation is not large, but we should have an estimation for the magnitude of the energy nonconservation. Especially, in some cases the magnitude of energy nonconservation might be huge. In some processes, such as stimulated emission, the energy conservation is a physical requirement, and thus in these cases we have to impose the energy conversation condition to the transition formula before calculations.

Comment 2: The energy nonconservation in transition processes do not violate the energy conservation proved in chapter 5. The latter says that if H has homogeneity (or uniformity) of time, then $\partial H/\partial t = 0$, and thus $dE/dt = 0$ from

$$\frac{d\langle H\rangle}{dt} = \frac{dE}{dt} = \left\langle \frac{i}{\hbar}[H, H] + \frac{\partial H}{\partial t} \right\rangle = 0. \tag{10.5.2.10}$$

The condition to obtain Eq. (10.5.2.10) is that the variation of state with time obeys the time-dependent Schrödinger equation. However, in the case time-dependent transition the system is in external time-dependent perturbation, the time is no longer of homogeneity. On the other hand, the transition between initial and final states is determined by the time-dependent transition probability instead of the time-dependent Schrödinger equation. More simply speaking, in quantum transition process, $\partial H/\partial t \neq 0$.

Comment 3: The energy sources of energy nonconservation come from vacuum space because the energy dependence of $|H'_{k'k}|^2 = |V_{k'k}|^2$ in Eq. (10.5.2.8) or $|H'_{k'k}\rho(k')|^2$ in Eq. (10.3.2.4) determines the energy nonconservation of transition process. The particular energy dependence of $|H'_{k'k}|^2\rho(k')$ is determined by the wavefunctions in states $|k>$ and $|k'>$, the form of the perturbation Hamiltonian H', and density of states, which are all dependent on the space structure such as dimensions, and so on. We name $|H_{k'k}|^2\rho(k')$ vacuum energy support ability.

At the end of this chapter 1, I would like to tell the readers that some references treat the quantum transition problems by using a method, which can be called "cutting up with one time knife". According to this method, all transition processes obey the Fermi golden rule, and, correspondingly, keep the energy conservation.

However, according to the theorem in this subsection, we should consider the energy dependence of $|H'_{k'k}|^2\rho(k')$ on k'. If the dependence of k' is weak, then we can obtain the time independence of the transition rate, i. e., Fermi golden rule. However, even in the case of approximate time independence of transition rate, the physics between our textbook and some references still has differences. Some references deduce the energy conservation under the hypotheses that duration time of perturbation $T = (t_\circ - 0) \to \infty$. On the contrary, this textbook deduces Fermi golden rule without using finite duration time of perturbation.

Theorem XXXXII: Zero initial transition rate: The initial transition rate of any nonstable system is certainly zero.

Theorem XXXXIII: Quantum Zeno effect: The survival probability of a nonstationary initial state increases with increase of observation times n in finite time interval $t = n\delta t$, where δt represents the time interval between two time observations (i. e., the time interval producing transition) and in continuous transition case, i. e., $n \to \infty$ and $\delta t \to 0$, the nonstationary state will stop decay or transition, i. e.,

$$P_{ex\to gr,n} \overset{\substack{n\to\infty \\ \delta t\to 0}}{\longrightarrow} 1.$$

Quantum Zeno effect is a present interesting subject. Readers can refer to Refs. [9, 10]. The quantum Zeno effect was proposed as a paradox by Refs. [11]. All until now references and textbooks like to connect the so-called measurement with Zeno effect. However, according to our statement and proof of Zeno effect, it is not necessary to connect measurements with Zeno effect.

1.8 Dissociated States

At first, we have to give a definition of dissociated (=unbound) state. In physics, the picture of dissociated state is a state which the particle can go to infinite far, the the energy spectrum, given by Schrödinger equation is continuous. However, in mathematics, this is not exact. The wavefunction of quantum mechanics as a mathematical function should have Fourier transform. Until now nobody knows that what are the exact conditions for a function $\Psi(x)$ to have Fourier transform. Ref. [3] uses a weak condition that if $\Psi(x) \in \mathcal{M}(\mathbb{R})$, then $\Psi(x)$ has Fourier transform.

Theorem XXXXIV: Energy nonconservation of dissociated states: Dissociated states do not satisfy energy conservation law.

proof:

At present, we have not yet give a general proof for this theorem. We give an example.

Assume that the mass of a particle is m, and the gravitation potential is $-mgx$ ($g = 9.81$ m/sec^2.). This potential is called linear potential, which can be generally expressed as Fx, due to that it is proportional to x. In many cases we meet with the linear potential. For example, an electron is emitted from metal surface, and then is accelerated by electric field $\mathcal{E}$ in x direction. The potential of this electron in the electric field is $-|e|\mathcal{E}x$.

The particle falls from $x = 0$ to ∞. Schrödinger equation is

$$\left(-\frac{\hbar^2}{2m}\frac{d^2}{dx^2} - mgx\right)\Psi = E\Psi, \tag{1.8.1}$$

where $x \geq 0$, g is the gravitation acceleration, and $E \geq 0$.

Taking natural units ($\hbar = m = mg = 1$), Eq. (1.8.1) becomes

$$\frac{d^2\Psi}{dx^2} + 2(E + x)\Psi = 0, \tag{1.8.2}$$

where $x \geq 0$. Set

$$\xi = 2^{1/2}(x + E) \geq 0, \tag{1.8.3}$$

then Eq. (1.8.2) becomes

$$\frac{d^2}{d\xi^2} + \xi\Psi = 0. \tag{1.8.4}$$

Set

$$z = \tfrac{2}{3}\xi^{3/2} \quad \Psi = \sqrt{\xi}u \ , \tag{1.8.5}$$

then

$$\frac{d^2u}{dz^2} + \frac{1}{z}\frac{du}{dz}\left[1 - \frac{(1/3)^2}{z^2}\right]u = 0. \tag{1.8.6}$$

Eq. (1.8.6) is Bessel equation with order of $\nu = 1/3$. Its general solution can be obtained by linear superposition of any two functions from $J_{1/3}$, $N_{1/3}$, $H^{(1)}_{1/3}$, and $H^{(2)}_{1/3}$. In physics, an unbound particle can move to infinity. Therefore, we need that the solution (wavefunction) of Eq. (1.8.6) is a traveling wave along the positive direction of x. Traveling wave means that the wave can go to ∞. Therefore, we have to take the first kind Hankel function $H^{(1)}_{1/3}(z)$, i. e.,

$$\Psi(\xi) = C\xi^{1/2}H^{(1)}_{1/3}\left(\frac{2}{3}\xi^{3/2}\right). \tag{1.8.7}$$

Eq. (1.8.7) tells us that the E is continuous. Let us at first introduce a wrong method to discuss the problem on the energy conservation (This wrong method can be found in many textbooks). Discuss the case $\xi \gg 1$. Using the expression of asymptotic expansion of Hankel function

$$H_{1/3}^{(1)}(z) = C_1\sqrt{\frac{2}{\pi z}}\exp\left[i\left(z - \frac{5\pi}{12}\right)\right], \quad |z| \to \infty, \tag{1.8.8}$$

we have

$$\Psi(\xi) = CC_1\xi^{-1/4}\exp\left(i\frac{2}{3}\xi^{3/2}\right), \quad \xi \to \infty. \tag{1.8.9}$$

The position probability current J and position probability density ρ are

$$J = \frac{\hbar}{2im}\left(\Psi^*\frac{d\Psi}{dx} - \Psi\frac{d\Psi^*}{dx}\right). \tag{1.8.10}$$

Using natural units (Note that $dx = 2^{-1/3}d\xi$) and substituting Eq. (1.8.9) into Eq. (1.8.10) yield

$$J = \frac{1}{2i}2^{1/3}\left(\Psi^*\frac{d\Psi}{d\xi} - \Psi\frac{d\Psi^*}{d\xi}\right) = 2^{1/3}. \tag{1.8.11a}$$

$$\rho = \Psi^*\Psi = \xi^{-1/2}. \tag{1.8.11b}$$

The Eq. (3.5.6.13) in chapter 3 of Ref. [3] shows that **J** corresponds to product of the velocity and the position probability density taken by single particle. Therefore, the velocity of the particle v is

$$v = \frac{J}{\rho} = 2^{1/2}\sqrt{\xi} = \sqrt{2(x+E)}. \tag{1.8.12}$$

In ordinary units, Eq. (1.8.12) becomes

$$v = \sqrt{2\left(gx + \frac{E}{m}\right)}. \tag{1.8.13}$$

Eq. (1.8.13) is the same as the result of classical mechanics. Because according to the energy conservation law in classical mechanics

$$\frac{1}{2}mv^2 - mgx = E. \tag{1.8.14}$$

According to Eq. (1.8.13), the unbound falling neutron at $x \to \infty$ will keep energy conservation. However, according to our exact derivations, we get Eq. (1.8.20) instead of Eq. (1.8.13). Eq. (1.8.20) indicates that the unbound falling neutron does not keep energy conservation although it is of energy continuous spectrum. The exercise (8) in the last section of chapter 4 of Ref. [3] requires the readers to point out the mistake in derivation of Eq. (1.8.11a).

Next we will see that the neutron falling processes do not obey the "energy conservation law". We do not assume $\xi \to \infty$. In this case we can still demonstrate the continuities of velocity v and energy spectrum E in case gravitational field. We have to use the exact expression of the first kind Hankel function $H_{1/3}^{(1)}(z)$ ($z \geq 0$) [12, 13]. Eq. (1.8.7) becomes

$$\begin{aligned}\Psi(\xi) &= C\xi^{1/2}H_{1/3}^{(1)}\left(\frac{2}{3}\xi^{3/2}\right) \\ &= \xi^{1/2}\frac{2e^{-i\pi/2\times 3}}{i\pi}\int_0^\infty dte^{i2\xi^{3/2}ch(t)/3}ch\left(\frac{t}{3}\right).\end{aligned} \tag{1.8.15}$$

In Eq. (1.8.15) the ξ and thus the E [See Eq. (1.8.3)] is continuous.

Substituting Eq. (1.8.15) into Eq. (1.8.11) yields

$$J = \frac{2\times 2^{1/3}\xi}{2i\pi^2}\left[\int_0^\infty dte^{-i2\xi^{3/2}ch(t)/3}ch\left(\frac{t}{3}\right)\frac{d}{d\xi}\int_0^\infty dte^{i2\xi^{3/2}ch(t)/3}ch\left(\frac{t}{3}\right)\right.$$
$$\left. - \int_0^\infty dte^{i2\xi^{3/2}ch(t)/3}ch\left(\frac{t}{3}\right)\frac{d}{d\xi}\int_0^\infty dte^{-i2\xi^{3/2}ch(t)/3}ch\left(\frac{t}{3}\right)\right]. \tag{1.8.16}$$

Substituting Eq. (1.8.15) into Eq. (1.8.11b) yields

$$\rho = \frac{4\xi}{\pi^2}\int_0^\infty dte^{-i2\xi^{3/2}ch(t)/3}ch\left(\frac{t}{3}\right)\int_0^\infty dte^{i2\xi^{3/2}ch(t)/3}ch\left(\frac{t}{3}\right). \tag{1.8.17}$$

From Eqs. (1.8.16) and (1.8.17) we have

$$\begin{aligned} v &= \frac{J}{\rho} = \frac{2^{1/3}}{2}\sqrt{\xi}\left[\frac{\int_0^\infty dte^{i\frac{2}{3}\xi^{3/2}ch(t)}ch\left(\frac{t}{3}\right)ch(t)}{\int_0^\infty dte^{i\frac{2}{3}\xi^{3/2}ch(t)}ch\left(\frac{t}{3}\right)} + \frac{\int_0^\infty dte^{-i\frac{2}{3}\xi^{3/2}ch(t)}ch\left(\frac{t}{3}\right)ch(t)}{\int_0^\infty dte^{-i\frac{2}{3}\xi^{3/2}ch(t)}ch\left(\frac{t}{3}\right)}\right] \\ &\equiv \frac{2^{1/3}\sqrt{\xi}}{2}\times I = \frac{1}{2}\sqrt{2(x+E)}\times I = \frac{1}{2}\sqrt{\left[\frac{2(mgx+E)}{m}\right]}\times I. \end{aligned} \tag{1.8.18}$$

Considering $ch(t) \geq 1$ for any value of t in the first equality of Eq. (1.8.18) yields

$$v \geq \sqrt{\left[\frac{2(mgx+E)}{m}\right]}. \tag{1.8.19}$$

From Eq. (1.8.19) we have

$$\frac{mv^2}{2} - mgx > E. \tag{1.8.20}$$

Eq. (1.8.20) indicates that even the particle can go to ∞ (i. e., unbound or dissociated particle) and the E is continuous, but the movement of particle in gravitational field does not satisfy energy conservation law.

1.9 Big Theorem of Energy

Combining Theorems XXXXIV, XXXI, XXXIX, XXIV, and corollary XVI yields the following theorem:

Theorem XXXXV: Big theorem of energy:
(1). The energy in an isolated (=closed) system is conservative.
(2). Although in an isolated system the system is in stationary state, if the system can be in distinct stationary states, then energy of the system can be nonconservative. This process does not be a typical quantum transition processes because it happens in an isolated system. Thus, we call it quasi-transition process.
(3). The energy in an open system is not conservative.

proof:
(1). The isolated system is of time uniformity, and thus theorem XXIV has proved this part of this theorem.

This theorem corresponds to the energy conservation law in classical thermodynamics and statistical mechanics. However, the quantum mechanics gives the proof, and thus becomes a theorem in quantum mechanics other than a law.

Next, let me give a more detail analyses. For an isolated system, the Hamiltonian is independent of time, nd thus the system is in stationary state. In this case we should understand that what does the average value of Hamiltonian mean?

Suppose that the system can be in many stationary states. The general form of this state is:

$$\Psi(\mathbf{r},t)=\sum_{n=1}^{N}\psi_{E_n}(\mathbf{r})e^{iE_nt/\hbar}. \qquad (1.9.1)$$

The average value of Hamiltonian is

$$<\Psi|H|\Psi>=(\mathbf{r},t)=\sum_{n=1}^{N}\psi_{E_n}(\mathbf{r})e^{iE_nt/\hbar}. \qquad (1.9.1)$$

(2). Suppose that there is a big system, and that this big system is an isolated system. However, if we separate this big system into two parts: one is still called system, and another part is called outside, which gives the system more or less perturbations, such as thermal conduction, light radiation, or any other action on the system, then the system will definitely produce quantum transition. Theorems XXXXI, XXXIX, and corollary XVI have proved that the energy in quantum transition process is more or less nonconservative.

However, the readers might feel that the energy nonconservation in the so-called quasi-transition processes should still be proved directly. Let me give two examples:

(i). The dissociated state or unbound state. Theorem XXXXIV had proved the energy nonconservation.

(ii). The bound state. Would readers please refer to the subsection 4.5.2 in Ref. [3]. This subsection gives a very detail calculations for the coldneutron on ground. The ground for the coldneutron is an infinitely large potential. Thus the coldneutron states are bound states. When the coldneutron makes falling motion. In this processes, the energy of coldneutron is not conservative. This should be a ordinary problem in quantum mechanics.

(3). If a system is an open system, then this means that this system suffers the perturbation from outside. This is typical quantum transition problem. Theorems XXXXI, XXXIX, and corollary XVI have proved that the energy in quantum transition process is more or less nonconservative.

1.10 Big Theorem of Wavepacket

Combining Theorems VI, IX-1, IX-2, IX-3, VII, X, and wavepacket law yields the following theorem:

Theorem XXXXVI: Big theorem of wavepacket:

(1). Any wavepacket, including free and bound, and massy and massless object, consists of three factors: guide-factor, front-factor, and body-factor.

(2). Quantum mechanics itself cannot decide if the wavepacket spread or not. Or, equivalently, spreading or non-spreading are determined by the inherent property of any object and/or its interaction with the environment.

proof:

(1). Theorems VI, IX-1, IX-2, IX-3 tell us that a free wavepacket consists of the three parts. Theorem X is Schrödinger equation theorem, which tells us that in a potential field the guide-factor will change, but the front-factor and body-factor do not change a wee bit.

(2). The guide-factor itself of an object is only a mathematical function. According to the wavepacket law, the action of a guide-factor is to guide the motion of the body-factor according to the spontaneous equiprobability symmetry breaking rule. However, the form of the guide-factor is determined by the interaction between the potential and the wavepacket.

We hope that readers do not think that the significance of the part 2 of theorem XXXXVI is equal to zero. Actually, until now all references and textbooks think that even a free electron will spread to infinite space in 10^{-35} second!

Chapter 2

Second Quantization Method

The best method to study the identical multiparticle system is to use second quantization method. In essence, the so-called second quantization is equivalent to occupation number representation (=occupation number space) instead of, for example, the momentum representation and coordinate representation. That is, the second quantization is not a new quantization, like to energy quantization from classical to quantum mechanics. If we treat a system with finite freedoms, then the second quantization does not bring big convenience to us. If there are infinite number of particles, then the system is of enumerable infinite dimensions. In nonrelativistic quantum mechanics the second quantization is already a much convenient method to treat the identical multiparticle system with infinite number of particles. Note that the particle creation and annihilation operators, introduced in second quantization method, do not have any essential action. In quantum field theory, the second quantization method is extended to continuous infinite dimensional system, which is called field, and is called field quantization.

Due to considering special theory of relativity, the particle creation and annihilation operators, introduced in second quantization method, do have essential action. Readers can find from appendix that even relativistic classical mechanics requires existence of the antiparticles.

The aims of this chapter are to introduce second quantization method in nonrelativistic quantum mechanics, and to make preparations for interaction between electron and phonon in chapter 3, for the quantum relaxation in chapter 5, and for the quantum field theory in chapter 9.

For our aims, we just explain the basic concepts on the second quantization, make some comments for some references and textbooks, give some basic formulas, and neglect tedious derivations.

2.1 Introduction

From Ref. [3] and chapter 1, we know that for an identical multiparticle system we have a theorem of identity. According to this theorem the important quantities of an identical multiparticle system are:

(A). The intrinsic properties of the particle, especially, the value of spin quantum number;

(B). The eigenvalues of permutation operator, which determine the symmetry of multiparticle wavefunction;

(C). How much particles are in every single state;

(D). On the identity of intrinsic properties of particles, we should note that in both classical and quantum mechanics the identical particles are distinguishable. Identity$\neq$Indistinguishability.

(For example, we can distinguish them by the locations in the position space.) For an identical multiparticle system it is not need to introduce the so-called postulate of indistinguishability of identical particles. However, AUNRTB (all until now references and text books) introduce the so-called principle of indistinguishability. In Ref. [3], we have negatived this principle in principle.

Based on the identity theorem, this chapter will introduce the second quantization method. To show that the second quantization method is equivalent to particle number representation, which is the same as any other representations, such as momentum or coordinate or energy representations, we just need to take one- and two-body operators as examples.

The theorem of identity tells us that to describe the multiparticle quantum state (Note that in our textbook the multiparticle quantum state, state, state vector, wavefunction, and guide-factor express the same thing) the most important two things are:

(A). Occupation numbers in every quantum state of single particle in the identical multiparticle system; Due to identity of particles, we do not need to indicate which particle occupies the single particle $|k_i>$, which one occupies the state $|k_j>$ and so on, although we can distinguish them. As soon as all the properties of the single particle states are known, the occupation number representation contains full information, what we need actually. In deed, any multiparticle state can be expanded by the complete set of these occupation number states with certain occupation numbers and then we can answer any question concerning the position distribution, momentum distribution, transition probabilities and so on. In an arbitrary multiparticle state expressed as a superposition of occupation number states, the average numbers of occupation of single states in general are non-integer.

(B). The symmetry of the multiparticle quantum state.

According to (A) and (B) we can use occupation number representation. For this purpose, we can use the creation and annihilation operators, used in harmonic oscillator problems (Refer to chapter 4 of Ref. [3].). We use k_i, where k denotes the complete quantum number set of the single particle state, for example $k = (\mathbf{p}, s_z)$, and we use the subscripts i to label the different eigenvalues of single particle state. We assume that here the single particle state vector $|k_i>$ are orthonormalized, i. e.,

$$<k_i|k_j>=\delta_{k_ik_j}\equiv\delta_{ij}. \tag{2.1.1}$$

Although this restriction can also be lifted because it is possible to work with nonorthogonal sets as well, but this textbook uses the condition Eq. (2.1.1). The Kronecker symbol in Eq. (2.1.1), as usual, has to be understood as the Dirac δ function for continuous numbers k_i. Of course, in practice, some sets k_i can be more convenient than others. Sometimes, especially in application to finite systems (atoms, molecules, nuclei), the single state $|k_i>$ will be called orbits or orbitals.

Define a multiparticle state vector as

$$|\{n_i\}>\equiv|n_1, n_2, \cdots, n_i, \cdots>, \tag{2.1.2}$$

where the n_i specify the occupation number of single particle wavefunctions with eigenvalue k_i. According to the type of symmetry of the multiparticle system, n_i could be only zero or one for antisymmetrical multiparticle wavefunction, and any nonnegative integers for symmetrical multiparticle wavefunction. We require that the new formalism, i. e., the occupation number representation, would automatically ensure this requirement.

According to the definition of n_i, one obtains the total particle number of the multiparticle system

$$N = \sum_i n_i. \tag{2.1.3}$$

The summation in Eq. (2.1.3) is for all the occupation numbers in all the possible single particle states. The only variables in the occupation number representation based on the single particle basis $|k_i>$ are the integers n_i. The quantummechanical operators have to be defined as acting on these "coordinates" (= particle's numbers in single particle eigenstate with eigenvalue k_i).

For this purpose, the appropriate tools are the creation and annihilation operators. We already have such an experience: for harmonic oscillation quanta, this apparatus was discussed in detail in chapter 4 in Ref. [3]. The creation and annihilation operators allow us to work in the occupation number representation where the states are characterized by the numbers of quanta in different modes, which are determined by single particle quantum state. In quantum mechanics, it is sufficient to understand the second quantization as a purely mathematical transformation from the representation where the basis vectors are the usual Schrödinger multiparticle wavefunction $\Psi(1, 2, \cdots, N)$ with the digital arguments combining all single particle variables to the basis of the multiparticle states in Eq. (2.1.2) in the occupation number representation. It is only in quantum field theory that the second quantization acquires deeper meaning.

To use occupation number representation (=second quantization representation) to describe the multiparticle wavefunction in fact begins to lay a foundation for quantum field theory, reexamines the way in which we define the multiparticle state. The "second quantization" can be called also "occupation number representation" or, simply, "particle number or Fock representation" or "Fock space" or "particle number or occupation number space".

In Fock space the particle numbers become variable. Therefore, Hilbert space for a given particle number is only a sector in a large Fock space that includes sectors with different particle numbers of all particle numbers of all types. This becomes absolutely necessary in quantum field theory, where the particles can be created and annihilated.

There are some viewpoints on the second quantization. For example, Ref. [14]: "A word of caution is in order. Our particle number notation for state vectors itself needs an important assumption, namely that there indeed exists a basis of noninteracting states. Interactions between particles can in fact, in principle, affect very nature. Whether or not, starting on this assumption, we can make a self-consistent theory that itself accurately describes nature can only be tested by experiment."

However, our wavepacket-only theory does not think that the second quantization needs the "assumption" mentioned in Ref. [14]. A identical particle system consists of many single wavepackets such as electron wavepackets. Although every wavepacket has its own guide-factor, but the collection of all single guide-factors needs to obey the identity theorem, and the identical multiparticle system needs to obey the guide of the multiparticle wavefunction (i. e., multiparticle guide-factor). The multiparticle wavefunction can be described by momentum representation, and, equivalently, by particle number representation.

2.2 Bosons

The procedure of the second quantization depends on the type of symmetry of identical multiparticle system. For bosons, the formalism practically coincides with what we have used for harmonic oscillators.

Let us now build a framework for a theory of identical multiboson systems using states in occupation number representation. The first of these is

$$|0, 0, \cdots, 0, \cdots >\equiv |\mathbf{0}>, \tag{2.2.1}$$

for which there are no particles in any single particle states. This state is called vacuum and is, as usual, normalized to unity. The second special case is

$$|0, 0, \cdots, n_i = 1, 0, \cdots >\equiv |k_i>, \tag{2.2.2}$$

which is the state in which there is exactly one particle in the state with eigenvalue k_i. Of course, this is just the single particle state that has dominated our discussion of quantum mechanics.

Now we need to learn how to build an identical multiboson states and then make sure that this building process respects permutation symmetry. In an obvious nod to the creation and annihilation operators that we first encountered in subsection 4.4.4 of chapter 4 in Ref. [3], we define a creation operator a_i^+ that increases by one the number of particles in the state with eigenvalue k_i, i. e.,

$$a_i^+|n_1, n_2, \cdots, n_i, \cdots >= \sqrt{(n_i+1)}|n_1, n_2, \cdots, n_i+1, \cdots > . \tag{2.2.3}$$

We define an annihilation operator a_i that annihilates by one the number of particles in the state with eigenvalue k_i, i. e.,

$$a_i|n_1, n_2, \cdots, n_i, \cdots >= \sqrt{n_i}|n_1, n_2, \cdots, n_i-1, \cdots > . \tag{2.2.4}$$

We can also define the occupation number operator

$$\widehat{n}_i = a_i^+ a_i, \tag{2.2.5}$$

for each single particle state k_i. The numbers n_i in Eq. (2.1.2) are the eigenvalues of the operators $\widehat{n}_i$ in Eq. (2.2.5). The eigenvalues of $\widehat{n}_i$ are nonnegative integers.

Theorem XXXXVII: Boson commutation relation: Boson creation and annihilation operators obey commutation relation.

Proof:

Using the method of action of operator on the state vector, one can obtain the following commutation rules:

$$[a_i, a_j] = [a_i^+, a_j^+] = 0, \quad [a_i, a_j^+] = \delta_{ij} \ . \tag{2.2.6}$$

The commutation relations in Eq. (2.2.6) are called commutation relation. QED.

The total particle number operator is

$$\widehat{N} = \sum_i \widehat{n}_i. \tag{2.2.7}$$

Using Eq. (2.2.4), one can show

$$|\{n_i\}>= \prod_i \frac{1}{\sqrt{n_i!}}(a_i^+)^{n_i}|\mathbf{0}> . \tag{2.2.8}$$

All annihilation operators give zero acting onto the vacuum state. By virtue of the commutation relation in Eq. (2.2.6), all creation operators in Eq. (2.2.8) are permutable. This means that the multiparticle wavefunction is symmetric with respect to the permutations.

The state vector in Eq. (2.2.8) is the eigenstate of occupation number operator $\widehat{n}_i$, and the eigenvalue is n_i. The state vector in Eq. (2.2.8) is also the eigenstate of multiparticle number operator $\widehat{N}$, and the eigenvalue is $\sum_i n_i = N$. The state vectors in Eq. (2.2.8) are the orthonormalized complete basis in occupation number representation.

2.3 Fermions

Theorem XXXXVIII: Fermion anticommutation relation: Fermion creation and annihilation operators obey anticommutation relation.

Proof:
Pauli exclusion theorem in chapter 1 tells us that the fermion particle number in any single particle state is 1 or zero. The multifermion state in occupation number representation is formally still given by Eq. (2.1.2). Introducing the creation operators, Eq. (2.1.2) can be written as

$$|\{n_i\}>=a_1^+a_2^+\cdots|\mathbf{0}> . \tag{2.3.1}$$

Considering the permutation antisymmetry, if $n_1=n_2=1$, then we have

$$|n_1,n_2,n_3,\cdots,n_i,\cdots>=-|n_2,n_1,n_3,\cdots,n_i,\cdots> \qquad (2.3.2),$$

that is,

$$a_1^+a_2^+|n_3,\cdots,n_i,\cdots>=-a_2^+a_1^+|n_3,\cdots,n_i,\cdots>$$

or

$$(a_1^+a_2^+ + a_2^+a_1^+)|n_3,\cdots,n_i,\cdots>=0. \tag{2.3.3}$$

Because $|n_3,\cdots,n_i,\cdots>$ is arbitrary, we have generally

$$[a_1^+,a_2^+]_+=0$$

or, more generally $i\neq j$,

$$[a_i^+,a_j^+]_+=0. \tag{2.3.4}$$

If $i=j$, then

$$a_i^+a_j^+=0. \tag{2.3.5}$$

Eq. (2.3.5) indicates that two fermions cannot be in the same single particle state, which is the mathematical expression of Pauli exclusion theorem. (Note that Pauli exclusion theorem has been proved by the antisymmetry of multiparticle wavefunction of fermion system.) The adjoint state of Eq. (2.3.1) is

$$<\{n_i\}|=<\mathbf{0}|\cdots a_2a_1. \tag{2.3.6}$$

Similarly to Eq. (2.3.4), we have

$$[a_i,a_j]_+\equiv a_ia_j+a_ja_i=0. \tag{2.3.7}$$

Next, let us prove

$$[a_i,a_j^+]_+=\delta_{ij}. \tag{2.3.8}$$

Ar first, $i=1,j=2$ and $n_2=1\neq 0$.

$$\begin{aligned} a_2|\{n_i\}>=a_2a_1^+a_2^+a_3^+a_4^+\cdots|\mathbf{0}> &= -a_2a_2^+a_1^+a_3^+a_4^+\cdots|\mathbf{0}>=-a_1^+a_3^+a_4^+\cdots|\mathbf{0}> \\ &= -a_1^+a_2a_2^+a_3^+a_4^+\cdots|\mathbf{0}> . \end{aligned}$$

Therefore,

$$[a_2,a_1^+]_+=0. \tag{2.3.9}$$

In case of $i=1,j=2$ and $n_2=0$, we can use the similar method to obtain Eq. (2.3.8).

In case of $i=1,j=1$ and $n_1=1$, using Eq. (2.3.4), we have

$$a_1a_1^+|\{n_i\}>=a_1a_1^+a_1^+a_2^+a_3^+a_4^+\cdots|\mathbf{0}>=0. \tag{2.3.10}$$

$$a_1^+ a_1|\{n_i\} >= a_1^+ a_1 a_1^+ a_2^+ a_3^+ a_4^+ \cdots |\mathbf{0}>= a_1^+ a_2^+ a_3^+ a_4^+ \cdots |\mathbf{0}> . \quad (2.3.11)$$

From Eqs. (2.3.10) and (2.3.11) we have

$$(a_1 a_1^+ + a_1^+ a_1)|\{n_i\} >= |\{n_i\}> . \quad (2.3.12)$$

In case of $i = 1, j = 1$ and $n_1 = 0$, similarly, we have Eq. (2.3.12) as well.

In summary, we have

$$[a_i, a_j]_+ = [a_i^+, a_j^+]_+ = 0, \quad [a_i, a_j^+]_+ = \delta_{ij} \ . \quad (2.3.13)$$

Quantities satisfying the anticommutation relation, Eq. (2.3.13), are called Grassman variables. Eq. (2.3.13) constructs a Grassman algebra, which contains all the algebraic properties of the creation and annihilation operators of the fermions. QED.

Similarly to the case of bosons in Eq. (2.2.3) and (2.2.4), for fermions, we have if $n_i = 0, 1$, then

$$a_i^+|n_1, \cdots, n_i, \cdots >= (-1)^{\sum_{\nu=1}^{i-1} n_\nu}\sqrt{1-n_i}|n_1, \cdots, n_i+1, \cdots > . \quad (2.3.14)$$

Because that: (a). a_i^+ needs to at first go through the operators $a_1^+ a_2^+ \cdots a_{i-1}^+$, then a_i^+ can act on the n_i; (b). The creation operator of Fermions obeys anticommutation relation, there appears a factor

$$(-1)^{n_1+n_2+\cdots+n_{i-1}} = (-1)^{\sum_{\nu=1}^{i-1} n_\nu}. \quad (2.3.15)$$

The Eq. (2.3.14) can be rewritten as

$$a_i^+|n_1, \cdots, n_i, \cdots >= (-1)^{\sum_{\nu=1}^{i-1} n_\nu}\sqrt{1-n_i}|n_1, \cdots, 1_i, \cdots > \delta_{n_i 0}. \quad (2.3.16)$$

Similarly,

$$a_i|n_1, \cdots, n_i, \cdots >= (-1)^{\sum_{\nu=1}^{i-1} n_\nu}\sqrt{n_i}|n_1, \cdots, 0_i, \cdots > \delta_{n_i 1}. \quad (2.3.17)$$

Owing to the importance of the Fermion operator description in many double-energy-level system studies, we treat here this problem.

Consider a two-energy-level system, the Hamiltonian eigenvalue equation of which is

$$H|i>= E_i|i> \quad with \ \ i = 0, 1 \quad (2.3.18)$$

which $|1>$ (single particle occupation) and $|0>$ (zero particle occupation) represent the ground and excited state, respectively. We have

$$\hat{N}|1>= a^+a|1>= 1|1> \quad and \quad \hat{N}|0>= a^+a|0>= 0|0>, \quad (2.3.19)$$

which can be viewed as eigenvalue equation.

It is now of interest to give matrix representations of the Fermion (creation and annihilation) operators. For this purpose, it is convenient to represent the two orthogonal kets $|1>$ and $|0>$, by two orthonarmalized column vectors of dimension 2 according to

$$|0>= \begin{pmatrix} 0 \\ 1 \end{pmatrix} \quad and \quad |1>= \begin{pmatrix} 1 \\ 0 \end{pmatrix}, \quad (2.3.20)$$

which satisfy the orthonormality properties, since one obtains, respectively,

$$<1|1>= (1,0)\begin{pmatrix} 1 \\ 0 \end{pmatrix} = 1$$

$$< 0|0 >= (0,1)\begin{pmatrix} 0 \\ 1 \end{pmatrix} = 1$$

$$< 1|0 >= (1,0)\begin{pmatrix} 0 \\ 1 \end{pmatrix} = 0$$

If we choose the matrix representations of a and a^+ as

$$a = \begin{pmatrix} 0 & 0 \\ 1 & 0 \end{pmatrix} \quad and \quad a^+ = \begin{pmatrix} 0 & 1 \\ 0 & 0 \end{pmatrix}, \tag{2.3.21}$$

then we can verify that these matrix represents have all the properties of operators a and a^+. We list as follows:

$$\begin{aligned} [a, a^+] &= \begin{pmatrix} 0 & 0 \\ 1 & 0 \end{pmatrix}\begin{pmatrix} 0 & 1 \\ 0 & 0 \end{pmatrix} + \begin{pmatrix} 0 & 1 \\ 0 & 0 \end{pmatrix}\begin{pmatrix} 0 & 0 \\ 1 & 0 \end{pmatrix} \\ &= \begin{pmatrix} 0 & 0 \\ 0 & 1 \end{pmatrix} + \begin{pmatrix} 1 & 0 \\ 0 & 0 \end{pmatrix} = \begin{pmatrix} 1 & 0 \\ 0 & 1 \end{pmatrix} \end{aligned}$$

$$[a, a] = \begin{pmatrix} 0 & 0 \\ 1 & 0 \end{pmatrix}\begin{pmatrix} 0 & 0 \\ 1 & 0 \end{pmatrix} + \begin{pmatrix} 0 & 0 \\ 1 & 0 \end{pmatrix}\begin{pmatrix} 0 & 0 \\ 1 & 0 \end{pmatrix} = \begin{pmatrix} 0 & 0 \\ 0 & 0 \end{pmatrix}$$

$$[a^+, a^+] = \begin{pmatrix} 0 & 1 \\ 0 & 0 \end{pmatrix}\begin{pmatrix} 0 & 1 \\ 0 & 0 \end{pmatrix} + \begin{pmatrix} 0 & 1 \\ 0 & 0 \end{pmatrix}\begin{pmatrix} 0 & 1 \\ 0 & 0 \end{pmatrix} = \begin{pmatrix} 0 & 0 \\ 0 & 0 \end{pmatrix}$$

$$\begin{aligned} \hat{N}|1> &= \begin{pmatrix} 0 & 1 \\ 0 & 0 \end{pmatrix}\begin{pmatrix} 0 & 0 \\ 1 & 0 \end{pmatrix}\begin{pmatrix} 1 \\ 0 \end{pmatrix} \\ &= \begin{pmatrix} 1 & 0 \\ 0 & 0 \end{pmatrix}\begin{pmatrix} 1 \\ 0 \end{pmatrix} = \begin{pmatrix} 1 \\ 0 \end{pmatrix} = |1> \end{aligned}$$

$$\hat{N}|0 >= \begin{pmatrix} 1 & 0 \\ 0 & 0 \end{pmatrix}\begin{pmatrix} 0 \\ 1 \end{pmatrix} = \begin{pmatrix} 0 \\ 0 \end{pmatrix} = 0|0 >$$

$$a|1 >= \begin{pmatrix} 0 & 0 \\ 1 & 0 \end{pmatrix}\begin{pmatrix} 1 \\ 0 \end{pmatrix} = \begin{pmatrix} 0 \\ 1 \end{pmatrix} = |0 >$$

$$a|0 >= \begin{pmatrix} 0 & 0 \\ 1 & 0 \end{pmatrix}\begin{pmatrix} 0 \\ 1 \end{pmatrix} = \begin{pmatrix} 0 \\ 0 \end{pmatrix} = 0$$

$$a^+|1 >= \begin{pmatrix} 0 & 1 \\ 0 & 0 \end{pmatrix}\begin{pmatrix} 1 \\ 0 \end{pmatrix} = \begin{pmatrix} 0 \\ 0 \end{pmatrix} = 0$$

$$a^+|0 >= \begin{pmatrix} 0 & 1 \\ 0 & 0 \end{pmatrix}\begin{pmatrix} 0 \\ 1 \end{pmatrix} = \begin{pmatrix} 1 \\ 0 \end{pmatrix} = |1 >$$

2.4 One- and Two-Body Operators of Bosons

To apply the developed second quantization formalism to physical problems, we need to express all operators of observables in the second quantization form. We often meet in practice with the one- and two-body operators.

2.4.1 One-Body Operators of Bosons

The general form of one-body operator is

$$\widehat{F}^{(1)} = \sum_{n'=1}^{N} \widehat{f}^{(1)}(q_{n'}), \tag{2.4.1.1}$$

where N denotes total number of particles, $n' = 1, 2, 3, \cdots$ are the label of particles, $\widehat{f}^{(1)}(q_{n'})$ denotes an one-body operator, and $\widehat{F}^{(1)}$ can be the system's total momentum, angular momentum, kinetic energy, magnetic moment etc.

We can prove that the second quantization form of $\widehat{F}^{(1)}$ is

$$\widehat{F}^{(1)} = \sum_{ij} f_{ij}^{(1)} a_i^+ a_j, \tag{2.4.1.2}$$

where

$$f_{ij}^{(1)} \equiv (\psi_{k_i}(q), \widehat{f}^{(1)}(q)\psi_{k_j}(q)). \tag{2.4.1.3}$$

$f_{ij}^{(1)}$ is the matrix element of one-body operator between single particle states ψ_{k_i} and ψ_{k_j}. The method to prove Eq. (2.4.1.2) is as follows:

Multiparticle wavefunction of N identical bosons, which is

$$\Psi^S = \Psi^S_{n_1 n_2 \cdots n_N}(q_1, q_2, \cdots, q_N) = \sqrt{\frac{\prod_{i=1}^{N} n_i!}{N!}} \sum_{\nu} P_\nu[\psi_{k_1}(q_1)\psi_{k_2}(q_2), \cdots, \psi_{k_N}(q_N)], \tag{2.4.1.4}$$

where q_i represents both the i-th boson's position space coordinates and spin.

In the occupation number representation (second quantization representation) Ψ^S is a state vector:

$$|n_1 n_2 \cdots n_i \cdots n_j \cdots >, \tag{2.4.1.5}$$

where the subscript i in n_i denotes k_i. We can demonstrate that matrix elements of both the $\widehat{F}^{(1)}$ in Eq. (2.4.1.2) on the state vector in Eq. (2.4.1.5) and the $\widehat{F}^{(1)}$ in Eq. (2.4.1.1) on the total wavefunction in Eq. (2.4.1.4) are the same. From this demonstration, we can confirm the equivalence of the q representations [for example, $q = (\mathbf{r}, s_z)$] and the occupation representation. The equivalence of the two representations is in the sense of matrix elements including the average values. Next, we neglect the derivation processes, and just give the mathematical relations to express the equivalence of both approaches.

(A). The non-diagonal matrix element of $\widehat{F}^{(1)}$ in the two representations:

$$\begin{aligned} & (\Psi^S_{\cdots(n_i+1)\cdots(n_j-1)\cdots}, \widehat{F}^{(1)}\Psi^S_{\cdots(n_i)\cdots(n_j)\cdots}) \\ = \; & < \cdots (n_j - 1) \cdots (n_i + 1) \cdots |\widehat{F}^{(1)}| \cdots (n_i) \cdots n_j \cdots > \\ = \; & \sqrt{(n_i + 1)n_j} f_{ij}^{(1)}. \end{aligned} \tag{2.4.1.6}$$

(B). The diagonal matrix element of $\widehat{F}^{(1)}$, i. e., the average value of $\widehat{F}^{(1)}$, in the two representation:

$$(\Psi^S_{n_1 n_2 \cdots}, \widehat{F}^{(1)}\Psi^S_{n_1 n_2 \cdots}) = < \cdots n_j \cdots n_i \cdots |\widehat{F}^{(1)}| \cdots n_i \cdots n_j \cdots > = \sum_i n_i f_{ii}^{(1)}. \tag{2.4.1.7}$$

2.4.2 Two-Body Operators of Bosons

The general form of two-body operator is

$$\widehat{F}^{(2)} = \sum_{i<j}^{N} \widehat{f}^{(2)}(q_i, q_j), \tag{2.4.2.1}$$

where N denotes total number of particles, $i, j = 1, 2, 3, \cdots$ are the labels of two particles, $\widehat{f}^{(2)}(q_i, q_j)$ denotes a two particle operator, such as the interaction between two particles $n' = i$ and $n' = j$. Generally, $\widehat{f}^{(2)}(q_i, q_j) = \widehat{f}^{(2)}(q_j, q_i)$. $\widehat{F}^{(2)}$ are the sum of all the two-body operators in boson system.

We can prove that the second quantization form of $\widehat{F}^{(2)}$ is

$$\widehat{F}^{(2)} = \frac{1}{2} \sum_{i'j'ij} f^{(2)}_{i'j',ji} a^+_{i'} a^+_{j'} a_j a_i, \tag{2.4.2.2}$$

where the summations are over the single particle's states.

$$f^{(2)}_{i'j',ji} \equiv (\psi_{k_{i'}}(q_1)\psi_{k_{j'}}(q_2), \widehat{f}^{(2)}(q_1, q_2)\psi_{k_j}(q_2)\psi_{k_i}(q_1)). \tag{2.4.2.3}$$

$f^{(2)}_{i'j',ji}$ is the matrix element of two-body operator between single particle state products $\psi_{k_{i'}}(q_1)\psi_{k_{j'}}(q_2)$ and $\psi_{k_j}(q_2)\psi_{k_i}(q_1)$.

We can demonstrate that matrix elements of both the $\widehat{F}^{(2)}$ in Eq. (2.4.2.2) on the state vector in Eq. (2.4.1.5) and the $\widehat{F}^{(2)}$ in Eq. (2.4.2.1) on the total wavefunction in Eq. (2.4.1.4) are the same. From this demonstration, we can conclude that only in the sense of the matrix elements including average value of the total two-body operator, the q representation (for example,$q = (\mathbf{r}, s_z)$) and the occupation number representation are equivalent.

For the detail derivations refer to Ref. [15, 9, 16]. In actual applications, the derivations are not important. The important thing to use is the formula directly. Here we would like to emphasize a concept problem. Some references think that the success of second quantization demonstrates the importance of the so called principle of indistinguishability of identical particles. On the contrary, this textbook thinks that the demonstration for the equivalence of the occupation number representations and, for example, momentum representation are just based on the comparisons between matrix elements including the average value of two-body operators. Therefore, the concept on the so-called indistinguishability of identical particles, accepted by some references, is not necessary. This book still insists the viewpoint that the so-called indistinguishability purely belongs to an utterly unjustifiable.

2.5 One- and Two-Body Operators of Fermions

2.5.1 One-Body Operators of Fermions

The general form of one-body operator is

$$\widehat{F}^{(1)} = \sum_{n'=1}^{N} \widehat{f}^{(1)}(q_{n'}), \tag{2.5.1.1}$$

where N denotes total number of particles, $n' = 1, 2, 3, \cdots$ are the label of particles, $\widehat{f}^{(1)}(q_{n'})$ denotes an one-body operator, and $\widehat{F}^{(1)}$ can be the system's total momentum, angular momentum, kinetic energy, magnetic moment etc.

We can prove that the second quantization form of $\widehat{F}^{(1)}$ is the same as that of bosons.

$$\widehat{F}^{(1)} = \sum_{ij} f_{ij}^{(1)} a_i^+ a_j, \tag{2.5.1.2}$$

where

$$f_{ij}^{(1)} \equiv (\psi_{k_i}(q), \widehat{f}^{(1)}(q)\psi_{k_j}(q)). \tag{2.5.1.3}$$

$f_{ij}^{(1)}$ is the matrix element of one-body operator between single particle states ψ_{k_i} and ψ_{k_j}. The method to prove Eq. (2.5.1.3) is as follows:

The multiparticle wavefunction of N identical fermions, which is

$$\Psi^A = \Psi^A_{k_1k_2\cdots k_N}(q_1, q_2, \cdots) = \frac{1}{\sqrt{N!}} \begin{vmatrix} \psi_{k_1}(q_1) & \cdots & \psi_{k_1}(q_N) \\ \psi_{k_2}(q_1) & \cdots & \psi_{k_2}(q_N) \\ \cdots & \cdots & \cdots \\ \psi_{k_N}(q_1) & \cdots & \psi_{k_N}(q_N) \end{vmatrix}, \tag{2.5.1.4}$$

where the change in sign of the multiparticle wavefunction under permutation of any two fermions follows immediately from the change of sign of a determinant when two of its columns are interchanged. In Eq. (2.5.1.4) q_i represents the i-th fermion's position space coordinates and spin.

In the occupation number representation (second quantization representation) Ψ^A is a state vector, and does not need to point out $q_1, q_2, \cdots$. Considering the Pauli exclusion theorem, every single state can only occupy one or zero particle, the state vector of an identical fermion system will just write $n_i = 1$ etc. Thus we have

$$\begin{aligned} & |n_1 = 1, n_2 = 1, \cdots n_i = 1 \cdots n_j = 1 \cdots > \\ \equiv\ & |1_1 1_2 \cdots 1_i \cdots 1_j \cdots > \\ \equiv\ & |11 \cdots 1 \cdots 1 \cdots > \equiv a_1^+ a_2^+ \cdots a_i^+ \cdots a_j^+ \cdots |\mathbf{0}>, \end{aligned} \tag{2.5.1.5}$$

where the subscript i in n_i denotes k_i, which is the eigenvalue of single particle's state.

We can demonstrate that matrix elements of both the $\widehat{F}^{(1)}$ in Eq. (2.5.1.2) on the state vector in Eq. (2.5.1.5) and the $\widehat{F}^{(1)}$ in Eq. (2.5.1.1) on the total wavefunction in Eq. (2.5.1.4) are the same. From this demonstration, we can confirm the equivalence of the q representations [for example, $q = (\mathbf{r}, s_z)$] and the occupation representation. The equivalence of the two representations is only in the sense of average values and matrix elements. Next, we neglect the derivation processes, and just give the mathematical relations to express the equivalence of two representations. Note that we can only prove that in the sense of average value and matrix elements the second quantization representation is equivalent to the other representation.

(A). The diagonal matrix element of $\widehat{F}^{(1)}$, which is the average value of $\widehat{F}^{(1)}$:

$$(\Psi^A_{k_1k_2\cdots}, \widehat{F}^{(1)}\Psi^A_{k_1k_2\cdots}) = < \cdots 1_j \cdots 1_i \cdots |\widehat{F}^{(1)}| \cdots 1_i \cdots 1_j \cdots > = \sum_i n_i f_{ii}^{(1)}. \tag{2.5.1.6}$$

(B). The non-diagonal matrix element of $\widehat{F}^{(1)}$ (0_j represents the occupation number in single particle state $\psi_{k_j}(q_{j'})$) $n_j = 0$) ($i < j$):

$$\begin{aligned} & (\Psi^A_{\cdots 1_i \cdots 0_j)\cdots}, \widehat{F}^{(1)}\Psi^A_{\cdots 0_i \cdots 1_j \cdots}) \\ =\ & < \cdots 0_j \cdots 1_i \cdots |\widehat{F}^{(1)}| \cdots 0_i \cdots 1_j \cdots > \\ =\ & (-1)^{\sum_{i'=i+1}^{j-1} n_{n'}} f_{ij}^{(1)}. \quad (i < j) \end{aligned} \tag{2.5.1.7}$$

2.5.2 Two-Body Operators

The general form of two-body operator is

$$\widehat{F}^{(2)} = \sum_{n'_i<n'_j}^{N} \widehat{f}^{(2)}(q_{n'_i}, q_{n'_j}), \tag{2.5.2.1}$$

where N denotes total number of particles, $n'_i, n'_j = 1, 2, 3, \cdots$ are the labels of two particles, $\widehat{f}^{(2)}(q_{n'_i}, q_{n'_j})$ denotes an two particle operator, such as the interaction between two particles n'_i and n'_j. Generally, $\widehat{f}^{(2)}(q_i, q_j) = \widehat{f}^{(2)}(q_j, q_i)$. $\widehat{F}^{(2)}$ are the sum of all the two-body operators in fermion system.

We can prove that the second quantization form of $\widehat{F}^{(2)}$ is

$$\widehat{F}^{(2)} = \frac{1}{2}\sum_{i'j'ij} f^{(2)}_{i'j',ji} a^{+}_{i'} a^{+}_{j'} a_j a_i, \tag{2.5.2.2}$$

where the summations are over the single particle's states. Apparently, Eq. (2.5.2.2) is the same as Eq. (2.4.2.2). However, the fermion's anticommutation relations of creation and annihilation operators distinguish from the boson's.

$$f^{(2)}_{i'j',ji} \equiv (\psi_{k_{i'}}(q_1)\psi_{k_{j'}}(q_2), \widehat{f}^{(2)}(q_1, q_2)\psi_{k_j}(q_2)\psi_{k_i}(q_1)). \tag{2.5.2.3}$$

$f^{(2)}_{i'j',ij}$ is the matrix element of two-body operator between single particle state products $\psi_{k_{i'}}(q_1)\psi_{k_{j'}}(q_2)$ and $\psi_{k_j}(q_2)\psi_{k_i}(q_1)$.

We can demonstrate that matrix elements of both the $\widehat{F}^{(2)}$ in Eq. (2.5.2.2) on the state vector in Eq. (2.4.1.5) and the $\widehat{F}^{(2)}$ in Eq. (2.5.2.1) on the total wavefunction in Eq. (2.5.1.4) are the same. From this demonstration, we can conclude that in the sense of the average values and matrix elements the q representation (for example,$q = (\mathbf{r}, s_z)$) and the occupation number representation are equivalent.

2.6 Field Operators

2.6.1 Commutation Relations

Suppose that there is a system, consisting of identical spin 1/2 fermions. The single particle states are the common eigenstates of momentum operator $\mathbf{p}$ and spin operator S_z. We use box normalization, and the volume of the box is V. The creation and annihilation operators are $a^{+}_{\mathbf{p}s_z}$ and $a_{\mathbf{p}s_z}$, respectively. Define the following two operators:

$$\Psi^{+}(\mathbf{r}, s_z) = \sum_{\mathbf{p}} \frac{1}{\sqrt{V}} e^{-i\mathbf{p}\cdot\mathbf{r}/\hbar} a^{+}_{\mathbf{p}s_z}. \tag{2.6.1.1.a}$$

$$\Psi(\mathbf{r}, s_z) = \sum_{\mathbf{p}} \frac{1}{\sqrt{V}} e^{i\mathbf{p}\cdot\mathbf{r}/\hbar} a_{\mathbf{p}s_z}. \tag{2.6.1.1.b}$$

We can show in the following that $\Psi^+(\mathbf{r}, s_z)$ $[\Psi(\mathbf{r}, s_z)]$ represents the creation (annihilation) operator of a particle at $\mathbf{r}$ and with spin 1/2.

$$\begin{aligned} < \mathbf{r}', s'_z|\Psi^+(\mathbf{r}, s_z)|\mathbf{0}> &= < \mathbf{r}', s'_z \left| \sum_{\mathbf{p}} \frac{1}{\sqrt{V}} e^{-i\mathbf{p}\cdot\mathbf{r}/\hbar} a^+_{\mathbf{p}s_z} \right| \mathbf{0} > \\ &= \sum_{\mathbf{p}} \frac{1}{\sqrt{V}} e^{-i\mathbf{p}\cdot\mathbf{r}/\hbar} < \mathbf{r}', s'_z|a^+_{\mathbf{p}s_z}> \\ &= \sum_{\mathbf{p}} \frac{1}{\sqrt{V}} e^{-i\mathbf{p}\cdot\mathbf{r}/\hbar} \frac{1}{\sqrt{V}} e^{i\mathbf{p}\cdot\mathbf{r}'/\hbar} \delta_{s_z s'_z} \\ &= \delta(\mathbf{r}-\mathbf{r}')\delta_{s_z s'_z}. \end{aligned} \quad (2.6.1.2)$$

Therefore, $\Psi^+(\mathbf{r}, s_z)$ acts on the vacuum state vector, produces a single particle at $(\mathbf{r}, s_z)$, and then this single particle state projects into a single particle state $|\mathbf{r}', s'_z>$ produces $\delta(\mathbf{r}-\mathbf{r}')\delta_{s_z s'_z}$. $\Psi^+(\mathbf{r}, s_z)$ $(\Psi(\mathbf{r}, s_z))$ represents the particle's creation (annihilation) operator in position space. The position space means field. Thus, we name the $\Psi^+(\mathbf{r}, s_z)$ and $\Psi(\mathbf{r}, s_z)$ field operators.

For the particle without spin, we define

$$\Psi^+(\mathbf{r}) = \sum_{\mathbf{p}} \frac{1}{\sqrt{V}} e^{-i\mathbf{p}\cdot\mathbf{r}/\hbar} a^+_{\mathbf{p}}. \quad (2.6.1.3.a)$$

$$\Psi(\mathbf{r}) = \sum_{\mathbf{p}} \frac{1}{\sqrt{V}} e^{i\mathbf{p}\cdot\mathbf{r}/\hbar} a_{\mathbf{p}}. \quad (2.6.1.3.b)$$

We show in the following that $\Psi^+(\mathbf{r})$ $[\Psi(\mathbf{r})]$ represents the creation (annihilation) operator of a particle at $\mathbf{r}$.

Utilizing the commutation relation $[a_{\mathbf{p}}, a^+_{\mathbf{p}'}] = \delta_{\mathbf{p}\mathbf{p}'}$ of bosons yields

$$\begin{aligned} [\Psi(\mathbf{r}), \Psi^+(\mathbf{r}')] &= \sum_{\mathbf{p}\mathbf{p}'} \frac{1}{V} e^{i(\mathbf{p}\cdot\mathbf{r}-\mathbf{p}'\cdot\mathbf{r}')/\hbar} [a_{\mathbf{p}}, a^+_{\mathbf{p}'}] \\ &= \sum_{\mathbf{p}} \frac{1}{V} e^{i\mathbf{p}\cdot(\mathbf{r}-\mathbf{r}')/\hbar} = \delta(\mathbf{r}-\mathbf{r}'). \end{aligned} \quad (2.6.1.4)$$

Utilizing the commutation relations $[a_{\mathbf{p}s_z}, a^+_{\mathbf{p}'s'_z}] = \delta_{\mathbf{p}\mathbf{p}'}\delta_{s_z s'_z}$ of spin 1/2 fermions yields

$$[\Psi(\mathbf{r}, s_z), \Psi^+(\mathbf{r}'s'_z)]_+ = \delta(\mathbf{r}-\mathbf{r}')\delta_{s_z s'_z}. \quad (2.6.1.5)$$

The inverse of Eq. (2.6.1.1) is

$$a^+_{\mathbf{p}s_z} = \int d\mathbf{r} \frac{1}{\sqrt{V}} e^{i\mathbf{p}\cdot\mathbf{r}/\hbar} \Psi^+(\mathbf{r}, s_z). \quad (2.6.1.6.a)$$

$$a_{\mathbf{p}s_z} = \int d\mathbf{r} \frac{1}{\sqrt{V}} e^{-i\mathbf{p}\cdot\mathbf{r}/\hbar} \Psi(\mathbf{r}, s_z). \quad (2.6.1.6.b)$$

The inverse of Eq. (2.6.1.3) is

$$a^+_{\mathbf{p}} = \int d\mathbf{r} \frac{1}{\sqrt{V}} e^{i\mathbf{p}\cdot\mathbf{r}/\hbar} \Psi^+(\mathbf{r}). \quad (2.6.1.7.a)$$

$$a_{\mathbf{p}} = \int d\mathbf{r} \frac{1}{\sqrt{V}} e^{-i\mathbf{p}\cdot\mathbf{r}/\hbar} \Psi(\mathbf{r}). \quad (2.6.1.7.b)$$

2.6.2 Time-Dependent Forms

For brevity, we discuss the case of time-independent potential. The stationary state Schrödinger equation is

$$\left(-\frac{\hbar^2}{2m}\nabla^2 + V(\mathbf{r})\right)\psi_k(\mathbf{r}) = E_k\psi_k(\mathbf{r}), \tag{2.6.2.1}$$

where k represents all the quantum number of a complete set of dynamical quantities. For convenience, in this subsection (2.6.2) we use k instead of the k_i in other sections of this chapter. In principle, k can contain momentum, z component of spin for a system of fermions or z component of isotopic spin for a system of nucleons. k can denote the state (Some books call the state the mode) of a single particle. The single particle states $\psi_k(\mathbf{r})$ construct an orthonormalized complete set:

$$\int d\mathbf{r}\psi_k^*(\mathbf{r})\psi_{k'}(\mathbf{r}) = \delta_{kk'}, \tag{2.6.2.2.a}$$

and

$$\sum_k \psi_k^*(\mathbf{r}')\psi_k(\mathbf{r}) = \delta(\mathbf{r}-\mathbf{r}'), \tag{2.6.2.2.b}$$

Define the following field operators:

$$\Psi^+(\mathbf{r},t) = \sum_k \psi_k^*(\mathbf{r})e^{iE_kt/\hbar}a_k^+. \tag{2.6.2.3.a}$$

$$\Psi(\mathbf{r},t) = \sum_k \psi_k(\mathbf{r})e^{-iE_kt/\hbar}a_k. \tag{2.6.2.3.b}$$

Using the orthonormality of $\{\psi_k\}$, one can show:

$$a_k = \int d\mathbf{r}\Psi(\mathbf{r},t)\psi_k^*(\mathbf{r})e^{iE_kt/\hbar}, \tag{2.6.2.4.a}$$

$$a_k^+ = \int d\mathbf{r}\Psi^+(\mathbf{r},t)\psi_k(\mathbf{r})e^{-iE_kt/\hbar}, \tag{2.6.2.4.b}$$

Note that according to these definitions a_k^+, a_k do not depend on time t.

From boson's commutation relation $[a_k, a_{k'}^+] = \delta_{kk'}$, we obtain the commutation relations of boson's field operators:

$$\begin{gathered}[\Psi(\mathbf{r},t), \Psi^+(\mathbf{r}',t)] = \delta(\mathbf{r}-\mathbf{r}') \\ [\Psi(\mathbf{r},t), \Psi(\mathbf{r}',t)] = 0 \\ [\Psi^+(\mathbf{r},t), \Psi^+(\mathbf{r}',t)] = 0.\end{gathered} \tag{2.6.2.5}$$

From fermion's anticommutation relation $[a_k, a_{k'}^+]_+ = \delta_{kk'}$ we obtain the commutation relations of fermion's field operators:

$$\begin{gathered}[\Psi(\mathbf{r},t), \Psi^+(\mathbf{r}',t)]_+ = \delta(\mathbf{r}-\mathbf{r}') \\ [\Psi(\mathbf{r},t), \Psi(\mathbf{r}',t)]_+ = [\Psi^+(\mathbf{r},t), \Psi^+(\mathbf{r}',t)]_+ = 0.\end{gathered} \tag{2.6.2.6}$$

Next, we can show that $\Psi^+(\mathbf{r},t)$ ($\Psi(\mathbf{r},t)$) represents creation (annihilation) of a particle at time t and spatial point $\mathbf{r}$.

At first we demonstrate that the action of $\Psi^+(\mathbf{r},t)$ on state vector in occupation number representation constructs a new state vector, which's total particle number is one more than the original state vector. Using the formula Eq. (2.6.3.2) in the next subsection that the total particle's operator $\widehat{N}$ satisfies $\widehat{N}|n_1n_2\cdots n_i\cdots>=\int d\mathbf{r}\Psi^+\Psi|n_1n_2\cdots n_i\cdots>=N|n_1n_2\cdots n_i\cdots>$, one obtains

$$\begin{aligned}
&\widehat{N}\left(\Psi^+(\mathbf{r},t)|n_1n_2\cdots n_i\cdots>\right)\\
=&\int d\mathbf{r}'\Psi^+(\mathbf{r}',t)\Psi(\mathbf{r}',t)\Psi^+(\mathbf{r},t)|n_1n_2\cdots n_i\cdots>\\
=&\int d\mathbf{r}'\Psi^+(\mathbf{r}',t)\left(\pm\Psi^+(\mathbf{r},t)\Psi(\mathbf{r}',t)+\delta(\mathbf{r}-\mathbf{r}')\right)|n_1n_2\cdots n_i\cdots>\\
=&\left(\Psi^+(\mathbf{r},t)\widehat{N}+\Psi^+(\mathbf{r},t)\right)|n_1n_2\cdots n_i\cdots>\\
=&(N+1)\left(\Psi^+(\mathbf{r},t)|n_1n_2\cdots n_i\cdots>\right),
\end{aligned}\tag{2.6.2.7}$$

where the upper + (lower −) of ± belongs to boson's (fermion's) case.

Second, we demonstrate that the created particle by $\Psi^+(\mathbf{r},t)$ is at time t at $\mathbf{r}$. We have known that the effect of a_k^+ is to produce a particle in single particle state $\psi_k(\mathbf{r})$. Thus, the probability amplitude of this added particle to appear at $\mathbf{r}$' is $\psi_k(\mathbf{r}')exp(-iE_kt/\hbar)$. We have defined in Eq. (2.6.2.3) that

$$\Psi^+(\mathbf{r},t)=\sum_k\psi_k^*(\mathbf{r})e^{iE_kt/\hbar}a_k^+,\tag{2.6.2.8}$$

and have known that $\Psi^+(\mathbf{r},t)$ adds a particle to the multiparticle state vector. The Eq. (2.6.2.8) tells us that the approach that $\Psi^+(\mathbf{r},t)$ adds a particle to the multiparticle state vector is as follows. It is needed to sum over all possible particles produced by a_k^+. Note that there is coefficient in front of a_k^+, $\psi_k^*(\mathbf{r})exp(iE_kt/\hbar)\neq 1$. Therefore, the probability amplitude of this added k state particle to appear at $\mathbf{r}'$ should be the product $\psi_k(\mathbf{r}')exp(-iE_kt/\hbar)\times\psi_k^*(\mathbf{r})exp(iE_kt/\hbar)$. So, the total probability amplitude of the added particle by $\Psi^+(\mathbf{r})$ at $\mathbf{r}'$ is

$$\sum_{k'}\psi_{k'}^*(\mathbf{r})e^{iE_{k'}t/\hbar}\psi_{k'}(\mathbf{r}')e^{-iE_{k'}t/\hbar}=\delta(\mathbf{r}-\mathbf{r}').\tag{2.6.2.9}$$

Eq. (2.6.2.9) tells us that the particle added by $\Psi^+(\mathbf{r})$ into the multiparticle state is definitely at $\mathbf{r}$, because the probability amplitude of the particle, added by $\Psi^+(\mathbf{r},t)$, is concentrated at point $\mathbf{r}$.

Similarly, the field operator $\Psi(\mathbf{r},t)$ annihilates a particle at spatial point $\mathbf{r}$ and at time t in the identical multiplicate system.

One can show that for N identical multiboson and multifermion system the orthonormalized basis in coordinate representation can be expressed by field operators as

$$|\mathbf{r}_1,\mathbf{r}_2,\cdots,\mathbf{r}_N>=\frac{1}{\sqrt{N!}}\Psi^+(\mathbf{r}_1,t)\Psi^+(\mathbf{r}_2,t)\cdots\Psi^+(\mathbf{r}_N,t)|\mathbf{0}>,\tag{2.6.2.10}$$

where $\Psi^+(\mathbf{r}_i,t)$ and $\Psi(\mathbf{r}_j,t)$ satisfy different commutation relations for bosons and fermions.

2.6.3 Forms of Field Operator of One- and Two-Body Operators

Utilizing Ψ^+ and Ψ can express the one- and two-body operators. We take that the single particle's eigenfunctions are plane wave. We give the following example:

(1). One-body operator:

(1.1). Kinetic energy operator T.

$$
\begin{aligned}
T &= \sum_{\mathbf{p}s_z} \frac{\mathbf{p}^2}{2m} a^+_{\mathbf{p}s_z} a_{\mathbf{p}s_z} \\
&= \frac{1}{2m} \sum_{\mathbf{p}s_z} \mathbf{p}^2 \int d\mathbf{r} \int d\mathbf{r}' \frac{1}{V} e^{i(\mathbf{p}\cdot\mathbf{r}-\mathbf{p}\cdot\mathbf{r}')/\hbar} \Psi^+(\mathbf{r}, s_z)\Psi(\mathbf{r}', s_z) \\
&= \frac{\hbar^2}{2m} \int d\mathbf{r} \int d\mathbf{r}' \sum_{\mathbf{p}s_z} \frac{1}{V} \nabla e^{i(\mathbf{p}\cdot\mathbf{r}/\hbar} \cdot \nabla' e^{i\mathbf{p}\cdot\mathbf{r}'/\hbar} \Psi^+(\mathbf{r}, s_z)\Psi(\mathbf{r}', s_z)
\end{aligned}
$$

After partial integrations one obtains

$$
\begin{aligned}
T &= \frac{\hbar^2}{2m} \int d\mathbf{r} \int d\mathbf{r}' \sum_{\mathbf{p}s_z} \frac{1}{V} e^{i(\mathbf{p}\cdot(\mathbf{r}-(\mathbf{r})')/\hbar} \nabla\Psi^+(\mathbf{r}, s_z) \cdot \nabla'\Psi(\mathbf{r}', s_z) \\
&= \frac{\hbar^2}{2m} \int d\mathbf{r} \int d\mathbf{r}' \sum_{s_z} \delta(\mathbf{r}-\mathbf{r}') \nabla\Psi^+(\mathbf{r}, s_z) \cdot \nabla'\Psi(\mathbf{r}', s_z) \\
&= \frac{\hbar^2}{2m} \sum_{s_z} \int d\mathbf{r} \nabla\Psi^+(\mathbf{r}, s_z) \cdot \nabla\Psi(\mathbf{r}, s_z), \qquad (2.6.3.1.a)
\end{aligned}
$$

Eq. (2.6.3.1.a) can also be expressed as

$$
T = \sum_{s_z} \int d\mathbf{r} \Psi^+(\mathbf{r}, s_z) \frac{(-i\hbar\nabla)^2}{2m} \Psi(\mathbf{r}, s_z). \qquad (2.6.3.1.b)
$$

Similarly, for particle without spin one obtains

$$
\begin{aligned}
T &= \sum_{\mathbf{p}} \frac{\mathbf{p}^2}{2m} a^+_{\mathbf{p}} a_{\mathbf{p}} \\
&= \sum_{\mathbf{p}} \frac{(-i\hbar\nabla)^2}{2m} a^+_{\mathbf{p}} a_{\mathbf{p}} \\
&= \frac{\hbar^2}{2m} \sum_{s_z} \int d\mathbf{r} \nabla\Psi^+(\mathbf{r}) \cdot \nabla\Psi(\mathbf{r}) \\
&= \int d\mathbf{r} \Psi^+(\mathbf{r}) \frac{\mathbf{p}^2}{2m} \Psi(\mathbf{r}) \\
&= \int d\mathbf{r} \Psi^+(\mathbf{r}) \frac{(-i\hbar\nabla)^2}{2m} \Psi(\mathbf{r}) \qquad (2.6.3.2)
\end{aligned}
$$

Apparently, if we think that the $\Psi(\mathbf{r})$ in Eq. (2.6.3.2) is a single particle wavefunction other than a field operator, then Eq. (2.6.3.2) gives the average value of kinetic energy of one particle. One can think that the introducing wavefunction and the kinetic energy operator corresponds to first quantization. However, in Eq. (2.6.3.2) wavefunction becomes operator apparently, thus that the wavefunction becomes operator should be called second quantization. Here, we have to emphasize that the statement that wavefunction becomes operator is not correct, it is just apparent.

(1.2). Particle number operator $\widehat{N}$:

Using Eq. (2.6.1.6), one can show that

$$\begin{aligned}\widehat{N} &= \sum_{\mathbf{p}s_z} a^+_{\mathbf{p}s_z} a_{\mathbf{p}s_z} \\ &= \sum_{s_z} \int d\mathbf{r} \Psi^+(\mathbf{r}, s_z)\Psi(\mathbf{r}, s_z) \\ &\equiv \int d\mathbf{r}\rho(\mathbf{r}), \qquad (2.6.3.3)\end{aligned}$$

where $\rho(\mathbf{r})$ represents the particle's number density at $\mathbf{r}$, and it is not a probability density in position space of one particle.

(1.3). Current density operator of a particle, $\mathbf{j}$:

$$\begin{aligned}\widehat{\mathbf{J}} &= \sum_{\mathbf{p}s_z} \frac{\mathbf{p}}{m} a^+_{\mathbf{p}s_z} a_{\mathbf{p}s_z} \\ &= \int d\mathbf{r} \frac{\hbar}{i2m} \sum_{s_z} [\Psi^+(\mathbf{r}, s_z)\nabla\Psi(\mathbf{r}, s_z) - (\nabla\Psi^+(\mathbf{r}, s_z))\Psi(\mathbf{r}, s_z)] \\ &\equiv \int d\mathbf{r}\mathbf{j}(\mathbf{r}), \qquad (2.6.3.4)\end{aligned}$$

where $\mathbf{j}(\mathbf{r})$ represents the current density operator of particle number at $\mathbf{r}$, and it is not a probability current density of one particle in position space.

(1.4). The general form of one-body operator is given by Eq. (2.4.1.1), which is as follows:

$$\widehat{F}^{(1)} = \sum_{n'=1}^{N} \widehat{f}^{(1)}(q_{n'}), \qquad (2.6.3.5)$$

where N denotes total number of particles, $n' = 1, 2, 3, \cdots$ are the labels of particles, $\widehat{f}^{(1)}(q_{n'})$ denotes an one-body operator, such as the kinetic energy plus the potential of the particle in an applied fields, and $\widehat{F}^{(1)}$ can be the system's total momentum, angular momentum, kinetic energy, magnetic moment etc.

The second quantization forms of $\widehat{F}^{(1)}$ for bosons and fermions are given by Eqs. (2.4.1.2) and (2.5.1.2),respectively. Formally, the two forms are the same, which is

$$\widehat{F}^{(1)} = \sum_{ij} f^{(1)}_{ij} a^+_i a_j. \qquad (2.6.3.6)$$

Using the equations between a_i, a^+_j and field operators and Eq. (2.6.3.6), one can demonstrate that, for both bosons and fermions

$$\widehat{F}^{(1)} = \sum_{s_z} \int d\mathbf{r}\Psi^+(\mathbf{r}, s_z)\widehat{f}^{(1)}(\mathbf{r})\Psi(\mathbf{r}, s_z). \qquad (2.6.3.7a)$$

(2). Two-body operator:

The general form of two-body operator is given by Eq. (2.4.2.1)

$$\widehat{F}^{(2)} = \sum_{i<j}^{N} \widehat{f}^{(2)}(q_i, q_j), \qquad (2.6.3.7b)$$

where N denotes total number of particles, $i, j = 1, 2, 3, \cdots$ are the labels of two particles, $\widehat{f}^{(2)}(q_i, q_j)$ denotes an two particle operator, such as the interaction between two particles

$n' = i$ and $n' = j$. Generally, $\widehat{f}^{(2)}(q_i, q_j) = \widehat{f}^{(2)}(q_j, q_i)$. $\widehat{F}^{(2)}$ are the sum of all the two-body operators in boson and fermion system.

We have proved that the $\widehat{F}^{(2)}$'s second quantization forms in terms of (a_i^+, a_j) for both bosons and fermions [See Eqs. (2.4.2.2) and (2.5.2.2)] are the same, formally,

$$\widehat{F}^{(2)} = \frac{1}{2} \sum_{i'j'ij} f^{(2)}_{i'j',ji} a^+_{i'} a^+_{j'} a_j a_i, \tag{2.6.3.8}$$

Using the equations between a_i, a_j^+ and field operators and Eq. (2.6.3.8), one can demonstrate that, for both bosons and fermions

$$\widehat{F}^{(2)} = \sum_{s_z s'_z} \int\int d\mathbf{r} d\mathbf{r}' \Psi^+(\mathbf{r}, s_z) \Psi^+(\mathbf{r}', s'_z) \widehat{f}^{(2)} \mathbf{r}, \mathbf{r}') \Psi(\mathbf{r}', s'_z) \Psi(\mathbf{r}, s_z). \tag{2.6.3.9}$$

Note particularly the ordering of the final two field annihilation operators, which is opposite to the two field creation operators.

2.6.4 Schrödinger Equation in Second Quantization Method

We start our discussion by merely reformulating the Schrödinger equation in the language of second quantization, or say, in occupation number representation. For detail of this subsection refer to Ref. [16].

In almost all the cases of interest, the Hamiltonian takes the form

$$H = \sum_{i=1}^{N} \left(\widehat{T}(q_i) + \widehat{V}^{(1)}(q_i) \right) + \frac{1}{2} \sum_{i \neq j = 1}^{N} \widehat{V}^{(2)}(q_i, q_j), \tag{2.6.4.1}$$

where q_i denotes the "coordinates" of the i-th particle, including the spatial coordinate $\mathbf{r}_i$ and any discrete variables such as the z component of spin for a system of fermions or the z component of isotropic spin for a system of nucleons; $\widehat{T}(q_i)$ represents kinetic energy operator of i-th particle in "coordinate" representation, $\widehat{V}^{(1)}(q_i)$ represents the potential energy operator of i-th particle in "coordinate" representation, and $\widehat{V}^{(2)}(q_i, q_j)$ represents the interaction energy operator between i-th anf j-th particles in "coordinate" representation, and the factor $1/2$ ensures that the interactions have to be counted once only when we make double sum runs over the indices i and j separately, excluding the value i equal to j. With this Hamiltonian, the time-dependent identical multiparticle Schrödinger equation in "coordinates" representation is given by

$$i\hbar \frac{\partial}{\partial t} \Psi(q_1, q_2, \cdots, q_N, t) = H \Psi(q_1, q_2, \cdots, q_N, t), \tag{2.6.4.2}$$

together with an appropriate set of boundary conditions for the total wavefunction Ψ. Notice that we use Ψ to indicate both field operator and wavefunctions.

If we use operators a_i, a_j^+, then either for bosons or for fermions the Schrödinger equation in occupation number representation is given by

$$i\hbar \frac{\partial}{\partial t} |\Psi(t)> = H |\Psi(t)>, \tag{2.6.4.3}$$

where either for bosons or or fermions, from Eqs. (2.6.3.6) and (2.6.3.8) the Hamiltonian is formally

$$H = \sum_{ij} (T^{(1)} + V^{(1)})_{ij} a_i^+ a_j + \frac{1}{2} \sum_{i'j'ij} V^{(2)}_{i'j',ji} a^+_{i'} a^+_{j'} a_j a_i, \tag{2.6.4.4}$$

and the identical multiparticle state vector $|\Psi(t)>$ is

$$|\Psi(t)>=\sum_{n_1n_2\cdots n_\infty} f(n_1n_2\cdots n_\infty,t)|n_1n_2\cdots n_\infty>. \qquad (2.6.4.5)$$

It is important to distinguish between the operator and numbers in Eq. (2.6.4.3). The H is an operator in this abstract occupation number representation because it depends on the creation and annihilation operators. In contrast, in Eq. (2.4.6.4), the matrix elements of the kinetic and potential energy taken between the single particle eigenstates of the Schrödibger equation in the non-occupation number representation are merely complex numbers multiplying by the operators. Eqs. (2.6.4.3-5) restate the Schrödibger equation in the occupation number representation, and all of the statistics and operator properties are contained in the creation and annihilation operators. The physical problem is clearly unchanged by the new formulation. In particular the coefficients f specify the connection between occupation and non-occupation number representation.

For every solution to the original time-dependent multiparticle Schrödibger equation there exists a set of expansion coefficient f. Given this set of expansion coefficients f, it is possible to construct a solution to the problem in occupation number representation, as shown above. Conversely, if the problem is solved in occupation number representation, we can determine a set of expansion coefficients f, which then yield a solution to the original time-dependent multiparticle Schrödibger equation.

The identical multiparticle system's Hamiltonian can also be expressed by the field operators. From Eq. (2.6.4.4) and the relations between (a_i, a_j^+) and field operators $(\Psi(\mathbf{r}), \Psi^+(\mathbf{r}))$ one can show that (For brevity, neglect spin)

$$\begin{aligned} H &= \int d\mathbf{r}\Psi^+(\mathbf{r})[T^{(1)}(\mathbf{r})+V^{(1)}(\mathbf{r})]\Psi(\mathbf{r}) \\ &+ \frac{1}{2}\int\int d\mathbf{r}d\mathbf{r}'\Psi^+(\mathbf{r})\Psi^+(\mathbf{r}')V^{(2)}(\mathbf{r},\mathbf{r}')\Psi^+(\mathbf{r}')\Psi^+(\mathbf{r}), \end{aligned} \qquad (2.6.4.6)$$

If we use field operators, then either for bosons or for fermions the identical N multiparticle Schrödinger equation in occupation number representation is given by

$$i\hbar\frac{\partial}{\partial t}|\Psi(t)>=H|\Psi(t)>, \qquad (2.6.4.7)$$

and the identical multiparticle state vector $|\Psi(t)>$ is

$$|\Psi(t)>=\int d\mathbf{r}_1\int d\mathbf{r}_2\cdots\int d\mathbf{r}_N f(\mathbf{r}_1\mathbf{r}_2\cdots\mathbf{r}_N,t)|\mathbf{r}_1\mathbf{r}_2\cdots\mathbf{r}_N)>. \qquad (2.6.4.8)$$

Chapter 3

Electron and Phonon in Condensed Matter

3.1 Introduction

A condensed matter system in its most basic form can be viewed in terms of two components.

One corresponds to electrons. The electron is of spin 1/2. Because the electron is a spin 1/2 particle, it belongs to fermion, and obeys Pauli exclusion theorem.

The second component corresponds to ionic or atomic lattices. It is easy to see that the most simple motion of the ions is the harmonic vibration around the equilibrium position. The individual ion looks like a harmonic oscillator. However, the ions in the condensed matter are not independent, and actually are correlated each other. Therefore, the vibration of ions in condensed matter is collective vibration. This chapter will prove that this collective vibration can be quantized. The quantum of this collective vibration is called phonon. The phonon can belong to different normal modes. The phonon connects with the collective oscillation of ions, and thus belongs to collective excitation. That some books call the energy quantum of single ion in harmonic oscillation the phonon or the quasi-phonon is a wrong concept.

After introducing the concept of phonon, we treat the collective oscillation of ions, and the interaction between electron and ion by the electron-phonon interaction. The application of quantum mechanics in condensed matter refers to one of the eminently successful applications.

The basic Hamiltonian of the condensed matter is

$$H = H_{el} + H_b + H_{el-b}, \tag{3.1.1}$$

where H_{el} includes the kinetic energy of electrons and the interaction between electrons, H_b represents the energy of ions, and H_{el-b} represents the interaction between electrons and phonons (=interaction between electrons and ions).

3.2 Free Fermi Gas

3.2.1 One-Dimensional Free Electron Gas

Let us at the first consider an one dimensional system consisting of free electrons (=free electron gas). Suppose that the system is in ground state, and has N noninteracting electrons confined in one dimension of length L. Each electron is described by a free wavefunction. Eq. (8.2.11) tells us that this wavefunction should be

$$\Psi_{k\lambda}(x, s_z) = \frac{1}{\sqrt{L}} e^{ikx} \Upsilon_\lambda(s_z), \tag{3.2.1.1}$$

where $\Upsilon_\lambda(s_z)$ designate the spin up and spin down states:

$$\Upsilon_{1/2}(s_z) = \begin{bmatrix} 1 \\ 0 \end{bmatrix}, \quad \Upsilon_{-1/2}(s_z) = \begin{bmatrix} 0 \\ 1 \end{bmatrix}, \tag{3.2.1.2}$$

which are normalized according to

$$\Upsilon^+_{\lambda_1} \Upsilon_{\lambda_2} = \delta_{\lambda_1 \lambda_2}. \tag{3.2.1.3}$$

We take, as we have done in chapter 3, that the wavefunction satisfies the periodic boundary condition:

$$\Psi_{k\lambda}(x + L) = \Psi_{k\lambda}(x). \tag{3.2.1.4}$$

Substituting Eq. (3.2.1.1) into Eq. (3.2.1.4) yields

$$k_n L = 2n\pi, \quad n = 0, \pm 1, \pm 2, \cdots, . \tag{3.2.1.5}$$

which implies that the momentum eigenvalues $\hbar k_n = p_n$ can only take discrete values.

The energy eigenvalues will then be

$$E_n = \frac{\hbar^2 k_n^2}{2m} = \frac{2n^2\pi^2\hbar^2}{mL^2}. \tag{3.2.1.6}$$

Because of the Pauli exclusion theorem, at most one electron can be placed in each level with a given value of k_n. Since an electron has spin which can take two values, spin up and spin down, we are allowed to put no more than two electrons in a given level as long as their spins are in opposite directions.

Thus N electrons will fill up levels with $n = 0, 1, 2, \cdots, N/2$. The last level will consist of either one or two electrons depending on whether N is odd or even. The energy of the highest level can then be obtained by substituting $n = N/2$ in Eq. (3.2.1.6), which gives

$$E_F = \frac{N^2\pi^2\hbar^2}{2mL^2}. \tag{3.2.1.7}$$

E_F is the so called Fermi energy in one dimensional case. The Fermi energy is a very important concept in condensed matter system. In metals, for example, even near absolute zero temperature the electrons continue to have kinetic energy, with the highest value given by the Fermi energy. The so-called chemical potential is the same as Fermi energy at zero temperature.

The total energy of the N electron system is then

$$E_{e-tot} = 2\sum_{n=0}^{N/2} \frac{2n^2\pi^2\hbar^2}{mL^2}, \tag{3.2.1.8}$$

where the factor 2 corresponds to the two spin values. Since N is very large in condensed matter, the above sum can be converted to an integral:

$$\begin{aligned} E_{e-tot} &= \sum_{n=0}^{N/2} n^2 \frac{4\pi^2\hbar^2}{mL^2} \\ &\approx \int_0^{N/2} dnn^2 \frac{4\pi^2\hbar^2}{mL^2} \\ &= \frac{N^3}{24} \frac{4\pi^2\hbar^2}{mL^2} \\ &= \frac{N^3\pi^2\hbar^2}{6mL^2}. \end{aligned} \tag{3.2.1.9}$$

The energy per electron is

$$E_{e-single} = \frac{E_{e-tot}}{N} = \frac{N^2\pi^2\hbar^2}{6mL^2}, \tag{3.2.1.10}$$

which implies that the energy of an individual free electron increases as N^2.

Let us compare this result with a system consisting of N bosons. Since there is no Pauli exclusion theorem preventing the bosons from occupying the same state, their ground state will consist of N particles occupying the same state, $n = 1$. Thus the total energy of this system will be N times the ground state energy. Hence,

$$E_{boson-tot} = N\frac{2\pi^2\hbar^2}{mL^2}, \tag{3.2.1.11}$$

and the energy of a single boson is

$$E_{boson-single} = \frac{E_{boson-tot}}{N} = \frac{\pi^2\hbar^2}{2mL^2}, \tag{3.2.1.12}$$

which remains a constant independent of N. This result is in sharp contrast to the case of electrons.

3.2.2 Three-Dimensional Free Electron Gas

We now go to the more realistic case of the three dimensions. We consider a cube of length L, and volume, $V = L^3$, inside which the wavefunction of a free single electron is given by

$$\Psi_{\mathbf{k}\lambda}(\mathbf{r}, s_z) = \frac{1}{\sqrt{V}} e^{i\mathbf{k}\cdot\mathbf{r}} \Upsilon_\lambda(s_z). \tag{3.2.2.1}$$

We take, as we have done in chapter 2, that the wavefunction satisfies the periodic boundary condition:

$$\Psi_{\mathbf{k}\lambda}(x, y, z) = \Psi_{\mathbf{k}\lambda}(x + L, y, z) = \Psi_{\mathbf{k}\lambda}(x, y + L, z) = \Psi_{\mathbf{k}\lambda}(x, y, z + L). \tag{3.2.2.2}$$

Substituting Eq. (3.2.2.1) into Eq. (3.2.2.2) yields

$$k_\alpha = \tfrac{2\pi n_\alpha}{L}, \quad \alpha = x, y, x, \quad n_\alpha = 0, \pm 1 \pm 2, \cdots, \tag{3.2.2.3}$$

which implies that the momentum eigenvalues $\hbar k_\alpha = p_\alpha$ can only take discrete values. The $\Upsilon_\lambda(s_z)$'s, as one dimension, designate the spin up and spin down states. $\Psi_{\mathbf{k}_\alpha\lambda_\beta}(\mathbf{r})$ construct an orthonormalized complete set.

We will discuss this problem by using the occupation number representation that we plan to use throughout the rest of this chapter. Using one body operator formula Eqs. (2.5.1.2-3) in occupation number representation, the total kinetic energy operator, which is equivalent to an unperturbed Hamiltonian, $H_\circ$, as follows:

$$\begin{aligned}
H_\circ &\equiv \widehat{H}_\circ = \widehat{F}^{(1)} = \sum_{\mathbf{k}_i\mathbf{k}_{i'}\lambda_j\lambda_{j'}} f^{(1)}_{\mathbf{k}_i\mathbf{k}_{i'}} a^+_{\mathbf{k}_i\lambda_j} a_{\mathbf{k}'_i\lambda_{j'}} \\
&= \sum_{\mathbf{k}_i\mathbf{k}_{i'}\lambda_j\lambda_{j'}} \int d\mathbf{r}\Psi^*_{\mathbf{k}_i\lambda_j}(\mathbf{r})\frac{(-i\hbar\nabla)^2}{2m}\Psi_{\mathbf{k}_{i'}\lambda_{j'}}(\mathbf{r})a^+_{\mathbf{k}_i\lambda_j}a_{\mathbf{k}'_i\lambda_{j'}} \\
&= \sum_{\mathbf{k}_i\mathbf{k}_{i'}\lambda_j\lambda_{j'}} \int d\mathbf{r}\Psi^*_{\mathbf{k}_i\lambda_j}(\mathbf{r})\frac{(\hbar|\mathbf{k}|)^2}{2m}\Psi_{\mathbf{k}_{i'}\lambda_{j'}}(\mathbf{r})a^+_{\mathbf{k}_i\lambda_j}a_{\mathbf{k}'_i\lambda_{j'}} \\
&= \sum_{\mathbf{k}_i\mathbf{k}_{i'}\lambda_j\lambda_{j'}} \delta_{\mathbf{k}_i\mathbf{k}_{i'}}\delta_{\lambda_j\lambda_{j'}}\frac{(\hbar k)^2}{2m}a^+_{\mathbf{k}_i\lambda_j}a_{\mathbf{k}'_i\lambda_{j'}} \\
&= \sum_{\mathbf{k}_i\lambda_j}\frac{(\hbar k)^2}{2m}a^+_{\mathbf{k}_i\lambda_j}a_{\mathbf{k}_i\lambda_j} \\
&\equiv \sum_{\mathbf{k}\lambda}\frac{(\hbar k)^2}{2m}a^+_{\mathbf{k}\lambda}a_{\mathbf{k}\lambda}, \qquad (3.2.2.4)
\end{aligned}$$

where $a^+_{\mathbf{k}\lambda}$ and $a_{\mathbf{k}\lambda}$ are the usual creation and annihilation operators for the electrons with the momentum's eigenvalue $\mathbf{k}$ and spin λ, respectively. If $n_{\mathbf{k}\lambda}$ is the election number operator, then

$$n_{\mathbf{k}\lambda} = a^+_{\mathbf{k}\lambda}a_{\mathbf{k}\lambda}, \qquad (3.2.2.5)$$

and

$$H_\circ = \sum_{\mathbf{k}\lambda}\frac{\hbar^2k^2}{2m}n_{\mathbf{k}\lambda}, \qquad (3.2.2.6)$$

which implies a sum of the kinetic energy of each mode, characterized by $\mathbf{k}$ and λ, multiplied by the number of particles in that mode. If we denote by $|F>$ the multiparticle state corresponding to the ground state of the N electrons, then the energy eigenvalue $E_\circ$ is given by

$$E_\circ =< F|H_\circ|F >= 2\int d^3n\frac{\hbar^2k^2}{2m}, \qquad (3.2.2.7)$$

where d^3n is the number of states available and the factor 2 accounts for the two spin states for each $\mathbf{k}$. From Eq. (3.2.2.3) we obtain d^3n as

$$d^3n = dn_xdn_ydn_z = \frac{Vd\mathbf{k}}{(2\pi)^3}. \qquad (3.2.2.8)$$

Since

$$d\mathbf{k} = 4\pi k^2dk, \qquad (3.2.2.9)$$

where 4π corresponds to the solid angle in the $\mathbf{k}$ space, we can evaluate the integral in Eq. (3.2.2.7) to obtain the total energy of the electron system. We find

$$E_\circ = \frac{V\hbar^2k_F^5}{10\pi^2m}, \qquad (3.2.2.10)$$

where the integration in k is carried out from 0 to k_F, the Fermi wavenumber. $\hbar k_F$ is the maximum momentum value for the ground state.

We can write the above relation in terms of the total number of electrons, N. We note that

$$N = \int d^3n = \int d\mathbf{k}\frac{V}{(2\pi)^3}, \tag{3.2.2.11}$$

which after integration gives

$$N = \frac{Vk_F^3}{3\pi^2}. \tag{3.2.2.12}$$

Thus, $E_\circ$ can be expressed as

$$E_\circ = \frac{3}{5}\frac{\hbar^2 k_F^2}{2m}N. \tag{3.2.2.13}$$

The energy per electron is

$$\frac{E_\circ}{N} = \frac{3}{5}E_F, \tag{3.2.2.14}$$

where E_F is the Fermi energy, which is related to k_F by

$$E_F = \frac{\hbar^2 k_F^2}{2m}. \tag{3.2.2.15}$$

Substituting the value of k_F from relation Eq. (3.2.2.12), we obtain E_F in terms of N with $V = L^3$.

$$E_F = \frac{\pi^2\hbar^2}{2mL^2}\left(\frac{3N}{\pi}\right)^{2/3}. \tag{3.2.2.16}$$

We can also write E_F in terms of the particle number density

$$n = \frac{N}{V}, \tag{3.2.2.17}$$

as

$$E_F = \frac{\hbar^2}{2m}(3\pi^2 n)^{2/3}. \tag{3.2.2.18}$$

In summary, we note that the ground state of the many electron system at low temperature consists of completely filled energy levels up to $E = E_F$, which is typically 5 - 10 eV at room temperature. Excitations of the levels are difficult to achieve at low temperature except near the Fermi surface: a particle inside the Fermi sea has nowhere to go since all available states are already occupied. These observations play an essential role in determining the properties of metal and other solids.

3.3 Interacting Electron Gas

To further illustrate the utility of second quantization (or, say, the occupation number representation), we consider a simple metal that provides a first approximation to a metal or a plasma. This system is an interacting electron gas placed in a uniformly distributed positive background chosen to ensure that the total system is neutral. In real metal or plasma, of course, the positive charge is localized in the ionic cores, whose dynamical motion must also be included in the calculation. These positive ions are much heavier than the electrons, however, and it is permissible to neglect the ionic motion entirely. In contrast, the assumption of a uniform background is more drastic; for this reason, the present model can provide only a qualitative account of real metal.

We are interested in the properties of the bulk medium. It therefore is convenient to enclose the system in a large cubic box with sides of length L; the limits $L \to \infty$ and total electron number $N \to \infty$ will be taken at the end of the calculation. However, the electron the total particle's operator $\widehat{N}$ satisfies $\widehat{N}|n_1n_2\cdots n_i\cdots>=\int d\mathbf{r}\Psi^+\Psi|n_1n_2\cdots n_i\cdots>=N|n_1n_2\cdots n_i\cdots>$, one obtains

$$\begin{aligned}
&\widehat{N}\left(\Psi^+(\mathbf{r},t)|n_1n_2\cdots n_i\cdots>\right)\\
=&\int d\mathbf{r}'\Psi^+(\mathbf{r}',t)\Psi(\mathbf{r}',t)\Psi^+(\mathbf{r},t)|n_1n_2\cdots n_i\cdots>\\
=&\int d\mathbf{r}'\Psi^+(\mathbf{r}',t)\left(\pm\Psi^+(\mathbf{r},t)\Psi(\mathbf{r}',t)+\delta(\mathbf{r}-\mathbf{r}')\right)|n_1n_2\cdots n_i\cdots>\\
=&\left(\Psi^+(\mathbf{r},t)\widehat{N}+\Psi^+(\mathbf{r},t)\right)|n_1n_2\cdots n_i\cdots>\\
=&(N+1)\left(\Psi^+(\mathbf{r},t)|n_1n_2\cdots n_i\cdots>\right),
\end{aligned}\tag{2.6.2.7}$$

where the upper + (lower −) of ± belongs to boson's (fermion's) case.

Second, we demonstrate that the created particle by $\Psi^+(\mathbf{r},t)$ is at time t at $\mathbf{r}$. We have known that the effect of a_k^+ is to produce a particle in single particle state $\psi_k(\mathbf{r})$. Thus, the probability amplitude of this added particle to appear at **r**' is $\psi_k(\mathbf{r}')exp(-iE_kt/\hbar)$. We have defined in Eq. (2.6.2.3) that

$$\Psi^+(\mathbf{r},t)=\sum_k\psi_k^*(\mathbf{r})e^{iE_kt/\hbar}a_k^+,\tag{2.6.2.8}$$

and have known that $\Psi^+(\mathbf{r},t)$ adds a particle to the multiparticle state vector. The Eq. (2.6.2.8) tells us that the approach that $\Psi^+(\mathbf{r},t)$ adds a particle to the multiparticle state vector is as follows. It is needed to sum over all possible particles produced by a_k^+. Note that there is coefficient in front of a_k^+, $\psi_k^*(\mathbf{r})exp(iE_kt/\hbar)\neq 1$. Therefore, the probability amplitude of this added k state particle to appear at $\mathbf{r}'$ should be the product $\psi_k(\mathbf{r}')exp(-iE_kt/\hbar)\times\psi_k^*(\mathbf{r})exp(iE_kt/\hbar)$. So, the total probability amplitude of the added particle by $\Psi^+(\mathbf{r})$ at $\mathbf{r}'$ is

$$\sum_{k'}\psi_{k'}^*(\mathbf{r})e^{iE_{k'}t/\hbar}\psi_{k'}(\mathbf{r}')e^{-iE_{k'}t/\hbar}=\delta(\mathbf{r}-\mathbf{r}').\tag{2.6.2.9}$$

Eq. (2.6.2.9) tells us that the particle added by $\Psi^+(\mathbf{r})$ into the multiparticle state is definitely at $\mathbf{r}$, because the probability amplitude of the particle, added by $\Psi^+(\mathbf{r},t)$, is concentrated at point $\mathbf{r}$.

Similarly, the field operator $\Psi(\mathbf{r},t)$ annihilates a particle at spatial point $\mathbf{r}$ and at time t in the identical multiplicate system.

One can show that for N identical multiboson and multifermion system the orthonormalized basis in coordinate representation can be expressed by field operators as

$$|\mathbf{r}_1,\mathbf{r}_2,\cdots,\mathbf{r}_N>=\frac{1}{\sqrt{N!}}\Psi^+(\mathbf{r}_1,t)\Psi^+(\mathbf{r}_2,t)\cdots\Psi^+(\mathbf{r}_N,t)|\mathbf{0}>,\tag{2.6.2.10}$$

where $\Psi^+(\mathbf{r}_i,t)$ and $\Psi(\mathbf{r}_j,t)$ satisfy different commutation relations for bosons and fermions.

2.6.3 Forms of Field Operator 0f One- and Two-Body Operators

Utilizing Ψ^+ and Ψ can express the one- and two-body operators. We take that the single particle's eigenfunctions are plane wave. We give the following example:

(1). One-body operator:

(1.1). Kinetic energy operator T.

In Eqs. (3.3.2-4) the only dynamical variables are those referring to the electrons, because the positive background is inert. Thus H_b is a pure c number and is readily evaluated for a uniform distribution $\rho_b(\mathbf{r}) = N/V$:

Hence, H_b in Eq. (3.3.3) is found to be

$$\begin{aligned} H_b &= \frac{1}{2}e^2\left(\frac{N}{V}\right)^2 \int\int d\mathbf{r}d\mathbf{r}'\frac{e^{-\mu|\mathbf{r}-\mathbf{r}'|}}{|\mathbf{r}-\mathbf{r}'|} \\ &= \frac{1}{2}e^2\left(\frac{N}{V}\right)^2 \int\int d\mathbf{r}d\mathbf{z}\frac{e^{-\mu z}}{z} = \frac{1}{2}e^2\frac{N^2}{V}\frac{4\pi}{\mu^2}. \end{aligned} \tag{3.3.5}$$

Here the translational invariance has been used to shift the origin of integration in the second line. The quantity $N^{-1}H_b$ diverges in the limit $\mu \to 0$, because the long range of the Coulomb potential allows every element of charge to interact with every other one.

In principle, the H_{el-b} is an one-body operator since it acts on each electron. For the present system, however, we may again use the translational invariance to write

$$\begin{aligned} H_{el-b} &= -e^2\sum_{i=1}^{N}\frac{N}{V}\int d\mathbf{r}\frac{e^{-\mu|\mathbf{r}-\mathbf{r}_i|}}{|\mathbf{r}-\mathbf{r}_i|} \\ &= -e^2\sum_{i=1}^{N}\frac{N}{V}\int d\mathbf{z}\frac{e^{-\mu z}}{z} = -e^2\frac{N^2}{V}\frac{4\pi}{\mu^2}. \end{aligned} \tag{3.3.6}$$

The sum of the two Hamiltonians is then given by

$$H_b + H_{el-b} = -\frac{e^2}{2}\frac{N^2}{V}\frac{4\pi}{\mu^2}. \tag{3.3.7}$$

We note that, in the language of second quantization, this is actually a c number (i. e., proportional to a unit operator in the multiparticle Hilbert space or, say, Fock space) since N corresponds to the eigenvalue of the number operator $\widehat{N}$.

The total Hamiltonian in Eq. (3.1.1) thus reduces to

$$H = H_b + H_{el-b} + H_{el} = -\frac{e^2}{2}\frac{N^2}{V}\frac{4\pi}{\mu^2} + H_{el}, \tag{3.3.8}$$

and all of the physical effects are contained in H_{el}.

Eq. (3.3.2) define $H_{el} = H_\circ + H'$. $H_\circ$ has been calculated in subsection 3.2.2, and the form of $H_\circ$ in occupation number representation has been given by Eq. (3.2.2.4). Now let us obtain the form of the perturbed Hamiltonian H' in occupation number representation. H' is a two-body operator. Eq. (2.4.2.2) is the general form of a two-body operator in occupation representation:

$$\widehat{F}^{(2)} = \frac{1}{2}\sum_{i'j'ij} f^{(2)}_{i'j',ji}a^+_{i'}a^+_{j'}a_j a_i, \tag{3.3.9}$$

where the summations are over the single particle's states.

$$f^{(2)}_{i'j',ji} \equiv (\psi_{k_{i'}}(q_1)\psi_{k_{j'}}(q_2), \widehat{f}^{(2)}(q_1,q_2)\psi_{k_j}(q_2)\psi_{k_i}(q_1)). \tag{3.3.10}$$

$f^{(2)}_{i'j',ji}$ is the matrix element of two-body operator between single particle state products $\psi_{k_{i'}}(q_1)\psi_{k_{j'}}(q_2)$ and $\psi_{k_j}(q_2)\psi_{k_i}(q_1)$. Using Eqs. (3.3.9-10), the form of H' in occupation number representation is

$$\begin{aligned}
H' &= \sum_{\substack{\mathbf{k}_{i'}\lambda_{i'}\mathbf{k}_{j'}\lambda_{j'}\\ \mathbf{k}_i\lambda_i\mathbf{k}_j\lambda_j}} \frac{1}{2}\frac{e^2}{V^2}\int\int d\mathbf{r}_1 d\mathbf{r}_2 \left[e^{-i\mathbf{k}_{i'}\cdot\mathbf{r}_1}\Upsilon^{(1)+}_{\lambda_{i'}}\right]\left[e^{-i\mathbf{k}_{j'}\cdot\mathbf{r}_2}\Upsilon^{(2)+}_{\lambda_{j'}}\right] \\
&\quad \frac{e^{-\mu|\mathbf{r}_1-\mathbf{r}_2|}}{|\mathbf{r}_1-\mathbf{r}_2|}\left[e^{i\mathbf{k}_j\cdot\mathbf{r}_2}\Upsilon^{(2)+}_{\lambda_j}\right]\left[e^{i\mathbf{k}_i\cdot\mathbf{r}_1}\Upsilon^{(1)+}_{\lambda_i}\right] a^+_{\mathbf{k}_{i'}\lambda_{i'}} a^+_{\mathbf{k}_{j'}} a_{\mathbf{k}_j\lambda_j} a_{\mathbf{k}_i\lambda_i} \\
&= \frac{1}{2}\frac{e^2}{V^2}\sum_{\substack{\mathbf{k}_{i'}\lambda_{i'}\mathbf{k}_{j'}\lambda_{j'}\\ \mathbf{k}_i\lambda_i\mathbf{k}_j\lambda_j}} \delta_{\lambda_{i'i}}\delta_{\lambda_{j'j}}\int d\mathbf{r}_2 e^{-i(\mathbf{k}_{i'}+\mathbf{k}_{j'}-\mathbf{k}_i-\mathbf{k}_j)\cdot\mathbf{r}_2} \\
&\quad \int d\mathbf{r} e^{i(\mathbf{k}_i-\mathbf{k}_{i'})\cdot\mathbf{r}}\frac{e^{-\mu r}}{r} a^+_{\mathbf{k}_{i'}\lambda_{i'}} a^+_{\mathbf{k}_{j'}} a_{\mathbf{k}_j\lambda_j} a_{\mathbf{k}_i\lambda_i} \\
&= \frac{1}{2}\frac{e^2}{V}\sum_{\substack{\mathbf{k}_{i'}\lambda_{i'}\mathbf{k}_{j'}\lambda_{j'}\\ \mathbf{k}_i\lambda_i\mathbf{k}_j\lambda_j}} \delta_{\lambda_{i'i}}\delta_{\lambda_{j'j}}\delta_{\mathbf{k}_{i'}+\mathbf{k}_{j'},\mathbf{k}_i+\mathbf{k}_j} \\
&\quad \frac{4\pi}{(\mathbf{k}_i-\mathbf{k}_{i'})^2+\mu^2} a^+_{\mathbf{k}_{i'}\lambda_{i'}} a^+_{\mathbf{k}_{j'}\lambda_{j'}} a_{\mathbf{k}_j\lambda_j} a_{\mathbf{k}_i\lambda_i}. \qquad (3.3.11)
\end{aligned}$$

In deriving Eq. (3.3.11) we have used the following relations:

$$\mathbf{r}_1-\mathbf{r}_2=\mathbf{r},\quad \Upsilon^{(2)+}_{\lambda_{j'}}\Upsilon^{(2)}_{\lambda_j}=\delta_{\lambda_j\lambda_{j'}},\ \Upsilon^{(1)+}_{\lambda_i}\Upsilon^{(1)}_{\lambda_{i'}}=\delta_{\lambda_i\lambda_{i'}}\ , \qquad (3.3.12)$$

$$\frac{1}{V}\int d\mathbf{r}_2 e^{-i(\mathbf{k}_{i'}+\mathbf{k}_{j'}-\mathbf{k}_i-\mathbf{k}_j)\cdot\mathbf{r}_2}=\delta_{\mathbf{k}_{i'}+\mathbf{k}_{j'},\mathbf{k}_i+\mathbf{k}_j}, \qquad (3.3.13)$$

$$\int d\mathbf{r} e^{i(\mathbf{k}_i-\mathbf{k}_{i'})\cdot\mathbf{r}}\frac{e^{-\mu r}}{r}=\frac{4\pi}{(\mathbf{k}_i-\mathbf{k}_{i'})^2+\mu^2}. \qquad (3.3.14)$$

We change variables and write

$$\mathbf{k}_{i'}-\mathbf{k}_i=\mathbf{k}_{j'}-\mathbf{k}_j=\mathbf{q},\quad \mathbf{k}_{i'}=\mathbf{k},\quad \mathbf{k}_j=\mathbf{p}\ . \qquad (3.3.15)$$

By summing over λ_i and λ_j and using Kronecker δ, H' can be simplified to

$$\begin{aligned}
H' &= \frac{e^2}{2V}\sum_{\mathbf{kpq}}\sum_{\lambda_1\lambda_2}\frac{4\pi}{q^2+\mu^2}a^+_{\mathbf{k+q},\lambda_1}a^+_{\mathbf{p-q},\lambda_2}a_{\mathbf{p}\lambda_2}a_{\mathbf{k}\lambda_1} \\
&= \frac{e^2}{2V}\sum_{\mathbf{kpq}}{}'\sum_{\lambda_1\lambda_2}\frac{4\pi}{q^2+\mu^2}a^+_{\mathbf{k+q},\lambda_1}a^+_{\mathbf{p-q},\lambda_2}a_{\mathbf{p}\lambda_2}a_{\mathbf{k}\lambda_1} \\
&+ \frac{e^2}{2V}\sum_{\mathbf{kp}}\sum_{\lambda_1\lambda_2}\frac{4\pi}{\mu^2}a^+_{\mathbf{k}\lambda_1}a^+_{\mathbf{p}\lambda_2}a_{\mathbf{p}\lambda_2}a_{\mathbf{k}\lambda_1}\equiv I+II. \qquad (3.3.16)
\end{aligned}$$

The prime in the summation sign in the first term implies that we exclude terms with $\mathbf{q}=0$. The second term contains only the terms with $\mathbf{q}=0$. To simplify the second term we now use the following anticommutation relations for the fermion operators,

$$[a_{\mathbf{p}\lambda_2},a_{\mathbf{k}\lambda_1}]_+=0,\quad and\quad [a^+_{\mathbf{p}\lambda_2},a_{\mathbf{k}\lambda_1}]_+=\delta_{\mathbf{kp}}\delta_{\lambda_1\lambda_2}\ , \qquad (3.3.17)$$

and obtain

$$a^+_{\mathbf{k}\lambda_1}a^+_{\mathbf{p}\lambda_2}a_{\mathbf{p}\lambda_2}a_{\mathbf{k}\lambda_1}=-a^+_{\mathbf{k}\lambda_1}a^+_{\mathbf{p}\lambda_2}a_{\mathbf{k}\lambda_1}a_{\mathbf{p}\lambda_2}. \qquad (3.3.18)$$

$$a^+_{\mathbf{p}\lambda_2}a_{\mathbf{k}\lambda_1} = \delta_{\mathbf{kp}}\delta_{\lambda_1\lambda_2} - a_{\mathbf{k}\lambda_1}a_{\mathbf{p}\lambda_2}. \tag{3.3.19}$$

Substituting Eqs. (3.3.17-19) into the second term II $(q = 0)$ of Eq. (3.3.16) yields

$$\begin{aligned} II &= \frac{e^2}{2V}\sum_{\mathbf{kp}}\sum_{\lambda_1\lambda_2}\frac{4\pi}{\mu^2}a^+_{\mathbf{k}\lambda_1}a_{\mathbf{k}\lambda_1}(a^+_{\mathbf{p}\lambda_2}a_{\mathbf{p}\lambda_2} - \delta_{\mathbf{kp}}\delta_{\lambda_1\lambda_2}) \\ &= \frac{e^2}{2V}\sum_{\mathbf{kp}}\sum_{\lambda_1\lambda_2}\frac{4\pi}{\mu^2}\widehat{n}_{\mathbf{k}\lambda_1}(\widehat{n}_{\mathbf{p}\lambda_2} - \delta_{\mathbf{kp}}\delta_{\lambda_1\lambda_2}) \\ &= \frac{e^2}{2V}\frac{4\pi}{\mu^2}(\widehat{N}^2 - \widehat{N}) \\ &= \frac{e^2}{2V}\frac{4\pi}{\mu^2}N^2 - \frac{e^2}{2V}\frac{4\pi}{\mu^2}N. \end{aligned} \tag{3.3.20}$$

Considering Eqs. (3.3.7-8), (3.3.2), (3.3.16) and (3.3.20), Hamiltonian written in occupation number representation is given by

$$\begin{aligned} H &= H_\circ + H' \\ &= \sum_{\mathbf{k}\lambda}\frac{(\hbar k)^2}{2m}a^+_{\mathbf{k}\lambda}a_{\mathbf{k}\lambda} + \frac{e^2}{2V}{\sum_{\mathbf{kpq}}}'\sum_{\lambda_1\lambda_2}\frac{4\pi}{q^2+\mu^2}a^+_{\mathbf{k+q},\lambda_1}a^+_{\mathbf{p-q},\lambda_2}a_{\mathbf{p}\lambda_2}a_{\mathbf{k}\lambda_1} \\ &- \frac{e^2}{2V}\frac{4\pi}{\mu^2}N \\ &\approx \sum_{\mathbf{k}\lambda}\frac{(\hbar k)^2}{2m}a^+_{\mathbf{k}\lambda}a_{\mathbf{k}\lambda} + \frac{e^2}{2V}{\sum_{\mathbf{kpq}}}'\sum_{\lambda_1\lambda_2}\frac{4\pi}{q^2+\mu^2}a^+_{\mathbf{k+q},\lambda_1}a^+_{\mathbf{p-q},\lambda_2}a_{\mathbf{p}\lambda_2}a_{\mathbf{k}\lambda_1}, \end{aligned} \tag{3.3.21}$$

where the sum in the momentum goes up to the Fermi wavenumber k_F, and the approximation in the last equality is as follows:

$$\frac{e^2}{2V}\frac{4\pi}{\mu^2}N = N\frac{4\pi e^2}{2L(L\mu)^2}. \tag{3.3.22}$$

In Eq. (3.3.22) μ represents a Coulomb screen factor, therefore, $L\mu >> 1$. In the large volume limit, the expression in Eq. (3.3.22) is zero.

The second term in Eq. (3.3.21) comes from the Coulomb interaction between electrons, and can also be evaluated. If we call this E'', then after some lengthy calculations (about 1.5 pages. See Ref. [16]), one obtains

$$E'' = -\frac{3e^2Nk_F}{4\pi}. \tag{3.3.23}$$

3.4 Phonon

We will show that the collective vibration of an one dimensional atom chain can be described by using phonon, which is a quantum of collective vibration of all ions in condensed matter.

Suppose that the one dimensional atom (or ion) chain consists of $N+1$ atoms, labeled by $l = 0, 1, 2, \cdots, N$. The spacing between two nearest neighbor atoms is a. Every atom is an harmonic oscillator (harmonic approximation). Let us explain that this one dimensional atom chain is a coupling vibrational chain. Because there are interactions between nearest neighbor atoms, the single atom's oscillation around its equilibrium position suffers the influence of other atom. For a coupling oscillator chain the atom itself's equilibrium position is not important, we should consider the relative displacement instead. Let x_l be the displacement of the l-th single oscillator (=atom) from its equilibrium position. For a coupling oscillators the displacement of l-th oscillator should be relative to x_{l+1}. Referring to the Hamiltonian of individual oscillator in such as Eq. (4.4.4.1) of Ref. [3], for the $N+1$ atom linear chain we have

$$\begin{aligned} H &= \sum_{l=0}^{N} \frac{p_l^2}{2M} + \frac{1}{2}M\omega^2 \sum_{l=1}^{N} (x_l - x_{l+1})^2 \\ &= \sum_{l=1}^{N} \left[\frac{p_l^2}{2M} + M\omega^2 x_l^2\right] - M\omega^2 \sum_{l=1}^{N} x_l x_{l+1}, \end{aligned} \tag{3.4.1}$$

where we have neglected the end effect because N is a very large number (for example, $N = 10^7$). The first term in Eq. (3.4.1) describes a set of N independent oscillators, while the second term in a sense corresponds to the interaction, or coupling, between an oscillator and its nearest neighbor. In order to obtain the energy eigenvalues of this system we need to diagonalize this Hamiltonian and write it in terms of normal modes so that H can be expressed in terms of a new set of independent oscillators.

The classical equation of motion for the above Hamiltonian is given by the differential equation

$$\ddot{x}_l = -\omega^2(2x_l - x_{l-1} - x_{l+1}). \tag{3.4.2}$$

The normal modes are obtained in the standard manner by writing

$$x_l = x_\circ e^{ikal} e^{-i\omega(k)t}, \tag{3.4.3}$$

where k is wavenumber and $\omega(k)$ the angular frequency. Eq. (3.4.3) then corresponds to the displacement of the oscillator l at time t. Because the vibration direction is in the direction of propagation of wave, we call it longitudinal vibration, which corresponds to sound wave of wavelength $2\pi/k$ and angular frequency $\omega(k)$.

Substituting Eq. (3.4.3) into Eq. (3.4.2) yields after canceling the common factors

$$\omega(k) = \sqrt{\omega^2[2 - e^{-ika} - e^{ika}]}, \tag{3.4.4a}$$

$$= 2\omega \sin\left(\frac{ka}{2}\right). \tag{3.4.4b}$$

Eq. (3.4.4) relates the normal mode's frequency to the wavenumber. It is referred to as a dispersion relation.

Therefore, at small k (wavelength larger than the period of the lattice a).

$$\begin{aligned} \omega(k) &\approx \left(\frac{d\omega(k)}{dk}\right) \\ &\equiv vk. \end{aligned} \tag{3.4.5}$$

By interpreting the wave running along this chain as sound waves, we obtain that v in Eq. (3.4.5) is the sound speed. From Eqs. (3.4.4-5) it is obtained that at low energies ($ka \approx 0$) $v = \omega a$. Thus, the long wavelength collective motion is similar to the density wave in a continuous medium where the discreteness of atomic structure is not important.

Let us now express the displacement x_l at time $t = 0$ in the form of a Fourier expansion

$$x_l = \frac{1}{\sqrt{N}} \sum_{k=1} e^{ikal} X_k, \tag{3.4.6}$$

with periodic boundary conditions. Its inverse is given by

$$X_k = \frac{1}{\sqrt{N}} \sum_{l=1}^{N} e^{-ikal} x_l = X_{-k}^{+}, \tag{3.4.7}$$

where we have used the orthogonality relation

$$\frac{1}{N} \sum_{l=1}^{N} e^{ial(k'-k)} = \delta_{k'k}. \tag{3.4.8}$$

Similar relations can be obtained for the momentum variables.

$$p_l = \frac{1}{\sqrt{N}} \sum_{k=1} e^{-ikal} P_k, \tag{3.4.9}$$

Its inverse is given by

$$P_k = \frac{1}{\sqrt{N}} \sum_{l=1}^{N} e^{ikal} p_l = P_{-k}^{+}. \tag{3.4.10}$$

Now let us go to quantum mechanics. We have

$$[x_l, p_{l'}] = i\hbar\delta_{ll'}. \tag{3.4.11}$$

From Eqs. (3.4.6-10) one obtains

$$\begin{aligned} [X_k, P_{k'}] &= \frac{1}{N} \sum_{ll'} e^{-ikal} e^{ik'al'} [x_l, p_{l'}] \\ &= \frac{i\hbar}{N} \sum_{l} e^{-ial(k-k')} = i\hbar\delta_{kk'}. \end{aligned} \tag{3.4.12}$$

Thus the amplitudes X_k and $P_{k'}$ themselves act as canonical operators.

Now let us express the Hamiltonian in terms of $X_k, P_{k'}$. Substituting Eqs. (3.4.6) and (3.4.9) into Eq. (3.4.1), utilizing Eqs. (3.4.9), (3.4.4.a), and considering the commutation relations of X_k's, one obtains

$$
\begin{aligned}
H &= \sum_{l=1}^{N} \frac{1}{2m} \frac{1}{\sqrt{N}} \sum_{k=1}^{N} e^{ikal} P_k \frac{1}{\sqrt{N}} \sum_{k'=1}^{N} e^{ik'al} P_{k'} \\
&+ \sum_{l=1}^{N} \frac{1}{2} M\omega^2 \frac{1}{\sqrt{N}} \sum_{k=1}^{N} e^{-ikal} X_k \sum_{k'=1}^{N} e^{-ik'al} X_{k'} \\
&- M\omega^2 \sum_{l=1}^{N} \frac{1}{\sqrt{N}} \sum_{k=1}^{N} e^{-ikal} X_k \frac{1}{\sqrt{N}} \sum_{k'=1}^{N} e^{-ik'a(l+1)} X_{k'} \\
&= \frac{1}{2M} \sum_{k=1}^{N} \sum_{k'=1}^{N} \delta_{k,-k'} P_k P_{k'} + \frac{1}{2} M\omega^2 \sum_{k=1}^{N} \sum_{k'=1}^{N} \delta_{k,-k'} X_k X_{k'} \\
&- M\omega^2 \sum_{k=1}^{N} \sum_{k'=1}^{N} e^{-ik'a} \delta_{k,-k'} X_k X_{k'} \\
&= \frac{1}{2M} \sum_{k=1}^{N} P_k P_{-k} + \frac{1}{2} M\omega^2 \sum_{k=1}^{N} X_k X_{-k} - m\omega^2 \sum_{k=1}^{N} e^{-ika} X_k X_{-k} \\
&= \frac{1}{2M} \sum_{k=1}^{N} P_k P_{-k} + \frac{1}{2} M\omega^2 \sum_{k=1}^{N} (2 - e^{-ika} - e^{ika}) X_k X_{-k} \\
&= \frac{1}{2M} \sum_{k=1}^{N} P_k P_{-k} + \frac{M}{2} \sum_{k=1}^{N} \omega^2(k) X_k X_{-k}. \qquad (3.4.13)
\end{aligned}
$$

$(X_k, P_{k'})$ are the normal (or collective) modes of the coupling oscillators.

Referring to, for example, Eqs. (4.4.4.4.a-b) in Ref. [3], we define the creation and annihilation operators as follows:

$$b_k = \sqrt{\frac{M\omega(k)}{2\hbar}} \left(X_k + \frac{i}{M\omega(k)} P_{-k} \right). \qquad (3.4.14)$$

$$b_k^+ = \sqrt{\frac{M\omega(k)}{2\hbar}} \left(X_{-k} - \frac{i}{M\omega(k)} P_k \right). \qquad (3.4.15)$$

Utilizing Egs. (3.4.12), one can demonstrate that

$$[b_k, b_{k'}^+] = \delta_{kk'}, \quad [b_k, b_{k'}] = [b_k^+, b_{k'}^+] = 0. \qquad (3.4.16)$$

These are the same type of commutation relations that the a's in Eq. (4.4.4.0) satisfy. The Hamiltonian is then found to be [Refer to the derivation for Eq. (4.4.4.11) in Ref. [3]]

$$
\begin{aligned}
H &= \sum_{k=1} \hbar\omega(k) \left(b_k^+ b_k + \frac{1}{2} \right) & (3.4.17.a) \\
&\equiv \sum_{k=1} \hbar\omega(k) \left(n_k + \frac{1}{2} \right) & (3.4.17.b) \\
&= \frac{1}{2} \sum_{k=1} \hbar\omega(k) + \sum_{k=1} \hbar\omega(k) n_k \\
&\equiv E_\circ + \sum_{k=1} \hbar\omega(k) n_k, & (3.4.17.c)
\end{aligned}
$$

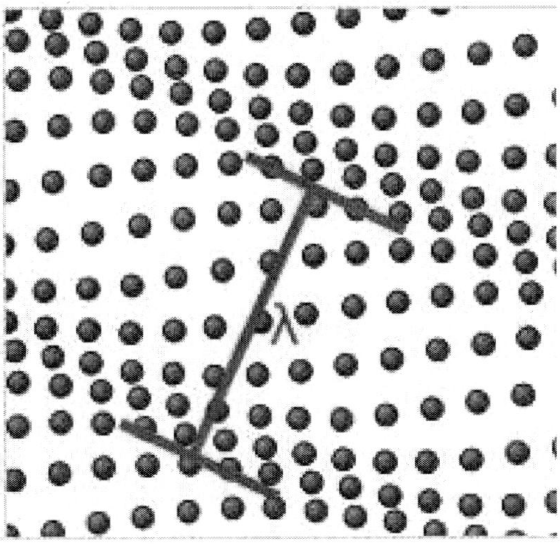

Figure 3.1: Lattice wave propagating through a square lattice (atom displacements greatly exaggerated). Due to the connections between atoms, the displacement of one or more atoms from their equilibrium positions will give rise to a set of vibration waves propagating through the lattice. One such wave is shown in the Fig. 3.1. The amplitude of the wave is given by the displacements of the atoms from their equilibrium positions. The wavelength λ is marked. There is a minimum possible wavelength, given by twice the equilibrium separation a between atoms. Solid physics tells us that any wavelength shorter than this can be mapped onto a wavelength longer than 2a, due to the periodicity of the lattice. Not every possible lattice vibration has a well-defined wavelength and frequency. However, the normal modes do possess well-defined wavelengths and frequencies. Note that many references think that this diagram represents the phonon propagation. This textbook thins that it is not. The phonon is a quantum of collective oscillation of lattice. Fig. 3.1 is just one picture of one in infinite different lattice waves. If we want to draw a picture for phonon, then we have to draw many small spheres, which represent many phonon wavepackets.

where $E_\circ$ is identified as ground state energy. The second term of Eq. (3.4.17.c) is identified as a sum corresponding to n_k collective (elementary) excitation, each with energy $\hbar\omega(k)$, which is now called quasiparticles, or simply phonons. As we know, these phonons connect with longitudinal vibration of atoms (i. e., along the direction of linear chain direction), which is just the direction of propagation of the wave, therefore, the phonon is polarized in longitudinal direction. The longitudinally polarized phonon is called acoustic phonon or acoustic branch phonon. n_k denotes the phonon number of wavenumber k or energy $\hbar\omega(k)$. Because $b_k, b^+_{k'}$ obey Eq. (3.4.16), phonons are bosons, regardless the statistics of original atoms (ions).

The oscillation of the atoms on lattice points in a crystal is a collective oscillation, which is called lattice wave. After the quantization of lattice wave we have phonons. Fig. 3.1 gives a image of the lattice wave or, say, the propagation of phonon in a two dimensional lattice.

After introducing phonon and its creation and annihilation operators, we can can express a state corresponding to the atomic system in Fock's space (=multiparticle Hilbert space) by writing

$$|\psi>=|n_1, n_2, \cdot, n_k, \cdot> . \tag{3.4.18}$$

Some textbooks call the quantization of the vibration of a single oscillator first quantization, and the quantization after introducing collective (or general canonical or normal) coordinates and momentums for the multioscillator system second quantization.

Using the phonon language, the displacement of l-th atom from its equilibrium, x_l, can be expressed as a Hermitian superposition of normal modes.

$$\begin{aligned} x_l &= \frac{1}{\sqrt{N}} \sum_{k=1} e^{ikal} X_k \\ &= \frac{1}{\sqrt{N}} \sum_{k=1} e^{ikal} \sqrt{\frac{\hbar}{2M\omega(k)}} (b_k + b^+_{-k}) \\ &= \sum_{k=1} \sqrt{\frac{\hbar}{2NM\omega(k)}} \left(e^{ikal} b_k + e^{-ikal} b^+_k\right), \end{aligned} \quad (3.4.19.a)$$

or

$$x_l(t) = \sum_{k=1} \sqrt{\frac{\hbar}{2NM\omega(k)}} \left(e^{ikal - i\omega(k)t} b_k + e^{-ikal + i\omega(k)t} b^+_k\right). \quad (3.4.19.b)$$

Solid state theory tells us that for a three dimensional solid, if j labels cell of the lattice of solid, α labels the atom in a cell, λ denotes the polarization of lattice wave (the two components of transverse wave, and the longitudinal wave), M_α denotes the mass of atom α, and $\mathbf{e}^{(\lambda)}_{\alpha\mathbf{k}}$ denotes polarization vector. The displacement of the α atom in cell j, $x_{j\alpha}$ is (Refer Ref. [17] for proof.)

$$x_{j\alpha} \equiv u_{j\alpha} = \sum_{\lambda\mathbf{k}} \sqrt{\frac{\hbar}{2NM_\alpha \omega_\lambda(\mathbf{k})}} \mathbf{e}^{(\lambda)}_{\alpha\mathbf{k}} \left(e^{i(\mathbf{k}\cdot\mathbf{R_j})} b_{\lambda\mathbf{k}} + e^{-i(\mathbf{k}\cdot\mathbf{R_j})} b^+_{\lambda\mathbf{k}}\right). \quad (3.4.19.c)$$

The readers do not need to derive this equation, and just know that in a more general cases, for example, one cell contains many atoms, Eq. (3.4.19.a) can be extended.

3.5 Electron-Phonon Interaction

Let us consider an one dimensional ion chain. Many electrons move in this chain. We use $U(r_i - R_l)$ to denote the interaction potential between electron at r_i and ion at R_l. If the chain is perfect, namely exactly periodic, then the electron's motion is free, namely there is no interaction between the chain and the electron. If there is a displacement from the l-th ion equilibrium position, x_l, then there is interaction between electrons and the ion chain. The interaction between the i-th electron and l-th ion is

$$H'_{i-l} = -\frac{\partial U(r_i - R_l)}{\partial R_l} x_l. \quad (3.5.1)$$

Due to the interactions between ions, the electron has actually interaction with the whole chain. Therefore, it is better to use the collective excitation of lattice (the chain), namely phonon, to study the interaction between electrons and ions in this multiparticle system. Note that if we consider electron-phonon interaction, then we cannot consider electron-ion interaction. The interaction Hamiltonian between all electrons and all atoms (or, say, ions) is

$$
\begin{aligned}
H_{el-ph} &= \sum_{i=1}^{N_e}\sum_{l=1}^{N} H'_{i-l} \\
&= \sum_{i=1}^{N_e}\left[\sum_{l=1}^{N}(-1)\frac{\partial U(r_i - R_l)}{\partial R_l}x_l\right] \\
&\equiv \widehat{F}^{(1)} \equiv \sum_{i=1}^{N_e}\widehat{f}^{(1)}(q_i),
\end{aligned}
\tag{3.5.2}
$$

where N_e denotes the total electron's number in the system. H_{el-ph} is a one-body operator for the electrons. Using the electron's field operators $\Psi^+(r, s_z)$ and $\Psi(r, s_z)$, considering the spin-independence of electron-phonon interaction, utilizing Eqs. (2.6.3.5), (2.6.3.7), (3.4.19.a), (2.6.1.1.a-b), defining

$$
D(k) = -\sqrt{\frac{\hbar}{2NM\omega(k)}}, \tag{3.5.3}
$$

making variable transformation $r - R_l = r - al \rightarrow r'$, using the following summation formula

$$
\sum_{l=1}^{N} e^{-i(p/\hbar - k - p'/\hbar)al} = N\delta_{p/\hbar - k - p'/\hbar, 2\pi w/a}, \quad w = 0, \pm 1, \pm 2, \cdots, \tag{3.5.4}
$$

using long wave approximation of phonon (namely, $k << 1/a$), and making integration by parts for the variable r', we obtain that H_{el-ph} can be transformed into an expression using the phonon language:

$$
\begin{aligned}
H_{el-ph} &= \int dr \Psi^+(r) \left[\sum_{l=1}^{N} \frac{\partial U(r-R_l)}{\partial R_l} x_l \right] \Psi(r) \\
&= \int dr \sum_p \frac{1}{\sqrt{L}} e^{-ipr/\hbar} a_p^+ \left[\sum_{l=1}^{N} \frac{\partial U(r-R_l)}{\partial R_l} x_l \right] \sum_{p'} \frac{1}{\sqrt{L}} e^{ip'r/\hbar} a_{p'} \\
&= \int dr \sum_p \frac{1}{\sqrt{L}} e^{-ipr/\hbar} a_p^+ \left[\sum_{l=1}^{N} \frac{\partial U(r-R_l)}{\partial R_l} \right. \\
&\quad \left. \sum_{k=1} D(k) \left(e^{ikal} b_k + e^{-ikal} b_k^+ \right) \right] \sum_{p'} \frac{1}{\sqrt{L}} e^{ip'r/\hbar} a_{p'} \\
&= \int dr \sum_p \frac{1}{L} e^{-ipr/\hbar} a_p^+ \sum_{l=1}^{N} \frac{\partial U(r-R_l)}{\partial R_l} \sum_{k=1} D(k) e^{ikal} b_k \sum_{p'} e^{ip'r/\hbar} a_{p'} \\
&+ \int dr \sum_p \frac{1}{L} e^{-ipr/\hbar} a_p^+ \sum_{l=1}^{N} \frac{\partial U(r-R_l)}{\partial R_l} \sum_{k=1} D(k) e^{-ikal} b_k^+ \sum_{p'} e^{ip'r/\hbar} a_{p'} \\
&= \sum_{pkp'} \frac{1}{L} D(k) a_p^+ b_k a_{p'} \sum_{l=1}^{N} e^{-i(p/\hbar - k - p'/\hbar)al} \int dr' e^{-i(p-p')r'/\hbar} \frac{dU(r')}{dr'} \\
&+ \sum_{pkp'} \frac{1}{L} D(k) a_p^+ b_k^+ a_{p'} \sum_{l=1}^{N} e^{-i(p/\hbar + k - p'/\hbar)al} \int dr' e^{-i(p-p')r'/\hbar} \frac{dU(r')}{dr'} \\
&= \sum_{pkp'} \frac{1}{L} D(k) a_p^+ b_k a_{p'} N \delta_{p/\hbar, k+p'/\hbar} i \frac{p-p'}{\hbar} a U(p-p') \\
&+ \sum_{pkp'} \frac{1}{L} D(k) a_p^+ b_k^+ a_{p'} N \delta_{p/\hbar+k, p'/\hbar} i \frac{p-p'}{\hbar} a U(p-p') \\
&= \sum_{pk} ikU(k) D(k) (a_p^+ b_k a_{p-\hbar k} + a_p^+ b_k^+ a_{p+\hbar k}), \qquad (3.5.5)
\end{aligned}
$$

where a is the length between two nearest neighbour ions, i. e., the lattice constant of this one dimensional ion chain, $L = Na$, and

$$U(k) = \frac{1}{a} \int dr U(r) e^{ikr}$$

Noting that a_p^+, a_p, b_k^+, b_k represent creation and annihilation operators of electron and phonon, respectively, the physical meanings of the two terms in Eq. (3.5.5) are clear. The first term contains $a_p^+ b_k a_{p-\hbar k}$, which means that the electron with momentum $p - \hbar k$ absorbs an acoustic phonon with momentum $\hbar k$, and becomes an electron with momentum p. In this process the momentum is conservative. The second term contains $a_p^+ b_k^+ a_{p+\hbar k}$, which means that the electron with momentum $p + \hbar k$ emits an acoustic phonon with momentum $\hbar k$, and becomes an electron with momentum p. In this process the momentum is conservative as well.

3.6 Exercises and Solutions

(1). Determine the fermi energy E_F and the Fermi velocity v_F, the speed of an electron at Fermi energy, for sodium. The molecular weight for sodium is 23 and the density of metallic sodium is 0.97 g/cm^3. There is one valence electron per atom for sodium.

Solution:

$$\frac{N}{V} = \frac{0.97\ g/cm^3}{23\ g/mole}(6.02 \times 10^{23}\ atoms/mple) = 2.5 \times 10^{22}\ \frac{atoms}{cm^3}.$$

Substituting this result into Eq. (3.2.2.18), we obtain $E_F = 3.1$ eV for sodium. Since this is the kinetic energy of the most energetic electron in the box, the Fermi velocity is determined from

$$E_F = \frac{mv_F^2}{2}.$$

For sodium, $v_F = 1.0 \times 10^6$ m/sec. Thus the most energetic electrons are nonrelativistic, which is necessary for self consistency since we have used the nonrelativistic Schrödinger equation to determine the allowed energies. But the speed of these most energetic electrons is appreciable, namely 0.3/100 of the speed of light.

(2). If we take the white dwarfs as electron gas, then the electron number density is of order of 10^{27} electrons/cm^3. Seek the Fermi Energy.

Solution:
Plugging the number density into the Eq. (3.2.2.18) gives

$$E_F \approx 3 \times 10^5 (eV).$$

(3). The radius of the heavy nucleus is given by

$$R = 1.25 \times 10^{-12} A^{1/3} (cm)$$

where A is the number of nucleons (=neutrons+potons), while the particle number density is given by the usual relation

$$n = \frac{A}{(4\pi/3)R^3}.$$

Seek the Fermi energy.

Solution:
Substituting all the given relations into Eq. (3.2.2.18) yields

$$E_F \approx 3 \times 10^7 (eV).$$

(4). Calculate the de Bloglie wavelength λ_F of an electron in a metal at the Fermi energy. Compare this length with the spacing between electrons in the metal, l_a.

Solution:

$$\lambda_F = \frac{\hbar\pi}{p_F}$$

From Eq. (3.2.2.18) one obtains

$$E_F = \frac{p_F^2}{2m} = \frac{h^2}{2m\lambda_F} = \frac{\hbar^2}{2m}(3\pi^2 n)^{2/3}$$

The spacing between electrons is roughly

$$l_a \approx (n)^{1/3} = \frac{1}{2}\left(\frac{3}{\pi}\right)^{1/3} \lambda_F$$

Therefore, the spacing between atoms in the metal is roughly half the de Bloglie wavelength of the most energetic conduction electrons. The lower the energetic electron's energy is, the larger the λ_F. The quantum effect becomes pronounced in the multiparticle systems when $l_a \lesssim \lambda_F$.

Chapter 4

Scattering Theory

4.1 Introduction

One of the best ways to understand the structure of particles and the forces between them is to scatter them off each other. This is particularly true at the microscopic level where the systems cannot be seen directly by eye, and must be probed by indirect means. The scattering process gives us information about the projectile, the target, and the interaction between them. A natural way to get information is to consider cases where two of these are known and learn about the third. For example, the structure of the photon and its coupling to matter are well understood (According to Ref. [3], photon is a wavepacket in position position. In practical treatment, but, the structure of the wavepacket of photon is not important, and is thus taken approximately as a point particle without structure.) The scattering experiment that the neutron, which is electrically neutral, interacts with the photon tells us that the neutron is built out charged constituents (two quarks with $2e/3$ and -$2e/3$), spin 1/2, and that the interaction between the two quarks gets very weak as they get close.

The C-C quantum transition is called scattering process. The scattering theory is a very extensive subject, and is addressed in many textbooks [18, 19, 12, 13, 20, 21, 22, 14]. Considering that we have found some mistakes in scattering theory of some references, we have to write this chapter. We list the following mistakes in Refs. [18, 19, 12, 13, 20, 21, 22, 14], and the viewpoints of this textbook. We hope that after reading this chapter, readers knows clearly that what and where these wrong are.

(1). The title of the textbook Ref. [13] is "Scattering theory of waves and particles".

(2). Some references take $t \to \infty$ before integration.

(3). Some references obtain conclusion that scattering processes in first order perturbation keep energy conservation.

(4). Some references obtain conclusion that scattering processes in second perturbation do not keep energy conservation, but call these energy nonconservation processes the virtual processes.

(5). Some references think that scattering processes can be treated in principle by time-independent Schrödinger equation, and, correspondingly, can use Green's function method.

This textbook thinks that:

(1). A wave, for example, a plane wave is a wavefunction of free particle. The wavefunction and the guide-factor are the same thing. The guide-factor does not carry any mass, charge, spin, and so on. The action of a guide-factor is to guide the motion of the body-factor of a wavepacket (wavepacket = particle) according to the spontaneous equiprobability symmetry breaking rule. When we write the interaction Hamiltonian, we always consider the true interaction between two substances of the two particles. Therefore, the so-called collision between plane wave and particle likes "Put Alice's hat on Bob's head". The scattering occurs always between particle and particle, such as photon-neutron, neutron-nucleon, and so on, other than wave-particle.

(2). We have to understand physical meaning of the t (or, more exactly, $t - t_\circ$). The t represents the collision (=interaction) time duration, which's magnitude order is less than 10^{-6} second. Therefore, the t cannot be taken as ∞ without any proof in mathematics and without illustration in physics.

(3). The scattering processes, often appearing in all references and textbboks, are essentially a quantum transition of continuous spectra into continuous spectra (Briefly, C-C transition). Several theorems in chapter 1 prove that the transition processes are generally energy-nonconservative.

(4). One cannot call the second order perturbation processes the virtual processes due to the energy nonconservation. One cannot *a priori* believe energy conservation, because the system in transition process is an open other than an isolated, or say, a closed system. In fact, the meson has been observed in experiments, and the producing process of meson should connect with a real process other than the so-called "virtual process".

(5). The time-independent Schrödinger equation and Green function equation always keep energy conservation, and thus they are, in principle, wrong for transition processes. The origins of mistakes will be interpreted in the next.

4.2 Scattering Matrix

4.2.1 Introduction

The aim of section 4.2 is to restate the transition theory in chapter 1 by introducing time evolution operator and S matrix, which were introduced by any references.

The interaction picture in chapter 1 tells us that if the Hamiltonian $H = H_\circ + H'$, where $H_\circ$ is , for simplicity, a time-independent operator and just H' depends on time, then the perturbation Hamiltonian H' in interaction picture is

$$\begin{aligned} H_I'(t) &= U_0^+(t,t')H'(t)U_0(t,t') = U(t',t)H'(t)U(t,t') \\ &= e^{iH_\circ(t-t')/\hbar}H'(t)e^{-iH_\circ(t-t')/\hbar}. \end{aligned} \tag{4.2.1.1}$$

The composition property of the time evolution operator is

$$U_0(t,t')U_0(t',t_1) = U_0(t,t_1). \tag{4.2.1.2}$$

Substituting Eqs. (4.2.1.1-2) into Eq. (5.7.1.19) of chapter 1, the time evolution operator in interaction picture becomes

$$\begin{aligned} U_I(t,t_\circ) &= T\exp\left(-\frac{i}{\hbar}\int_{t_\circ}^{t} dt' H_I'(t')\right) \\ &= \sum_{n=0}^{\infty}\left(\frac{-i}{\hbar}\right)^n \int_{t_\circ}^{t} dt_1 H_I'(t_1)\int_{t_\circ}^{t_1} dt_2 H_I'(t_2)\cdots\int_{t_\circ}^{t_{n-1}} dt_n H_I'(t_n) \\ &= \sum_{n=0}^{\infty}\left(\frac{-i}{\hbar}\right)^n \int_{t_\circ}^{t} dt_1 \int_{t_\circ}^{t_1} dt_2 \cdots \int_{t_\circ}^{t_{n-1}} dt_n U_0(t',t_1)H'(t_1)U_0(t_1,t_2) \\ &\times\ H'(t_2)U_0(t_2,t_3)\cdots U_0(t_{n-1},t_n)H'(t_n)U_0(t_n,t'), \end{aligned} \tag{4.2.1.3}$$

where t' is an arbitrary time parameter.

If $H_\circ$ does not depend on time explicitly, then Eq. (4.2.1.3) becomes

$$\begin{aligned} U_I(t,t_\circ) &= T\exp\left(-\frac{i}{\hbar}\int_{t_\circ}^{t} dt' H_I'(t')\right) \\ &= \sum_{n=0}^{\infty}\left(\frac{-i}{\hbar}\right)^n \int_{t_\circ}^{t} dt_1 \int_{t_\circ}^{t_1} dt_2 \cdots \int_{t_\circ}^{t_{n-1}} dt_n e^{-iH_\circ(t'-t_1)/\hbar} H'(t_1) \\ &\times\ e^{-iH_\circ(t_1-t_2)/\hbar} H'(t_2) e^{-iH_\circ(t_2-t_3)/\hbar} \cdots \\ &\quad e^{-iH_\circ(t_{n-1}-t_n)/\hbar} H'(t_n) e^{-iH_\circ(t_n-t')/\hbar}, \end{aligned} \tag{4.2.1.4}$$

The time evolution operator $U_I(t,t_\circ)$ is called the scattering matrix or S matrix or collision operator [18, 22].

$$U_I(t,t_\circ) \equiv S(t,t_\circ). \tag{4.2.1.5}$$

The definition of scattering matrix is often with $t \to +\infty$ and $t_\circ \to -\infty$, and thus S means $S(-\infty,+\infty)$ [22]. This textbook does not use this default values. According to some theorems on transition processes in chapter 1, the limits $t \to +\infty$ and $t_\circ \to -\infty$ are wrong in both physics and mathematics, and lead to the wrong energy conservation in transition processes.

Let us prove that the transition probability amplitude between initial unperturbed state $|k>$ and final unperturbed state $|k'>$ (suppose that $|k>$ and $|k'>$ just represent one state, respectively.) is the matrix element of the S matrix (which is the time evolution operator in interaction picture), where the unperturbed states are taken to be at any arbitrary time.

For the calculation of transition probability, recall that the expansion of a state $|\phi>$ in terms of an orthogonal complete set of states, $|\psi_n>$, is

$$|\phi>=\sum_n <\psi_n|\phi>|\psi_n>, \tag{4.2.1.6}$$

and, therefore, according to the spontaneous equiprobability symmetry breaking, for the state $|\phi>$, the probability of finding the particle being in state $|\psi_n>$ is $|<\psi_n|\phi>|^2$. Now suppose that the state $|\phi>$ is a state $|\psi_i(t)>$, where the state at an earlier time t' was an unperturbed state $|\psi_i^{(0)}(t')>$, typically an eigenstate of $H_\circ$. Then we know that the state at time t is

$$|\psi_i(t)>=U(t,t')|\psi_i^{(0)}(t')>, \tag{4.2.1.7}$$

and $U(t,t')$ contains full Hamiltonian including the perturbation $H'(t)=V(t)$. The system will not be an unperturbed state any more, but a superposition of unperturbed states. For the state $|\psi_i(t)>$, the probability finding the particle in state $|\psi_f^{(0)}(t)>$ will be $|<\psi_f^{(0)}(t)|\psi_i(t)>|^2$. Therefore, the probability amplitude for transition from an unperturbed state $|\psi_i^{(0)}(t_\circ)>$ to an unperturbed state $|\psi_f^{(0)}(t)>$ between $t_\circ$ and t is the matrix element of the time evaluation operator in the interaction picture

$$\begin{aligned} <\psi_f^{(0)}(t)|\psi_i(t)> &= <\psi_f^{(0)}(t)|U(t,t_\circ)|\psi_i^{(0)}(t_\circ)> \\ &= <\psi_f^{(0)}(t')|U_0^+(t,t')U(t,t_\circ)U_0(t_\circ,t')|\psi_i^{(0)}(t')> \\ &= <\psi_f^{(0)}(t')|U_I(t,t_\circ)|\psi_i^{(0)}(t')>, \end{aligned} \tag{4.2.1.8}$$

where the unperturbed states are taken at some arbitrary fixed states. Ref. [18] proved that it can be proved that the t' in Eq. (4.2.1.8) is irrelevant for the transition matrix element.

This proof is as follows. The Schrödinger equations for the unperturbed states $|\psi_i^{(0)}(t')>$, $|\psi_f^{(0)}(t')>$, time evolution operator $U_0(t_\circ, t')$ and $U_0^+(t,t')$ imply

$$\frac{\partial}{\partial t'} < \psi_f^{(0)}(t')|U_0^+(t,t')U(t,t_\circ)U_0(t_\circ,t')|\psi_i^{(0)}(t') >= 0, \qquad (4.2.1.9)$$

i. e., the choice of the parameter t' is irrelevant for the transition matrix element. We set $t'=0$ in the following.

If we substitute the Eq. (4.2.1.3) into into Eq. (4.2.1.8), then we get a series ($H'=V$)

$$\begin{aligned} < \Psi_f^{(0)}(t)|\psi_i(t) > &= < \psi_f^{(0)}(t')|U_I(t,t_\circ)|\psi_i^{(0)}(t') >=< \psi_f^{(0)}(0)|U_I(t,t_\circ)|\psi_i^{(0)}(0) > \\ &= \sum_{n=0}^{\infty}\left(\frac{-i}{\hbar}\right)^n \int_{t_\circ}^{t} dt_1 \int_{t_\circ}^{t_1} dt_2 \cdots \int_{t_\circ}^{t_{n-1}} dt_n \\ &\times < \psi_f^{(0)}(0)|e^{iH_\circ t_1/\hbar}H'(t_1)e^{-iH_\circ(t_1-t_2)/\hbar}H'(t_2)e^{-iH_\circ(t_2-t_3)/\hbar} \\ &\times \cdots e^{-iH_\circ(t_{n-1}-t_n)/\hbar}H'(t_n)e^{-iH_\circ(t_n)/\hbar}|\psi_i^{(0)}(0) > . \end{aligned} \qquad (4.2.1.10)$$

Suppose that the unperturbed states are energy eigenstates of $H_\circ$

$$|\psi_f^{(0)}(0) >= |\psi_{k'}(0) >= |k' >, \qquad H_\circ|\psi_{k'}(0) >= E_{k'}|\psi_{k'}(0) >, \qquad (4.2.1.11)$$

$$|\psi_i^{(0)}(0) >= |\psi_k(0) >= |k >, \qquad H_\circ|\psi_k(0) >= E_k|\psi_k(0) >, \qquad (4.2.1.12)$$

Substituting Eq. (4.2.1.10) into Eq. (4.2.1.8) yields

$$\begin{aligned} &< \psi_f^{(0)}(t')|U_I(t,t_\circ)|\psi_i^{(0)}(t') >=< k'|U_I(t,t_\circ)|k > \\ &= \delta_{k',k} - \frac{i}{\hbar}\int_{t_\circ}^{t} dt_1 e^{i\omega_{k'k}t_1} < k'|H'(t_1)|k > \\ &- \frac{1}{\hbar^2}\sum_n \int_{t_\circ}^{t} dt_1 \int_{t_\circ}^{t_1} dt_2 e^{i\omega_{k'n}t_1} < k'|H'(t_1)|n > \\ &\times e^{i\omega_{nk}t_2} < n|H'(t_2)|k > + \cdots , \end{aligned} \qquad (4.2.1.13)$$

where $\omega_{k'k} = (E_{k'} - E_k)/\hbar$.

The transition probability from special initial state $|k>$ to a special finite state $|k'>$ is

$$P_{k\to k'}(t,t_\circ) = | < k'|U_I(t,t_\circ)|k > |^2. \qquad (4.2.1.14)$$

According to the derivation processes of Eq. (4.2.1.13), the initial and final states are the eigenstates of $H_\circ$. General speaking, the states $|k>$ and $|k'>$ in Eq. (4.2.1.14) can refer to different bases of states.

Due to the unitarity of $U_I(t,t_\circ)$, one can prove that $P_{k\to k'}(t,t_\circ)$ satisfies the normalization condition as a probability,

$$\begin{aligned} \sum_{k'} P_{k\to k'}(t,t_\circ) &= \sum_{k'} < k|U_I^+(t,t_\circ)|k' >< k'|U_I(t,t_\circ)|k > \\ &= < k|U_I(t_\circ,t)U_I(t,t_\circ)|k >=< k|k >= 1, \end{aligned} \qquad (4.2.1.15)$$

where the summation over k' includes the k.

The transition probability amplitude $< k'|U_I(t,t_\circ)|k >$ can also be denoted as scattering matrix element.

$$\begin{aligned} S_{k'k}(t,t_\circ) &= < k'|U_I(t,t_\circ)|k >=< k|U_I^+(t,t_\circ)|k' >^* \\ &= < k|U_I^-(t_\circ,t)|k' >^* = \left(S_{kk'}^{-1}(t,t_\circ)\right)^* = \left(S^{-1+}(t,t_\circ)\right)_{k'k}, \end{aligned} \qquad (4.2.1.16)$$

4.2.2 D-D Transition

D-D transition means the transition between the states in discrete spectra. Although the D-D transition does not belong to the scattering processes, due to that they belong to transition processes, we address here the D-D scattering. Using Eqs. (4.2.1.13-14) and (4.2.1.16), one can obtain the formula of first order transition probability for a discrete spectrum as follows. At first, we write the transition probability from a special initial state $|k>$ to a special state $|k'>$ in first order perturbation approximation

$$\begin{aligned} P_{k\to k'}(t,t_\circ) &= |S_{k'k}(t,t_\circ)|^2 = \left|-\frac{i}{\hbar}\int_{t_\circ}^{t} dt_1 e^{i\omega_{k'k}t_1} < k'|H'(t_1)|k>\right|^2 \\ &\neq \left|-\frac{i}{\hbar}\int_{-\infty}^{\infty} dt_1 e^{i\omega_{k'k}t_1} < k'|H'(t_1)|k>\right|^2 \\ &= \frac{2\pi}{\hbar^2}\left|< k'|H'(\omega_{k'k})|k>\right|^2. \end{aligned} \quad (4.2.2.1)$$

IIere we have to especially point out that the transition theorems in chapter 1 prove that we cannot take $t\to\infty$ and $t_\circ\to-\infty$. However, some references (such as [18, 20]) take this limits, and thus use the last expression of Eq. (4.2.2.1).

If $H'(t)$ is a monochromatic (=periodic=harmonic) perturbation,

$$H'(t) = V(t) = We^{\pm i\omega t}, \quad (4.2.2.2)$$

then Eq. (4.2.2.1) can be simplified as

$$\begin{aligned} P_{k\to k'}(t,t_\circ) &= |S_{k'k}(t,t_\circ)|^2 = \left|-\frac{i}{\hbar}\int_{t_\circ}^{t} dt_1 e^{i(\omega_{k'k}\pm\omega)t_1} < k'|W)|k>\right|^2 \\ &= \frac{1}{\hbar^2}| < k'|W|k > |^2 \frac{1}{(\omega_{k'k}\pm\omega)^2}\left|e^{i(\omega_{k'k}\pm\omega)t} - e^{i(\omega_{k'k}\pm\omega)t_\circ}\right|^2 \\ &= \frac{4}{\hbar^2}\left|< k'|W|k>\right|^2 \frac{1}{(\omega_{k'k}\pm\omega)^2}\sin^2[(\omega_{k'k}\pm\omega)(t-t_\circ)/2]. \end{aligned} \quad (4.2.2.3)$$

Some references (such as [18]) take the limits $t\to\infty$ and $t_\circ\to-\infty$, use the mathematical formula [20]

$$\lim_{\alpha\to\infty}\frac{\sin\alpha x}{x} = \pi\delta(x),$$

and thus obtain

$$P_{k\to k'}(t,t_\circ) = P_{k\to k'}(\infty,-\infty) = |S_{k'k}(\infty,-\infty)|^2 = \frac{(2\pi)^2}{\hbar^2}| < k'|W|k > |^2[\delta(\omega_{k'k}\pm\omega)]^2. \quad (4.2.2.4)$$

The following treatment of Ref. [18] is to use a mathematical formula

$$\delta(0) = \lim_{\omega\to0}\delta(\omega) = \lim_{\omega\to0}\lim_{T\to\infty}\int_{-T/2}^{T/2} dt e^{i\omega t} = \lim_{T\to\infty}\frac{T}{2\pi},$$

and obtain

$$P_{k\to k'}(\infty,-\infty) = |S_{k'k}(\infty,-\infty)|^2 = \frac{\pi}{\hbar^2}| < k'|W|k > |^2\delta(\omega_{k'k}\pm\omega)\lim_{T\to\infty}T. \quad (4.2.2.4)'$$

This textbook thinks that Eq. (4.2.2.4)′ is not reasonable obviously.

Using Eqs. (4.2.1.13-14) and (4.2.1.16), one can obtain the formula of total first order transition probability for a discrete spectrum from a special initial state $|k>$ to all possible finite states except the initial state $|k>$ as follows. At first, we write the transition probability from a special initial state $|k>$ to a special state $|k'>$ and then make summation over all possible final states.

$$\begin{aligned} P = \sum_{k' \neq k} P_{k \to k'}(t, t_\circ) &= \sum_{k' \neq k} |S_{k'k}(t, t_\circ)|^2 \\ &= \sum_{k' \neq k} \left| -\frac{i}{\hbar} \int_{t_\circ}^{t} dt_1 e^{i\omega_{k'k} t_1} < k'|H'(t_1)|k > \right|^2 . \end{aligned} \tag{4.2.2.5}$$

4.2.3 D-C Transition

D-C transition means a transition from discrete to continuous spectra. For example, we seek the probability for the system to go from a bound (discrete) state $|k>$ into a plane wave state $|\mathbf{k}'>=|k'>$ under the influence of a perturbation $H'(t) = V(t)$. Because $\mathbf{k}'$ is continuous, we should seek the differential transition probability amplitude into a volume element $d\mathbf{k}'$ around the wavevector $\mathbf{k}'$ instead of just one discrete states.

$$dP_{k \to \mathbf{k}'}(t, t_\circ) = d\mathbf{k}'| < \mathbf{k}'|U_I(t, t_\circ)|k > |^2. \tag{4.2.3.1}$$

Suppose that the volume of $\mathbf{k}'$ space is $\mathcal{K}$. The total transition probability from initial state $|k>$ to $\mathcal{K}$ space is

$$P_{k \to \mathcal{K}} = \int_{\mathcal{K}} d\mathbf{k}'| < \mathbf{k}'|U_I(t, t_\circ)|k > |^2. \tag{4.2.3.2}$$

If we define the transition probability density per $\mathbf{k}'$-space volume, $\mathcal{P}_{k \to \mathbf{k}'}$, instead of $dP_{k \to \mathbf{k}'}(t, t_\circ)$, then

$$\begin{aligned} \mathcal{P}_{k \to \mathbf{k}'}(t, t_\circ) &= \frac{dP_{k \to \mathbf{k}'}(t, t_\circ)}{d\mathbf{k}'} = | < \mathbf{k}'|U_I(t, t_\circ)|k) > |^2 \\ &= |S_{k'k}(t, t_\circ)|^2. \end{aligned} \tag{4.2.3.3}$$

Thus, the S matrix element is a transition probability density amplitude.

From the above discussions we can see that the S matrix theory is just a restatement of the transition theory in chapter 1. Therefore, if one takes the wrong limits $t, t_\circ \to \infty, -\infty$, one also will obtain the Fermi golden rule and energy conservation in transition processes. Let us use scattering theory to prove that the Fermi golden rule and energy conservation are generally not correct, and are approximate in some cases.

In D-C transition, we assume that the initial state $|k>$ is bound state with energy $E_k < 0$. The final state is an unbound state $|k'>$ of $H_\circ$ with $E' > 0$,

$$H_\circ|k'>= E'|k'> . \tag{4.2.3.4}$$

Note that sometimes, for brevity, we use $k' = \mathbf{k}'$ to represent all the quantum number of a state $|k'>$. The differential transition probability amplitude in first order perturbation approximation is

$$\begin{aligned} dP_{k \to k'} &= dE'\rho(E')| < k'|U_I(t, t_\circ)|k > |^2 \\ &= dE'\rho(E')|S_{k'k}(t, t_\circ)|^2 \\ &= dE'\rho(E') \left| -\frac{i}{\hbar} \int_{t_\circ}^{t} dt_1 e^{i\omega_{k'k} t_1} < k'|H'(t_1)|k > \right|^2 \\ &\neq dE'\rho(E') \left| -\frac{i}{\hbar} \int_{-\infty}^{\infty} dt_1 e^{i\omega_{k'k} t_1} < k'|H'(t_1)|k > \right|^2 \\ &= dE'\rho(E') \frac{2\pi}{\hbar^2} |< k'|H'(\omega_{k'k})|k >|^2 . \end{aligned} \tag{4.2.3.5}$$

Suppose that the perturbation in Eq. (4.2.3.4) is given by Eq. (4.2.2.2). This case corresponds to the ionization of hydrogen atom by absorption of one photon. In Ref. [3] there are many discussions on the ionizations. We calculate the first order approximation.

$$\begin{aligned} dP_{k\to k'} &= dE'\rho(E')|<k'|U_I(t,t_\circ)|k>|^2 \\ &= dE'\rho(E')|S_{k'k}(t,t_\circ)|^2 \\ &= dE'\rho(E')\frac{1}{\hbar^2}\frac{|<k'|W|k>|^2}{[(\omega_{k'k}-\omega)/2]^2} sin^2[(\omega_{k'k}-\omega)(t-t_\circ)/2]. \end{aligned} \quad (4.2.3.6)$$

Under first order approximation, the differential transition rate into a final state energy interval $E' \to E' + dE'$ is

$$\begin{aligned} dw_{k\to k'} &= \frac{\partial}{\partial t}dP_{k\to k'}(t,t_\circ) = dE'\rho(E')\frac{\partial}{\partial t}|<k'|U_I(t,t_\circ)|k>|^2 \\ &= dE'\rho(E')\frac{\partial}{\partial t}|S_{k'k}(t,t_\circ)|^2 \\ &= dE'\rho(E')\frac{1}{\hbar^2}\frac{|<k'|W|k>|^2}{[(\omega_{k'k}-\omega)/2]^2}\frac{\partial}{\partial t} sin^2[(\omega_{k'k}-\omega)(t-t_\circ)/2]. \end{aligned} \quad (4.2.3.7)$$

Integration over the final state energy E' then yields an expression for the transition rate of initial state $|k>$

$$\begin{aligned} w_{k\to all\ states\ of\ k'} &= \int_0^\infty dw_{k\to k'} \\ &= \int_0^\infty dE'\rho(E')\frac{1}{\hbar^2}\frac{|<k'|W|k>|^2}{[(\omega_{k'k}-\omega)/2]^2} \\ &\times \frac{\partial}{\partial t} sin^2[(\omega_{k'k}-\omega)(t-t_\circ)/2]. \end{aligned} \quad (4.2.3.8)$$

4.2.4 Transition Caused by Second Order Perturbation

To study the transition, caused just by second order perturbation, we take the third term in Eq. (4.2.1.13) here

$$
\begin{aligned}
S^{(2)}_{k'k}(t,t_\circ) &= -\frac{1}{\hbar^2}\sum_n \int_{t_\circ}^{t} dt_1 \int_{t_\circ}^{t_1} dt_2 \\
&\times \; e^{i\omega_{k'n}t_1} < k'|H'(t_1)|n > e^{i\omega_{nk}t_2+\epsilon} < n|H'(t_2)|k > \quad (\epsilon \to 0) \\
&= -\frac{1}{\hbar^2}\sum_n < k'|H'|n >< n|H'|k > \int_{t_\circ}^{t} dt_1 \\
&\times \; e^{i\omega_{k'n}t_1}\frac{1}{i\omega_{nk}+\epsilon}[e^{i\omega_{nk}t_1} - e^{i\omega_{nk}t_\circ}] \\
&= -\frac{1}{\hbar^2}\sum_n \frac{< k'|H'|n >< n|H'|k >}{i\omega_{nk}} \\
&\times \left\{\int_{t_\circ}^{t} dt_1 e^{i\omega_{k'k}t_1} - \int_{t_\circ}^{t} dt_1 e^{i\omega_{k'n}t_1} e^{i\omega_{nk}t_\circ}\right\} \\
&= -\frac{1}{\hbar^2}\sum_n \frac{< k'|H'|n >< n|H'|k >}{i\omega_{nk}+\epsilon}\left\{\frac{1}{i\omega_{k'k}}[e^{i\omega_{k'k}t} - e^{i\omega_{k'k}t_\circ}]\right. \\
&- \left.\frac{1}{i\omega_{k'n}}[e^{i\omega_{k'n}t} - e^{i\omega_{k'n}t_\circ}]e^{i\omega_{nk}t_\circ}\right\} \\
&= -\frac{1}{\hbar^2}\sum_n \frac{< k'|H'|n >< n|H'|k >}{i\omega_{nk}+\epsilon} \\
&\left\{\frac{1}{i\omega_{k'k}}e^{i\omega_{k'k}t_\circ}[e^{i\omega_{k'k}(t-t_\circ)} - 1] - \frac{1}{i\omega_{k'n}}e^{i\omega_{k'k}t_\circ}[e^{i\omega_{k'n}(t-t_\circ)} - 1]\right\}
\end{aligned}
$$

$$
\begin{aligned}
&= -\frac{1}{\hbar^2}\sum_n \frac{< k'|H'|n >< n|H'|k >}{i\omega_{nk}+\epsilon - i\epsilon} \\
&\times \left\{\frac{1}{i\omega_{k'k}}e^{i\omega_{k'k}t_\circ}e^{i\omega_{k'k}(t-t_\circ)/2}2isin[\omega_{k'k}(t-t_\circ)/2]\right. \\
&- \left.\frac{1}{i\omega_{k'n}}e^{i\omega_{k'k}t_\circ}e^{i\omega_{k'n}(t-t_\circ)/2}2isin[\omega_{k'n}(t-t_\circ)/2]\right\}. \qquad (4.2.4.1)
\end{aligned}
$$

In the first equality of Eq. (4.2.4.1), the $|n>$ represent an arbitrary complete set, which can contain both discrete and continuous spectrum. In the second equality of Eq. (4.2.4.1), for simplicity, suppose $H'(t) = H' = constant \neq 0, \; in \; time \; interval \; t_\circ \to t.$

$$
\begin{aligned}
\left|S^{(2)}_{k'k}(t,t_\circ)\right|^2 &= \frac{1}{\hbar^4}\left|\sum_n \frac{< k'|H'|n >< n|H'|k >}{\omega_{nk}}\right|^2 \\
&\times \left\{\frac{sin^2[\omega_{k'k}(t-t_\circ)/2]}{(\omega_{k'k}/2)^2} + \frac{sin^2[\omega_{k'n}(t-t_\circ)/2]}{(\omega_{k'n}/2)^2}\right. \\
&+ \left.\frac{sin[\omega_{k'k}(t-t_\circ)/2]sin[\omega_{k'n}(t-t_\circ)/2]}{\omega_{k'k}/2\omega_{k'n}/2}\right\}. \qquad (4.2.4.2)
\end{aligned}
$$

The total probability rate of transition, caused just by the second order perturbation, from an initial state $|k>$ to all possible final states is

$$\begin{aligned} w^{(2)}_{k\to all\ k'} &= \int_0^\infty dE'\rho(E')\frac{\partial}{\partial t}\left|S^{(2)}_{k'k}(t,t_\circ)\right|^2 \\ &\approx \frac{1}{\hbar^4}\int_0^\infty dE'\rho(E')\left|\sum_n \frac{<k'|H'|n><n|H'|k>}{\omega_{nk}+i\epsilon}\right|^2 \\ &\times \frac{\partial}{\partial t}\left\{\frac{sin^2[\omega_{k'k}(t-t_\circ)/2]}{(\omega_{k'k}/2)^2}+\frac{sin^2[\omega_{k'n}(t-t_\circ)/2]}{(\omega_{k'n}/2)^2}\right\}. \end{aligned} \tag{4.2.4.3}$$

In the second equality we neglect the third term of Eq. (4.2.4.2), because this term is oscillatory. Suppose that the states $|k'>$ are in a continuous spectrum. In Eq. (4.2.4.3) $E' = E'(k') = E'(\mathbf{k}')$.

Let us make further assumptions. Suppose that $\rho(E')\left|\sum_n \frac{<k'|H'|n><n|H'|k>}{\omega_{nk}}\right|^2$ is E'-independent. Thus, Eq. (4.2.4.3) becomes

$$\begin{aligned} w^{(2)}_{k\to all\ k'} &= \frac{1}{\hbar^4}\rho(E')\left|\sum_n \frac{<k'|H'|n><n|H'|k>}{\omega_{nk}-i\epsilon}\right|^2 \\ &\times \frac{\partial}{\partial t}\int_0^\infty dE'\left\{\frac{sin^2[\omega_{k'k}(t-t_\circ)/2]}{(\omega_{k'k}/2)^2}+\frac{sin^2[\omega_{k'n}(t-t_\circ)/2]}{(\omega_{k'n}/2)^2}\right\} \\ &\approx \frac{1}{\hbar^4}\rho(E')\left|\sum_n \frac{<k'|H'|n><n|H'|k>}{\omega_{nk}-i\epsilon}\right|^2 \\ &\times \frac{\partial}{\partial t}\left\{\hbar(t-t_\circ)\int_{-\infty}^\infty \frac{dE_{k'k}(t-t_\circ)}{2\hbar}\frac{sin^2[\omega_{k'k}(t-t_\circ)/2]}{(\omega_{k'k}(t-t_\circ)/2)^2}\right. \\ &+ \left.\hbar(t-t_\circ)\int_{-\infty}^\infty \frac{dE_{k'n}(t-t_\circ)}{2\hbar}\frac{sin^2[\omega_{k'n}(t-t_\circ)/2]}{(\omega_{k'n}/2)^2}\right\} \\ &= \frac{1}{\hbar^4}\rho(E')\left|\sum_n \frac{<k'|H'|n><n|H'|k>}{\omega_{nk}-i\epsilon}\right|^2 \hbar(\pi+\pi) \\ &= \frac{2\pi}{\hbar}\rho(E')\left|\sum_n \frac{<k'|H'|n><n|H'|k>}{E_{nk}-i\epsilon}\right|^2. \end{aligned} \tag{4.2.4.4}$$

Some references in the past about 80 years also obtained Eq. (4.2.4.4) [18]. However, we have to point out that the physical considerations in this textbook are completely different. In his famous lectures on nuclear physics at the University of Chigago in 1949, Fermi coined the phrase "Golden Rule #2" for the transition rate in first order perturbation approximation. (This textbook also calls "Golden Rule #2" the Fermi golden rule.) Fermi denoted the corresponding second order perturbation formula for transition rate

$$w^{(2)}_{k\to all\ k'} = \frac{2\pi}{\hbar}\rho(E')\left|\sum_n \frac{<k'|H'|n><n|H'|k>}{E_{nk}-i\epsilon}\right|^2\Bigg|_{E'=E(k')=E=E(k),\epsilon\to 0} \tag{4.2.4.5}$$

as "Golden Rule #1", because it is important for nuclear reactions through intermediate compound nuclei [23, 18].

Considering that Eq. (4.2.4.5) is adopted by some references, it is worth to address the derivation method of Eq. (4.2.4.5) by some references such as Ref. [18]. There are two conditions in the derivations of some references.

$$\begin{aligned}
w^{(2)}_{k\to all\ k'} &= \lim_{(t-t_\circ)\to\infty} \frac{1}{t-t_\circ}\int_0^\infty dE'\rho(E')\left|S^{(2)}_{k'k}(t,t_\circ)\right|^2 \\
&= \lim_{T\to\infty}\frac{1}{T}\int_0^\infty dE'\rho(E')\left|S^{(2)}_{k'k}(T/2,-T/2)\right|^2 \\
&= \int_0^\infty dE'\rho(E')\frac{1}{\hbar^4}\left|\sum_n <k'|H'|n><n|H'|k>\right|^2 \\
&\times \lim_{T\to\infty}\frac{1}{T}\left|\int_{-T/2}^{T/2}dt_1\int_{-T/2}^{t_1}dt_2 e^{i\omega_{k'n}t_1}e^{(i\omega_{nk}+\epsilon)t_2}\right|^2 \\
&= \frac{1}{\hbar^4}\int_0^\infty dE'\rho(E')\left|\sum_n <k'|H'|n><n|H'|k>\right|^2 \\
&\times \lim_{T\to\infty}\frac{1}{T}\left|\int_{-T/2}^{T/2}dt_1 e^{i\omega_{k'k}t_1}\frac{1}{i\omega_{nk}+\epsilon}\right|^2 \\
&= \frac{1}{\hbar^4}\rho(E')\left|\frac{\sum_n <k'|H'|n><n|H'|k>}{\omega_{nk}-i\epsilon}\right|^2\int_0^\infty dE' \\
&\times \lim_{T\to\infty}\frac{1}{T}\left|\frac{2isin(\omega_{k'k}T/2}{i\omega_{k'k}}\right|^2 \\
&= \frac{1}{\hbar^4}\rho(E')\left|\frac{\sum_n <k'|H'|n><n|H'|k>}{\omega_{nk}-i\epsilon}\right|^2\int_0^\infty dE' \\
&\times \lim_{T\to\infty}\left|\frac{sin^2[(\omega_{k'k}/2)T]}{T(\omega^2_{k'k}/2)^2}\right|^2
\end{aligned}$$

$$\begin{aligned}
&= \frac{1}{\hbar^4}\rho(E')\left|\frac{\sum_n <k'|H'|n><n|H'|k>}{\omega_{nk}-i\epsilon}\right|^2\int_0^\infty dE'\pi\delta(\omega_{k'k}/2) \\
&= \frac{1}{\hbar^4}\rho(E')\left|\frac{\sum_n <k'|H'|n><n|H'|k>}{\omega_{nk}-i\epsilon}\right|^2\int_0^\infty dE'\pi\hbar\delta(E_{k'k}) \\
&= \frac{2\pi}{\hbar}\rho(E')\left|\frac{\sum_n <k'|H'|n><n|H'|k>}{E_{nk}-i\epsilon}\right|^2\Bigg|_{E'=E(k')=E(k),\epsilon\to 0}. \qquad (4.2.4.6)
\end{aligned}$$

The first equality requires that the duration time is infinitely large, i. e., $t\to\infty$ and $t_\circ\to-\infty$. The third equality comes from the second equality of Eq. (4.2.4.1), and the ϵ is introduced to ensure the convergence of integration over t_2. The fifth equality requires an approximate condition, i. e., $\rho(E')\left|\frac{\sum_n <k'|H'|n><n|H'|k>}{\omega_{nk}-i\epsilon}\right|^2$ is independent of k'. The Eq. (4.2.4.6) is the typical derivation method adopted by some references [18].

Although the expressions, Eqs. (4.2.4.4) and (4.2.4.5), in this textbook and some references are nearly the same, but the physical ideas of the two methods are completely different. In the method of this textbook the integration over the final state's E' needs essentially to include the contributions coming from all the final states with even infinitely large energy.

On the contrary, in the method of some references the integration over the final state E' is formal, because the integrand includes an energy conservation factor $\delta(E' - E_k)$, i. e., the initial state can only transit into these final states which's energies are equal to the energy of the initial state. The question is that: "Does the energy in second order perturbation transition keep conservation or not?" To answer this question, we should note that the basic starting formulas of the two methods in this textbook and some references are exactly the same. But the method of this textbook needs only one approximate condition, and other steps of derivations are exact. In the derivation of Eq. (4.2.4.6), some references need not only the approximative condition but also the duration time infinitively large condition. In the derivations $H'(t) = H' = V = constant\ in\ time\ interval\ -\infty \to +\infty$. As is well known, if the perturbation is really a constant from $t = -\infty$ to $t = +\infty$, then the system is in certain stationary state, and thus there is no quantum transition. Moreover, the infinitely large condition of duration time is also not reasonable in magnitude order. As is well known, the duration time of perturbation should be nearly equal to the relaxation time. The magnitude order of the later is only about 10^{-6} *to* 10^{-15} second. Therefore, we conclude that the infinitely large condition of duration time is wrong both in physics and in mathematics.

The approximate condition, i. e., k'-independence of $\rho(E')\left|\frac{\sum_n <k'|H'|n><n|H'|k>}{E_{nk}-i\epsilon}\right|^2$, is very difficult to be satisfied. For example, generally, for a three dimensional system, $\rho(E') \propto \sqrt{E'}$ is already strongly energy-dependent. However, in the past nearly 80 years, the conclusions coming from Fermi golden rule were more than 10^3. This textbook estimates that more than 500 conclusions might have large errors, need to make exact calculations once again without using the two conditions, and might have some new discoveries.

4.3 Scattering Theory

In subsection 4.3.1 we address the general theory of the scattering off perturbation, in which the incident particle collides with a potential field and at the same time absorbs a photon. In subsection 4.3.2 we explain the physical meaning of the scattering amplitude (Some textbooks call it scattering length [24].). In subsection 4.3.3 we take the Rutherford scattering as an example to make comparisons of two scattering theories: this textbook's and some references' (such as Refs. [18, 20, 25, 24, 26]) theories. According to our analyses, some references' theories are wrong in principle, and are approximate only in some cases. As is well known, all the sizes of nuclei are determined by, for example, Rutherford scattering experiments. However, we would like to point out that all the so-called experimental measurements are connected tightly with formulas of scattering theory. If one uses some references' scattering theory to analyze Rutherford scattering experimental data, the radius of one nucleon is 1.2 fm [27]. For example, the radius of nucleus of ^{208}Pb is $(208\times1.2^3)^{1/3} \approx 7$ fm. However, if one uses this textbook's scattering theory to analyze Rutherford scattering experimental data, then the radius of one nucleon is 0.254×1.2 fm, and the radius of nucleus of ^{208}Pb is $0.254\times(208\times1.2^3)^{1/3} \approx 0.254\times 7$ fm. Note that this difference of 1 : 0.254 comes from $|<k'|H'|k>|^2\rho(k') = |<k'|W|k>|^2\rho(k') \propto \sqrt{E'} = \sqrt{mv^2/2}$. If $|<k'|H'|k>|^2\rho(k') = |<k'|W|k>|^2\rho(k') \propto E' = mv^2/2$, and one uses this textbook's scattering theory to analyze Rutherford scattering experimental data, then the radius of one nucleon is 0.023×1.2 fm, and the radius of nucleus of ^{208}Pb is $0.023 \times (208 \times 1.2^3)^{1/3} \approx 0.023 \times 7$ fm.

The above numerical examples show that this textbook's and some references' scattering theories have more or less numerical difference. From subsections 4.3.1-4 you will see that the scattering theories based on time-dependent perturbation theory or stationary state method (=Schrödinger equation+Green's function) require the energy conservation in scattering processes. On the contrary, the scattering theory of this textbook requires the energy nonconservation in the scattering processes. This textbook thinks that:

(i). Some references' scattering theories are wrong in principle, because they require the energy conservation in scattering processes;

(ii). In particular treatments of any scattering problems, some references give a Dirac delta function, which leads to simplification of numerical calculations. However, if $| < k'|H'|k > |^2\rho(k')$ (see the text) is strong k'-dependent, then the influence of energy nonconservation will be more stronger. In this case if one still use some references' theories, then it might be cause qualitative mistakes.

4.3.1 Scattering off a Periodic Perturbation

For transition between the states in continuous spectra, e. g., $|\mathbf{k}>\equiv|k>\rightarrow|\mathbf{k}'>\equiv|k'>$, the S matrix element is

$$S_{k',k} =< k'|U_I(t,t_\circ)|k >, \tag{4.3.1.1}$$

The differential scattering cross section $d\sigma_{k\rightarrow k'}$ is defined as follows:

$$j_{in}d\sigma_{k\rightarrow k'} = dw_{k\rightarrow k'} = \frac{dP_{k\rightarrow k'}}{t-t_\circ} = \frac{dn_{k'}|S_{k',k}|^2}{t-t_\circ}, \tag{4.3.1.2}$$

where $dn_{k'}$ represents the number of states in $d\mathbf{k}'$, j_{in} represents the the probability current density of one incident particle in state $|k>$, the dimension of which is $[1/(time\times area)]$, and $(t-t_\circ)\equiv T$ represents the duration time of scattering process. In physics, after scattering the incident particle in state $|k>$ can transit into many different final states with different values of $\mathbf{k}'$. We call $d\sigma_{k\rightarrow k'}$ the differential scattering cross section, and $dw_{k\rightarrow k'}$ represents the transition probability rate into a small volume element $d\mathbf{k}'$. The dimension of $d\sigma_{k\rightarrow k'}$ is *area*.

Our Hamiltonian is

$$H = \frac{\mathbf{p}^2}{2m} + We^{-i\omega t}, \tag{4.3.1.3}$$

and our unperturbed initial state is a plane wave $|k>$ (more exactly speaking, the guide-factor of particle is plane wave). The first order approximation of scattering matrix element is

$$\begin{aligned} S_{k',k} &= < k'|U_I^{(1)}(t,t_\circ)|k >= -\frac{i}{\hbar}\int_{t_\circ}^{t} dt_1 \exp\left[i\left(\hbar\frac{k'^2-k^2}{2m}-\omega\right)t\right] < k'|W|k > \\ &= -\frac{i}{\hbar} < k'|W|k > \frac{1}{i[\hbar(k'^2-k^2)/2m-\omega]} \\ &\times \left\{\exp\left[i\left(\hbar\frac{k'^2-k^2}{2m}-\omega\right)t\right] - \exp\left[i\left(\hbar\frac{k'^2-k^2}{2m}-\omega\right)t_\circ\right]\right\}. \end{aligned} \tag{4.3.1.4}$$

Substituting Eq. (4.3.1.4) into Eq. (4.3.1.2) yields

$$\begin{aligned} d\sigma_{k\rightarrow k'} &= \frac{dw_{k\rightarrow k'}}{j_{in}} \\ &= dn_{k'}\frac{| < k'|W|k > |^2}{j_{in}\hbar^2(t-t_\circ)}\frac{sin^2\{[\hbar(k'^2-k^2)/2m-\omega](t-t_\circ)/2\}}{\{[\hbar(k'^2-k^2)/2m-\omega]/2\}^2}, \end{aligned} \tag{4.3.1.5}$$

where the dimension of the probability current density $\vec{j}_{in}$ has to be $1/(time\times area)$. Take $\psi=\exp(i\mathbf{k}\cdot\mathbf{r})/\sqrt{V}$. Note that for $< k'|W|k >$ we have also to take $\psi=\exp(i\mathbf{k}\cdot\mathbf{r})/\sqrt{V}$.

$$\vec{j}_{in} = \frac{\hbar}{2im}(\psi^+\nabla\psi - \nabla\psi^+\cdot\psi) = \frac{\hbar\mathbf{k}}{mV} = \frac{\mathbf{v}}{V}, \tag{4.3.1.6}$$

where **v** represents the velocity of the incident particle. Substituting the Eq. (4.3.1.6) into Eq. (4.3.1.5), we obtain the total scattering cross section per solid angle

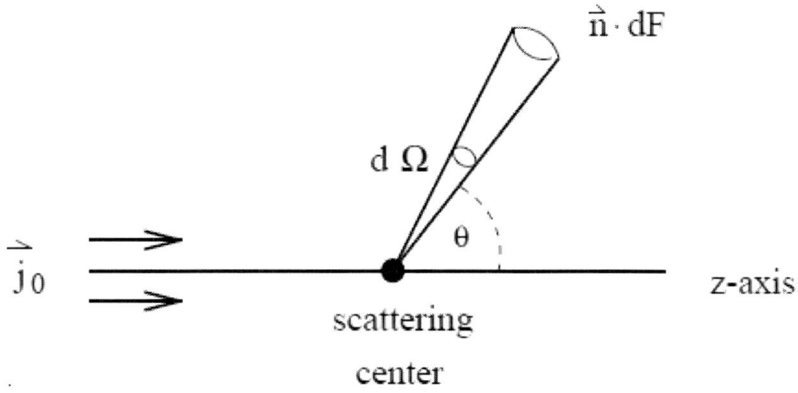

Figure 4.1: $\vec{j}_{\circ} \equiv \vec{j}_{in}$. $\vec{n}$ is an unit vector, and characterizes the direction of the areal element $dF = r^2 d\Omega$ through which the scattering particle penetrates.

$$\begin{aligned}
\frac{d\sigma}{d\Omega} &= \int \frac{d\sigma_{k\to k'}}{d\Omega} = \int \frac{dw_{k\to k'}}{d\Omega j_{in}} \\
&= \int \frac{dn_{k'}}{d\Omega} \frac{|<k'|W|k>|^2}{j_{in}\hbar^2(t-t_\circ)} \frac{sin^2\{[\hbar(k'^2-k^2)/2m-\omega](t-t_\circ)/(2\hbar)\}}{\{[\hbar(k'^2-k^2)/2m-\omega]/(2\hbar)\}^2} \\
&= \frac{V}{|\mathbf{v}|}\int_0^\infty dE'\rho(E',\Omega)\frac{|<k'|W|k>|^2}{\hbar^2(t-t_\circ)} \frac{sin^2[(E'-E-\hbar\omega)(t-t_\circ)/(2\hbar)]}{[(E'-E-\hbar\omega)/(2\hbar)]^2} \\
&= \frac{V}{|\mathbf{v}|}A\frac{|<k'|W|k>|^2}{\hbar^2(t-t_\circ)}\int_0^\infty dE'\sqrt{E'}\frac{sin^2[(E'-E-\hbar\omega)(t-t_\circ)/(2\hbar)]}{[(E'-E-\hbar\omega)/(2\hbar)]^2}
\end{aligned}$$

$$
\begin{aligned}
&\approx \frac{V}{|\mathbf{v}|} A \frac{|<k'|W|k>|^2}{\hbar^2} \frac{(2\hbar)^{3/2}}{\sqrt{t-t_\circ}} \int_0^\infty d\frac{(E'-E-\hbar\omega)(t-t_\circ)}{(2\hbar)} \\
&\times \sqrt{\frac{(E'-E-\hbar\omega)(t-t_\circ)}{(2\hbar)}} \frac{sin^2[(E'-E-\hbar\omega)(t-t_\circ)/(2\hbar)]}{[(E'-E-\hbar\omega)(t-t_\circ)/(2\hbar)]^2} \\
&= A\frac{Vm|<k'|W|k>|^2}{\hbar^3} \frac{(2\hbar)^{3/2}}{k\sqrt{t-t_\circ}} \int_{-(E+\hbar\omega)(t-t_\circ)/(2\hbar)}^\infty dx\sqrt{x}\frac{sin^2 x}{x^2} \\
&\equiv B\frac{(2\hbar)^{3/2}}{k\sqrt{t-t_\circ}} \int_{-(E+\hbar\omega)(t-t_\circ)/(2\hbar)}^\infty dx\sqrt{x}\frac{sin^2 x}{x^2} \\
&\equiv B\frac{(2\hbar)^{3/2}}{k\sqrt{t-t_\circ}} \int_{-\infty}^\infty dx\sqrt{x}\frac{sin^2 x}{x^2} \\
&= B\frac{(2\hbar)^{3/2}}{k\sqrt{t-t_\circ}}\sqrt{\pi} = \frac{V^2 m^{5/2}}{2\pi\hbar^{9/2}} \frac{|<k'|W|k>|^2}{k\sqrt{t-t_\circ}} \\
&= \frac{V^2 m^{5/2}}{2\pi\hbar^{9/2}} \frac{1}{k\sqrt{t-t_\circ}} \left| \int d\mathbf{r} \frac{1}{\sqrt{V}} e^{-i\mathbf{k}'\cdot\mathbf{r}} W(\mathbf{r}) \frac{1}{\sqrt{V}} e^{i\mathbf{k}\cdot\mathbf{r}} \right|^2 \\
&= \frac{(2\pi)^2 m^{5/2}}{\hbar^{9/2}} \frac{|W(\mathbf{k}'-\mathbf{k})|^2}{k\sqrt{t-t_\circ}} \\
&= \frac{8\pi m^{5/2}}{\hbar^{9/2}} \frac{1}{k\sqrt{t-t_\circ}} \frac{1}{|\mathbf{k}'-\mathbf{k}|} \left| \int_0^\infty dr r \sin[|\mathbf{k}'-\mathbf{k}|r] W(\mathbf{r}) \right|^2, \qquad (4.3.1.7)
\end{aligned}
$$

where $d\sigma$ is the differential scattering cross section per unit solid angle. In the third equality $E' = E_{k'}$, $E = E_k$, and $\rho(E', \Omega)$ represents the density of states per unit energy and per unit solid angle. In the fourth equality we use $\rho(E', \Omega) = A\sqrt{E'}$, *and* $A = Vm^{3/2}\sqrt{2}/(2\pi\hbar)^2$ [See the Eq. (10.4.4.3) in Ref. [3]]. In fifth equality the $< k'|W|k >$ is approximately taken as an energy independent of k', the aim of which is to see that in the simplest case how the dependence of $| < k'|W|k > |^2\rho(k')$ on k' to influence the values of the integral. In the twelfth equality, we introduce Fourier transform [3]. In the last equality we assume that the potential is of spherical symmetry.

$$
\frac{1}{(2\pi)^{3/2}} \int d\mathbf{r} e^{-i(\mathbf{k}'-\mathbf{k})\cdot\mathbf{r}} W(\mathbf{r}) \equiv W(\mathbf{k}'-\mathbf{k}). \qquad (4.3.1.8)
$$

If $|W(\mathbf{k}-\mathbf{k}')|^2$ is independent of Ω, then the total scattering cross section is

$$
\begin{aligned}
\sigma &= \int d\Omega \frac{d\sigma}{d\Omega} = \frac{(2\pi)^2 V^2 m^{5/2}}{\hbar^{9/2}} \frac{|<k'|W|k>|^2}{k\sqrt{t-t_\circ}} 4\pi \\
&= \frac{2(2\pi)^3 m^{5/2}}{\hbar^{9/2}} \frac{|W(\mathbf{k}'-\mathbf{k})|^2}{k\sqrt{t-t_\circ}}. \qquad (4.3.1.9)
\end{aligned}
$$

However, some references (such as Refs. [18, 20, 25, 24]) make the calculations quite different from our Eq. (4.3.1.7). It is necessary to introduce their method in detail at first, and then make comparisons. The starting point of some references is same as the fourth equality of Eq. (4.3.1.7), and then make the illegal limit, i. e., $(t - t_\circ) \to \infty$ in the following derivations.

$$\begin{aligned}
\frac{d\sigma}{d\Omega} &= \frac{V}{|\mathbf{v}|}A\frac{|<k'|W|k>|^2}{\hbar^2(t-t_\circ)}\int_0^\infty dE'\sqrt{E'}\frac{sin^2[(E'-E-\hbar\omega)(t-t_\circ)/(2\hbar)]}{[(E'-E-\hbar\omega)(t-t_\circ)/(2\hbar)]^2} \\
&= A\frac{Vm|<k'|W|k>|^2}{\hbar^3}\frac{\sqrt{E'}}{k}\int_0^\infty dE'\frac{sin^2[(E'-E-\hbar\omega)(t-t_\circ)/(2\hbar)]}{(t-t_\circ)[(E'-E-\hbar\omega)/(2\hbar)]^2} \\
&= \frac{Vm^2|<k'|W|k>|^2}{(2\pi)^3\hbar^5}\frac{k'}{k}\pi\int dE'\delta\left(\frac{E'-E-\hbar\omega}{2\hbar}\right) \\
&= \frac{V^2m^2|<k'|W|k>|^2}{(2\pi)^3\hbar^5}\frac{k'}{k}2\hbar\pi\Big|_{\frac{\hbar^2k'^2}{2m}-\frac{\hbar^2k^2}{2m}-\hbar\omega} \\
&= \frac{V^2m^2}{(2\pi)^3\hbar^5}\left|\int d\mathbf{r}\frac{1}{\sqrt{V}}e^{-i\mathbf{k}'\cdot\mathbf{r}}W(\mathbf{r})\frac{1}{\sqrt{V}}e^{i\mathbf{k}\cdot\mathbf{r}}\right|^2\frac{k'}{k}2\hbar\pi\Big|_{\frac{\hbar^2k'^2}{2m}-\frac{\hbar^2k^2}{2m}-\hbar\omega} \\
&= \left|\frac{m}{2\pi\hbar^2}\int d\mathbf{r}e^{-i(\mathbf{k}'-\mathbf{k})\cdot\mathbf{r}}W(\mathbf{r})\right|^2\frac{k'}{k}\Big|_{\frac{\hbar^2k'^2}{2m}-\frac{\hbar^2k^2}{2m}-\hbar\omega} \\
&= \left|\frac{\sqrt{2\pi}mW(\mathbf{k}'-\mathbf{k})}{\hbar^2}\right|^2\frac{k'}{k}\Big|_{k'=\sqrt{k^2+2m\omega/\hbar}} \\
&\equiv |-f(\mathbf{k}'-\mathbf{k})|^2\frac{k'}{k}\Big|_{k'=\sqrt{k^2+2m\omega/\hbar}}. \qquad (4.3.1.10)
\end{aligned}$$

If the incident particle just collides with the target, and does not absorbs the photon, then $\omega = 0$. In this condition, the sixth equality of Eq. (4.3.1.10) is same as the the Eq.(13.2.18) in Ref. [20]. The seventh equality of Eq. (4.3.1.10) is same as the Eq. (13.43) of Ref. [18].

To see the physical meaning of $f(\mathbf{k}'-\mathbf{k})$ in Eq. (4.3.1.10) please see the next subsection.

4.3.2 Physical Meaning of Scattering Amplitude

To see the physical meaning of scattering amplitude, let us at first consider a simplest case, i. e., suppose $k' = k$. For this aim we derive Eq. (4.3.1.10) with $\omega = 0$ by the time-independent Schrödinger equation [18].

The time-independent Schrödinger equation in potential $W(\mathbf{r})$ is

$$\left(-\frac{\hbar^2}{2m}\nabla^2 + W(\mathbf{r})\right)\psi(\mathbf{r}) = E\psi(\mathbf{r}) = \frac{\hbar^2k^2}{2m}\psi(\mathbf{r}). \qquad (4.3.2.1)$$

Define Green function $G(\mathbf{r},\mathbf{r}')$, which satisfies [20, 18, 25]

$$(\nabla^2 + k^2)G(\mathbf{r},\mathbf{r}') = -\delta(\mathbf{r}-\mathbf{r}'). \qquad (4.3.2.2)$$

One can verify both

$$\psi(\mathbf{r}) = \frac{2m}{\hbar^2}\int d\mathbf{r}'G(\mathbf{r},\mathbf{r}')W(\mathbf{r}')\psi(\mathbf{r}'). \qquad (4.3.2.3)$$

$$\begin{aligned}
\psi(\mathbf{r}) &= \psi^{(in)}(\mathbf{r}) + \psi^{(out)}(\mathbf{r}) = e^{i\mathbf{k}\cdot\mathbf{r}} + \psi^{(out)}(\mathbf{r}) \\
&= e^{i\mathbf{k}\cdot\mathbf{r}} + \frac{2m}{\hbar^2}\int d\mathbf{r}'G(\mathbf{r},\mathbf{r}')W(\mathbf{r}')\psi(\mathbf{r}') \qquad (4.3.2.4)
\end{aligned}$$

which is called Lippman-Schwinger equation [20].

Next we seek the Green function. Eq. (4.3.2.2) is of transitional invariance. The solution of Eq. (4.3.2.2) is [20, 18, 25]

$$G(\mathbf{r}) = \frac{1}{4\pi r}e^{ikr}. \qquad (4.3.2.5)$$

Substituting Eq. (4.3.2.5) into Eq. (4.3.2.4) yields

$$\begin{aligned}\psi(\mathbf{r}) &= e^{i\mathbf{k}\cdot\mathbf{r}} - \frac{m}{2\pi\hbar^2}\int d\mathbf{r}'\frac{W(\mathbf{r}')}{|\mathbf{r}-\mathbf{r}'|}e^{ik|\mathbf{r}-\mathbf{r}'|}\psi(\mathbf{r}') \\ &\approx e^{i\mathbf{k}\cdot\mathbf{r}} - \frac{m}{2\pi\hbar^2}\int d\mathbf{r}'\frac{W(\mathbf{r}')}{|\mathbf{r}-\mathbf{r}'|}e^{ik|\mathbf{r}-\mathbf{r}'|+i\mathbf{k}\cdot\mathbf{r}'}, \qquad (4.3.2.6)\end{aligned}$$

where $k = \mathbf{k}$. The first equality of Eq. (4.3.2.6) is an integral equation. To obtain the second equality of Eq. (4.3.2.6), we simply reinserted the leading plane wave term for $\psi(\mathbf{r}')$ in the integrand. According to the spirit of perturbation, this is the first order perturbation (W) approximation, which is called Born approximation.

For $r >> r'$, i. e., the perturbation is in a small region,

$$|\mathbf{r}-\mathbf{r}'| = \sqrt{r^2 - 2rr'\cos\theta + r'^2} \approx r - r'\cos\theta = r - \frac{1}{r}\mathbf{r}\cdot\mathbf{r}' = r - \overrightarrow{n}\cdot\mathbf{r}', \qquad (4.3.2.7)$$

where the definition of $\overrightarrow{n} = \mathbf{r}/r$ (See Fig. 4.1).

$$\frac{1}{|\mathbf{r}-\mathbf{r}'|} = \frac{1}{r(1 - r'\cos\theta/r)} \approx \frac{1}{r}. \qquad (4.3.2.8)$$

Substituting Eq. (4.3.2.7-8) into Eq. (4.3.2.6) yields the wavefunction of Born approximation

$$\begin{aligned}\psi(\mathbf{r}) &= e^{i\mathbf{k}\cdot\mathbf{r}} - \frac{m}{2\pi\hbar^2 r}e^{ikr}\int d\mathbf{r}'W(\mathbf{r}')e^{i(\mathbf{k}-k\overrightarrow{n})\cdot\mathbf{r}'} \\ &= e^{i\mathbf{k}\cdot\mathbf{r}} + \frac{1}{r}e^{ikr}(-)\frac{m}{2\pi\hbar^2}\int d\mathbf{r}'e^{-i(k\mathbf{n}-\mathbf{k})\cdot\mathbf{r}'} \\ &\equiv e^{i\mathbf{k}\cdot\mathbf{r}} + f(\Delta\mathbf{k})\frac{1}{r}e^{ikr} \\ &\equiv \psi^{(in)}(\mathbf{r}) + \psi^{(out)}(\mathbf{r}), \qquad (4.3.2.9)\end{aligned}$$

where

$$\begin{aligned}f(\Delta\mathbf{k}) &= f(\mathbf{k}'-\mathbf{k}) = -\frac{m}{2\pi\hbar}\int d\mathbf{r}'e^{-i(\mathbf{k}'-\mathbf{k})}W(\mathbf{r}') \\ &= -\sqrt{2\pi}\frac{m}{\hbar^2}W(\mathbf{k}'-\mathbf{k}) \\ &= -(2\pi)^2\frac{m}{\hbar^2} < k'|W|k > \\ &= f(2k\sin[(\theta)/2]) = -\frac{2m}{\hbar^2|\mathbf{k}'-\mathbf{k}|}\int_0^\infty dr r\sin(|\mathbf{k}'-\mathbf{k}|r)W(r) \qquad (4.3.2.10)\end{aligned}$$

is called scattering amplitude, which's physical meaning can be found in the third equality of Eq. (4.3.2.9). The scattering amplitude represents the amplitude of the scattering spherical wave. Note that in Eq. (4.3.2.10) $< k'|W|k >=< k\,\overrightarrow{n}\,|W|\mathbf{k} >$, and θ represents scattering angle. The last equality of Eq. (4.3.2.10) can be found in many references such as [18, 20].

The wavefunction in Eq. (4.3.2.9) is a superposition of wavefunctions (=guide-factors) both the incident and scattering particles simultaneously. $\psi^{(out)}$ is called outgoing spherical wave. The probability current density $\vec{j}_{out}$ of scattering particle is therefore

$$\begin{aligned}\vec{j}_{out} &= \frac{\hbar}{2im}(\psi^{(out)+}\nabla\psi^{(out)} - \nabla\psi^{(out)+}\cdot\psi)\Big|_{r\to\infty} \\ &= \frac{\hbar k\vec{n}}{mr^2}|f(\mathbf{k}'-\mathbf{k})|^2. \qquad (4.3.2.11)\end{aligned}$$

$$\vec{j}_{in} = \frac{\hbar}{2im}(\psi^{(in)+}\nabla\psi - \nabla\psi^{(in)+}\cdot\psi^{(in)}) = \frac{\hbar\mathbf{k}}{m}. \tag{4.3.2.12}$$

Note that because we use the delta function normalization in Eqs. (4.3.2.11-12) other than the box normalization, the dimensions of probability current density are *length/time* other than $1/(time \times area)$. However, the ratio of the two probability current density is independent of the approaches of normalization of the plane waves. Substituting Eqs. (4.3.2.11-12) into

$$\frac{d\sigma}{d\Omega} = \lim_{r\to\infty} r^2\frac{j_{out}}{j_{in}} = |f(\mathbf{k}'-\mathbf{k})|\,. \tag{4.3.2.13}$$

yields

$$\frac{d\sigma}{d\Omega} = |f(\mathbf{k}'-\mathbf{k})|\,, \tag{4.3.2.14}$$

which is same as the Eq. (4.3.1.10) in case of $\omega = 0$.

Referring to Eq. (4.3.2.14), we require that boundary condition ($r \to \infty$) of scattering waves is spherical waves. Under this boundary condition, we can derive an exact formula for the total scattering amplitude, which is a superposition of scattering amplitudes of all possible scattering spherical waves. Let us start from the exact formula of differential cross section of the second equality in Eq. (4.3.1.7).

$$\begin{aligned}\frac{d\sigma}{d\Omega} &= \int\frac{d\sigma_{k\to k'}}{d\Omega} = \int\frac{dw_{k\to k'}}{d\Omega j_{in}} \\ &= \int\frac{dn_{k'}}{d\Omega}\frac{|<k'|W|k>|^2}{j_{in}\hbar^2(t-t_\circ)}\frac{sin^2\{[\hbar(k'^2-k^2)/2m-\omega](t-t_\circ)/(2\hbar)\}}{\{[\hbar(k'^2-k^2)/2m-\omega]/(2\hbar)\}^2} \\ &= \int dn_{k'}\frac{1}{d\Omega j_{in}\hbar^2(t-t_\circ)}\frac{sin^2\{[\hbar(k'^2-k^2)/2m-\omega](t-t_\circ)/(2\hbar)\}}{\{[\hbar(k'^2-k^2)/2m-\omega]/(2\hbar)\}^2} \\ &\times \frac{(2\pi)^3}{V^2}|W(\mathbf{k}'-\mathbf{k})|^2 \\ &\equiv \int dn_{k'}\left|K[k',k,(t-t)_\circ,\Omega](2\pi)^{3/2}W(\mathbf{k}'-\mathbf{k}\right|^2 \\ &\equiv \int dn_{k'}]\,|f[k',k,(t-t_\circ),\Omega]|^2\,. \qquad (4.3.2.15)\end{aligned}$$

Next we prove that $f[k',k,(t-t_\circ),\Omega]$ represents the scattering amplitude of scattering spherical wave with wave $\mathbf{k}'$. We define a Schrödinger equation

$$(\nabla + k'^2)\psi^{(scat)}(\mathbf{r}) = 4\pi KW(\mathbf{r})\psi^{(scat)}(\mathbf{r}). \tag{4.3.2.16}$$

The $\Psi^{(scat)}(\mathbf{r})$ satisfies the following integral equation

$$\psi^{(scat)}(\mathbf{r}) = K4\pi\int d\mathbf{r}'G(\mathbf{r}-\mathbf{r}')W(\mathbf{r}')\psi^{(scat)}(\mathbf{r}'), \tag{4.3.2.17}$$

where $G(\mathbf{r}-\mathbf{r}')$ is called Green function, and satisfies

$$(\nabla + k'^2)G(\mathbf{r}-\mathbf{r}') = \delta(\mathbf{r}-\mathbf{r}). \tag{4.3.2.18}$$

It is easy to verify that [20, 18, 25]

$$G(\mathbf{r}-\mathbf{r}') = \frac{1}{4\pi} e^{ik|\mathbf{r}-\mathbf{r}'|}. \tag{4.3.2.19}$$

If you substitute the Eqs. (4.3.2.18-19) into Eq. (4.3.2.17), you will know that the $\psi^{(scat)}(\mathbf{r})$ is really the solution of Eq. (4.3.2.16). The $W(\mathbf{r}')$ is a perturbation. The $\psi^{(scat)}(\mathbf{r}')$ in the integrand of Eq. (4.3.2.17) is substituted by simple plane wave of free electron. (This is called Born approximation.) In Born approximation, Eq. (4.3.2.17) becomes

$$\begin{aligned} \psi^{(scat)}(\mathbf{r})\Big|_{r\to\infty} &\approx K4\pi \int d\mathbf{r}' G(\mathbf{r}-\mathbf{r}') W(\mathbf{r}') e^{i\mathbf{k}\cdot\mathbf{r}'} \\ &= K\frac{1}{r} e^{i\mathbf{k}'\cdot\mathbf{r}} \int d\mathbf{r}' e^{-i(\mathbf{k}'-\mathbf{k})} W(\mathbf{r}') \\ &= K\frac{1}{r} e^{i\mathbf{k}'\cdot\mathbf{r}} (2\pi)^{3/2} W(\mathbf{k}'-\mathbf{k}) \\ &= f[k', k, (t-t_\circ), \Omega] \frac{1}{r} e^{i\mathbf{k}'\cdot\mathbf{r}}. \end{aligned} \tag{4.3.2.20}$$

Our expression of scattering amplitude in Eq. (4.3.2.20) is more complex. But if you take $k' = k$, then you can obtain the result same as that from Lippman-Schwinger equation in the Born approximation and in the boundary condition that the scattering wave at $r \to \infty$ is a spherical wave.

4.3.3 Rutherford Scattering

Let us take the Rutherford scattering between electron and charged nucleus as an example of scattering process, and make more detail derivations and numerical calculations by using the two scattering theories: this textbook's and some references' theories.

There is a matrix element factor $< k'|W|k >$ in Eq. (4.3.1.7). The total wavefunctions of the electron-nucleus system are product of the wavefunctions of electron and nucleus. We use plane waves as an approximation for wavefunction of electron. Thus initial wavefunctions are

$$\Psi_k = \frac{1}{\sqrt{V}} e^{i\mathbf{p}\cdot\mathbf{r}/\hbar} \Phi_i(\mathbf{R}), \quad and \quad \Psi_{k'} = \frac{1}{\sqrt{V}} e^{i\mathbf{p}'\cdot\mathbf{r}/\hbar} \Phi_f(\mathbf{R}), \tag{4.3.3.1}$$

or, by using Dirac notation,

$$|\Psi_k >= |\mathbf{k} > |i >, \quad and \quad |\Psi_{k'} >= |\mathbf{k}' > |f > . \tag{4.3.3.2}$$

This can also be expressed as

$$< \mathbf{r}|\mathbf{k} >= \frac{1}{\sqrt{V}} e^{i\mathbf{p}\cdot\mathbf{r}/\hbar} = \frac{1}{\sqrt{V}} e^{i\mathbf{k}\cdot\mathbf{r}}, \tag{4.3.3.3a}$$

$$< \mathbf{r}|\mathbf{k}' >= \frac{1}{\sqrt{V}} e^{i\mathbf{p}'\cdot\mathbf{r}/\hbar} = \frac{1}{\sqrt{V}} e^{i\mathbf{k}'\cdot\mathbf{r}}, \tag{4.3.3.3b}$$

$$< \mathbf{R}|i >= \Phi_i(\mathbf{R}), \quad and \quad < \mathbf{R}|f >= \Phi_f(\mathbf{R}). \tag{4.3.3.3c}$$

The $\Phi(\mathbf{R})$ are the normalized wavefunctions of the nucleus. We do not normalized the plane waves in Eq. (4.3.3.1).

Let us now evaluate the matrix elements. Since the wavefunction factorizes into an electronic and a nuclear part, the volume element dV also implies integration over both volume dV_e (electronic space) and dV_n (nuclear space). The indices e and n stand for electron and nucleus, respectively. Thus

$$< \Psi_{k'}|H'|\Psi_k >= \frac{Ze^2}{V}\int \Phi_f^*(\mathbf{R})\Phi_i(\mathbf{R})\int \frac{\exp(-|\mathbf{r}-\mathbf{R}|/d)}{|\mathbf{r}-\mathbf{R}|}e^{i(\mathbf{k}-\mathbf{k}')\cdot\mathbf{r}}dV_e dV_n. \tag{4.3.3.4}$$

We first calculate the integration over the electronic coordinates and abbreviate $\mathbf{k}-\mathbf{k}'$ by the vector $\mathbf{s}$:

$$\mathbf{k}-\mathbf{k}' = \mathbf{s}. \tag{4.3.3.5}$$

Therefore, $\hbar\mathbf{s} = \mathbf{p}-\mathbf{k}'$ is momentum transfer from the electron to the nucleus during the scattering process. The integration over dV_e is evidently a function of $\mathbf{R}$:

$$J_e(\mathbf{R}) = \int \frac{\exp(-|\mathbf{r}-\mathbf{R}|/d)}{|\mathbf{r}-\mathbf{R}|}e^{i(\mathbf{k}-\mathbf{k}')\cdot\mathbf{r}}dV_e. \tag{4.3.3.6}$$

Since we integrate over the whole space, we can change our system of coordinates and integrate over the whole space again (the integration boundaries do not need be changed). We replace $\mathbf{r}$ by $\mathbf{r}'$=$\mathbf{r}$-$\mathbf{R}$ and introduce spherical coordinates with θ being the angle between $\mathbf{s}$ and $\mathbf{r}'$.

In terms of these coordinates we have

$$dV_c = r'^2 dr' \sin\theta d\theta d\phi, \tag{4.3.3.7}$$

and, from $\mathbf{r}'$=$\mathbf{r}$+$\mathbf{R}$, we get

$$\begin{aligned}(\mathbf{k}-\mathbf{k}')\cdot\mathbf{r} &= \mathbf{s}\cdot\mathbf{r} = \mathbf{s}\cdot\mathbf{r}' + \mathbf{s}\cdot\mathbf{R} \\ &= sr'\cos\theta + \mathbf{s}\cdot\mathbf{R}.\end{aligned} \tag{4.3.3.8}$$

Thus the integral becomes

$$\begin{aligned}J_e(\mathbf{R}) &= \int \frac{e^{-r'/d}}{r'}e^{i(sr'\cos\theta+\mathbf{s}\cdot\mathbf{R})}r'^2\sin\theta dr' d\theta d\phi \\ &= 2\pi e^{i\mathbf{s}\cdot\mathbf{R}}\int_0^\infty dr' \int_0^\pi d\theta r' e^{-r'/d}e^{isr'\cos\theta}\sin\theta \\ &= 2\pi e^{i\mathbf{s}\cdot\mathbf{R}}\int_0^\infty dr' r' e^{-r'/d}\frac{1}{isr'}(e^{isr'}-e^{-isr'}) \\ &= 2\pi e^{i\mathbf{s}\cdot\mathbf{R}}\frac{1}{is}\int_0^\infty dr'(e^{i(s-1/d)r'}-e^{-i(s-1/d)r'}) \\ &= -2\pi e^{i\mathbf{s}\cdot\mathbf{R}}\frac{1}{is}\left(\frac{1}{is-1/d}+\frac{1}{is+1/d}\right) \\ &= e^{i\mathbf{s}\cdot\mathbf{R}}\frac{4\pi}{s^2+1/d^2}.\end{aligned} \tag{4.3.3.9}$$

This result can be simplified: the term $1/d^2$ in the denominator can be neglected if $s^2d^2 >> 1$. This means that the momentum transfer must be not too small. Hence,

$$J_e(\mathbf{R}) \approx e^{i\mathbf{s}\cdot\mathbf{R}}\frac{4\pi}{s^2}. \tag{4.3.3.10}$$

Therefore the matrix takes the form

$$< \Psi_{k'}|W|\Psi_k >= \frac{Ze^2}{V}\frac{4\pi}{s^2}\int \Phi_f^*(\mathbf{R})e^{i\mathbf{s}\cdot\mathbf{R}}\Phi_i(\mathbf{R})dV_n. \tag{4.3.3.11}$$

For elastic scattering, the state of the nucleus is not changed. Let the nucleus be in its ground state Φ. Thus we have $\Phi = \Phi_f = \Phi_i$, and the product $Z\Phi^*(\mathbf{R})\Phi(\mathbf{R})$ is the density distribution of the protons in the nucleus. Instead of the wavefunctions, we can introduce the charge density $\rho_p(\mathbf{R})$ of the atomic nucleus [more accurately: $\rho_p(\mathbf{R})$ is the charge density without the factor e, which we have explicitly taken out]:

$$Z\Phi^*(\mathbf{R})\Phi(\mathbf{R}) = \rho_p(\mathbf{R}) \quad with \quad \int \rho_p(\mathbf{R})dV_n = Z. \tag{4.3.3.12}$$

For further simplification we assume spherical symmetry of the charge distribution:

$$\rho_p(\mathbf{R}) = \rho_p(R). \tag{4.3.3.13}$$

This assumption is only valid for atomic nuclei in the vicinity of the magic numbers. Therefore, we obtain for the matrix element

$$\begin{aligned} < \Psi_{k'}|H'|\Psi_k > &= \frac{4\pi e^2}{s^2}\int \rho_p(R)e^{is\cdot\mathbf{R}}\Phi_i(\mathbf{R})dV_n \\ &= \frac{4\pi e^2}{Vs^2}F(\mathbf{s}). \end{aligned} \tag{4.3.3.14}$$

The quantity $F(\mathbf{s})$ is called a form factor. It is the Fourier-transformed charge distribution and reflects the deviation of the nuclear charge distribution from point structure. Indeed, if the nucleus is assumed to be point-like, i. e., $\rho_p(\mathbf{R}) = \delta^3(\mathbf{R})$, we get $F = 1$.

The form factor can be further evaluated by again introducing spherical coordinates and using the axis defined by $\mathbf{s}$ as the polar axis. Hence, we get

$$dV_n = R^2 \sin\theta dRd\theta d\phi \quad and \quad \mathbf{s}\cdot\mathbf{R} = sR\cos\theta, \tag{4.3.3.15}$$

and therefore

$$\begin{aligned} F(\mathbf{s}) &= 2\pi\int_0^\infty\int_0^\pi \rho_p(R)e^{is\cos\theta R}R^2\sin\theta dRd\theta \\ &= 2\pi\int_0^\infty \rho_p(R)\frac{1}{isR}(e^{isR} - e^{-isR})R^2dR \\ &= \frac{4\pi}{s}\int_0^\infty \rho_p(R)\sin(sR)RdR \\ &= \frac{4\pi}{s^2}\rho_p\left[R_0\cos(sR_0) - \frac{(sR_0)}{s}\right]. \end{aligned} \tag{4.3.3.16}$$

The integral in the third equality can be calculated only if the charge distribution $\rho_p(R)$ is known. To continue our derivation, from the fourth equality is is supposed that $\rho_p(R)$ is a constant, and R_0 represents the radius of the nucleus. Our present result can be summarized as

$$< \Psi_{k'}|W|\Psi_k >= \frac{4\pi e^2}{Vs^2}F(s). \tag{4.3.3.17}$$

If $k' = k$, then

$$s^2 = k'^2 + k^2 - 2k'k\cos\theta \approx 2k^2 2\sin^2(\theta/2). \tag{4.3.3.18}$$

Substituting Eq. (4.3.3.17) into the third equality of Eq. (4.3.1.7) and setting $\omega = 0$ yield

$$
\begin{aligned}
\frac{d\sigma}{d\Omega} &= \frac{V}{|\mathbf{v}|}\int_0^\infty dE'\rho(E',\Omega)\frac{|<k'|W|k>|^2}{\hbar^2(t-t_\circ)}\frac{sin^2[(E'-E)(t-t_\circ)/(2\hbar)]}{[(E'-E)/(2\hbar)]^2} \\
&= \frac{AV^2m}{\hbar|\mathbf{k}|}\int_0^\infty dE'\sqrt{E'}\left|\frac{4\pi e^2}{Vs^2}F(s)\right|^2\frac{1}{\hbar^2(t-t_\circ)}\frac{sin^2[(E'-E)(t-t_\circ)/(2\hbar)]}{[(E'-E)/(2\hbar)]^2} \\
&= \frac{Am}{\hbar|\mathbf{k}|}\int_0^\infty dE'\sqrt{E'}\left|\frac{4\pi e^2}{s^2}\frac{4\pi}{s^2}\rho_p\left[R_0\cos(sR_0)-\frac{(sR_0)}{s}\right]\right|^2 \\
&\times \frac{1}{\hbar^2(t-t_\circ)}\frac{sin^2[(E'-E)(t-t_\circ)/(2\hbar)]}{[(E'-E)/(2\hbar)]^2}
\end{aligned}
$$

$$
\begin{aligned}
&= \frac{Am}{\hbar k}(4\pi e^2 4\pi\rho_p)^2\int_0^\infty dE'\sqrt{E'}\left|\frac{1}{s^4}\left[R_0\cos(sR_0)-\frac{(sR_0)}{s}\right]\right|^2 \\
&\times \frac{1}{\hbar^2(t-t_\circ)}\frac{sin^2[(E'-E)(t-t_\circ)/(2\hbar)]}{[(E'-E)/(2\hbar)]^2} \\
&\approx \frac{Am}{\hbar k}(4\pi e^2 4\pi\rho_p)^2\int_0^\infty dE'\sqrt{E'}\frac{4R_0^6}{9s^4}\frac{1}{\hbar^2(t-t_\circ)}\frac{sin^2[(E'-E)(t-t_\circ)/(2\hbar)]}{[(E'-E)/(2\hbar)]^2} \\
&\approx \frac{Am}{\hbar k}\frac{(\pi e^2 4\pi\rho_p)^2 4R_0^6}{[k^2\sin^2(\theta/2)]^2 9}\int_0^\infty dE'\sqrt{E'}\frac{1}{\hbar^2(t-t_\circ)}\frac{sin^2[(E'-E)(t-t_\circ)/(2\hbar)]}{[(E'-E)/(2\hbar)]^2} \\
&\approx \frac{Am}{\hbar k}\frac{(\pi e^2 4\pi\rho_p)^2 4R_0^6}{[k^2\sin^2(\theta/2)]^2 9}\frac{(2\hbar)^{3/2}}{\hbar^2\sqrt{t-t_\circ}}\int_0^\infty d\frac{(E'-E-\hbar\omega)(t-t_\circ)}{(2\hbar)} \\
&\times \sqrt{\frac{(E'-E-\hbar\omega)(t-t_\circ)}{(2\hbar)}}\frac{sin^2[(E'-E-\hbar\omega)(t-t_\circ)/(2\hbar)]}{[(E'-E-\hbar\omega)(t-t_\circ)/(2\hbar)]^2} \\
&= \frac{Am}{\hbar k}\frac{(\pi e^2 4\pi\rho_p)^2 4R_0^6}{[k^2\sin^2(\theta/2)]^2 9}\frac{(2\hbar)^{3/2}}{\hbar^2\sqrt{t-t_\circ}}\int_{-(E+\hbar\omega)(t-t_\circ)/(2\hbar)}^\infty dx\sqrt{x}\frac{sin^2x}{x^2} \\
&\approx \frac{Am}{\hbar k}\frac{(\pi e^2 4\pi\rho_p)^2 4R_0^6}{[k^2\sin^2(\theta/2)]^2 9}\frac{(2\hbar)^{3/2}}{\hbar^2\sqrt{t-t_\circ}}\int_{-\infty}^\infty dx\sqrt{x}\frac{sin^2x}{x^2} \\
&= \frac{Am}{\hbar k}\frac{(\pi e^2 4\pi\rho_p)^2 4R_0^6}{[k^2\sin^2(\theta/2)]^2 9}\frac{(2\hbar)^{3/2}}{\hbar^2\sqrt{t-t_\circ}}\sqrt{\pi}, \qquad (4.3.3.19)
\end{aligned}
$$

In the fifth equality, we assume $sR_0 < 1$, and expand the trigonometric functions. In the sixth equality, due to the factor $[(E'-E)/(2\hbar)]^2$ in the denominator of integrand, the contribution from $E'-E=0$ should be larger. Eq. (4.3.3.19) is the differential scattering cross section formula.

Next, we derive some references' differential scattering cross section formula. Some references' starting point is the same as the sixth equality of Eq. (4.3.3.19).

$$
\begin{aligned}
\left.\frac{d\sigma}{d\Omega}\right|_{some\ references} & \approx \frac{Am}{\hbar k}\frac{(\pi e^2 4\pi\rho_p)^2 4R_0^6}{[k^2\sin^2(\theta/2)]^2 9}\int_0^\infty dE'\sqrt{E'}\frac{1}{\hbar^2(t-t_\circ)} \\
& \times \frac{sin^2[(E'-E)(t-t_\circ)/(2\hbar)]}{[(E'-E)/(2\hbar)]^2} \\
& \approx \frac{Am}{\hbar k}\frac{(\pi e^2 4\pi\rho_p)^2 4R_0^6}{[k^2\sin^2(\theta/2)]^2 9}\frac{\sqrt{E'}2}{\hbar}\int_0^\infty d\frac{E'-E}{2\hbar} \\
& \times \lim_{(t-t_\circ)\to\infty}\frac{1}{(t-t_\circ)}\frac{sin^2[(E'-E)(t-t_\circ)/(2\hbar)]}{[(E'-E)/(2\hbar)]^2} \\
& = \frac{Am}{\hbar k}\frac{(\pi e^2 4\pi\rho_p)^2 4R_0^6}{[k^2\sin^2(\theta/2)]^2 9}\frac{\sqrt{E'}2}{\hbar}\int_0^\infty d\frac{E'-E}{2\hbar}\delta\left(\frac{E'-E}{2\hbar}\right) \\
& = \frac{Am}{\hbar k}\frac{(\pi e^2 4\pi\rho_p)^2 4R_0^6}{[k^2\sin^2(\theta/2)]^2 9}\frac{2\sqrt{E'}}{\hbar}\pi. \qquad (4.3.3.20)
\end{aligned}
$$

Although the basic starting point, i. e., the first equality, of this textbook and some references is exactly the same, in the second equality the differences begin to appear. The second equality makes an unproved approximation, i. e., taking out the $\sqrt{E'}$ from the integral sign from 0 to ∞. The third equality makes an even more serious assumption, i. e., $(t-t_\circ)\to\infty$. From exactly mathematical formula, this assumption leads to the Dirac delta function in the fourth equality. The delta function means that the scattering processes are energy conservation.

From Eqs. (4.3.3.19-20) we obtain

$$
\begin{aligned}
\frac{(d\sigma/d\Omega)_{this\ textbook}}{(d\sigma/d\Omega)_{some\ references}} & = \sqrt{\frac{2\hbar}{\pi(t-t_\circ)E}} \\
& = \sqrt{\frac{4\hbar}{\pi(t-t_\circ)mv^2}} = \sqrt{\frac{4\hbar v}{\pi l m v^2}} \\
& = \sqrt{\frac{4\hbar}{\pi l m v}} \\
& \approx \sqrt{\frac{1\times 1.05\times 10^{-27}}{\pi 10^{-12} 9.1\times 10^{-28}\times 1\times 10^5}} = 3.7\times 10^3, \qquad (4.3.3.21)
\end{aligned}
$$

where E is the energy of incident electron, $E = mv^2/2$, and $t - t_\circ$ is the duration time of the scattering of electron by the nucleus, $l = (t-t_\circ)v$ is the scattering length. In the last equality we take $l = 10^{-12}$ cm, $v = 10^5$ cm/second.

Let us estimate the influence of this difference to the so-called value of radius of a nucleus. Experimentalists do not know which formula is correct. Experimentalists only can give the experimental value of the differential scattering cross section. If the experimental value is $(d\sigma/d\Omega)_{experimental}$, then this textbook and some references will give different value of the radius of the nucleus at target. The ration is

$$
\frac{R_{0,this\ textbook}}{R_{0,some\ references}} = \left(\sqrt{\frac{\pi l m v}{4\hbar}}\right)^{1/6} = \left(\frac{1}{3.7\times 10^3}\right)^{1/6} = 0.254, \qquad (4.3.3.22)
$$

which means that from the same experimental value of the differential scattering cross section, the radius of the nucleus, determined by Rutherford scattering experiment and our scattering theory, will be less than the previous value, determined by Rutherford scattering experiment and some references' scattering theory.

Let us make a comparison for the case $| < k'|W|k > |^2\rho(k') \propto E'$ *other than* $\sqrt{E'}$. In this case, we have from Eq. (4.3.3.19)

$$\begin{aligned}
\frac{d\sigma}{d\Omega} &= \mathcal{N}\int_0^\infty dE'E'\frac{1}{(t-t_\circ)}\frac{sin^2[(E'-E)(t-t_\circ)/(2\hbar)]}{[(E'-E)/(2\hbar)]^2} \\
&= \mathcal{N}\frac{(2\hbar)^2}{t-t_\circ}\int_0^\infty d\frac{(E'-E)(t-t_\circ)}{2h}\left[\frac{(E'-E)(t-t_\circ)}{2\hbar}+\frac{E(t-t_\circ)}{2\hbar}\right] \\
&\times \frac{sin^2[(E'-E)(t-t_\circ)/(2\hbar)]}{[(E'-E)(t-t_\circ)/(2\hbar)]^2}
\end{aligned}$$

$$\begin{aligned}
&= \mathcal{N}\frac{(2\hbar)^2}{t-t_\circ}\int_{-E(t-t_\circ)}^\infty dx\frac{sin^2x}{x}+\mathcal{N}2\hbar E\int_0^\infty dx\frac{sin^2x}{x^2} \\
&= \mathcal{N}\frac{(2\hbar)^2}{t-t_\circ}\int_{-E(t-t_\circ)}^\infty dx\frac{sin^2x}{x}+\mathcal{N}2\hbar E\frac{\pi}{2}
\end{aligned}$$

$$\approx \mathcal{N}\left[\frac{(2\hbar)^2}{t-t_\circ}\times 500.635+2\pi\hbar E\right]. \tag{4.3.3.23}$$

For the lat equality we use $\approx$, because to avoid the divergence of integration at infinite energy we use $E^{0.999}$ other than E^1.

$$\begin{aligned}
\left.\frac{d\sigma}{d\Omega}\right|_{some\ refences} &= \mathcal{N}\int_0^\infty dE'E'\frac{1}{t-t_\circ}\frac{sin^2[(E'-E)(t-t_\circ)/(2\hbar)]}{[(E'-E)/(2\hbar)]^2} \\
&\approx \mathcal{N}E'\int_0^\infty dE'\frac{1}{t-t_\circ}\frac{sin^2[(E'-E)(t-t_\circ)/(2\hbar)]}{[(E'-E)/(2\hbar)]^2} \\
&= \mathcal{N}E'2\hbar\int_0^\infty d\frac{E'-E}{2\hbar}\lim_{(t-t_\circ)\to\infty}\frac{sin^2[(E'-E)(t-t_\circ)/(2\hbar)]}{[(E'-E)/(2\hbar)]^2} \\
&= \mathcal{N}E'2\hbar\pi\int_0^\infty d\frac{E'-E}{2\hbar}\pi\delta\left(\frac{E'-E}{2\hbar}\right) \\
&= \mathcal{N}E'2\hbar\pi. \qquad (4.3.3.24)
\end{aligned}$$

$$\begin{aligned}
\frac{(d\sigma/d\Omega)_{this\ textbook}}{(d\sigma/d\Omega)_{some\ references}} &= \left[\frac{(2\hbar)^2}{(t-t_\circ)}\times 500.835+2\pi\hbar E\right]/(2\pi\hbar E) \\
= \ &\frac{(2\hbar)^2}{(t-t_\circ)2\pi\hbar E}\times 500.835+1 \\
= \ &\frac{4\hbar}{\pi(t-t_\circ)mv^2}\times 500.635+1 \\
\approx \ &\frac{4\hbar}{\pi lmv}\times 500.635. \qquad (4.3.3.25)
\end{aligned}$$

Using the same data for Eq. (4.3.3.22) yields

$$\begin{aligned}
\frac{R_{0,this\ textbook}}{R_{0,some\ references}} &= \left(\frac{\pi lmv}{4\hbar 500.635}\right)^{1/6}\left(\frac{1}{(3.7\times 10^3)^2/500.635}\right)^{1/6} \\
&= (0.254)^2 0.355 = 0.023,
\end{aligned}$$

which means that from the same experimental value of the differential scattering cross section, the radius of the nucleus, determined by Rutherford scattering experiment and our scattering theory, will be much less than the previous value, determined by Rutherford scattering experiment and Some references' scattering theory.

4.4 Exercises and Solutions

(1). This exercise is for section 4.2.3.

Some references take the limits $t \to \infty$ and $t_\circ \to -\infty$, and obtain results different from the results in this textbook. Next, we introduce the method of Some references to obtain the $w_{k \to all\ k's}$. Would the readers please to make comparisons with our Eq. (4.2.3.8), and points out the mistakes in the derivations.

From the fourth equality in Eq. (4.2.3.5) and under the conditions of harmonic perturbation and first order approximation, Some references (such as [18]) make the following derivations:

$$\begin{aligned} dP_{k \to k'} &= dE'\rho(E')\left|-\frac{i}{\hbar}\int_{-\infty}^{\infty} dt_1 e^{i\omega_{k'k}t_1} < k'|H'(t_1)|k >\right|^2 \\ &= dE'\rho(E')\frac{1}{\hbar^2}| < k'|W|k > |^2(2\pi)^2[\delta(\omega_{k'k} - \omega)]^2 \\ &= dE'\rho(E')\frac{1}{\hbar^2}| < k'|W|k > |^2(2\pi)^2\delta(\omega_{k'k} - \omega) \\ &\times \lim_{(\omega_{k'k}-\omega)\to 0}\lim_{T\to\infty}\frac{1}{2\pi}\int_{T/2}^{T/2} dt e^{i(\omega_{k'k}-\omega)t} \\ &= dE'\rho(E')\frac{1}{\hbar^2}| < k'|W|k > |^2(2\pi)^2\delta(\omega_{k'k} - \omega)\lim_{T\to\infty}\frac{T}{2\pi}, \end{aligned} \tag{1.1}$$

where the first equality assumes the the duration time infinitely large condition, and the second equality assume the k'-independence of $\rho(E')\frac{1}{\hbar^2}| < k'|W|k > |^2(2\pi)^2$. Therefore,

$$\begin{aligned} w_{k\to all\ k's} &= \int_0^\infty dw_{k\to k'}\int_0^\infty \frac{1}{T}dP_{k\to k'} \\ &= \int_0^\infty dE'\rho(E')\frac{1}{\hbar^2}| < k'|W|k > |^2(2\pi)^2\delta[(\omega_{k'k} - \omega)\hbar]\hbar \\ &\times \lim_{T\to\infty}\frac{1}{T}\frac{T}{2\pi} = \frac{2\pi}{\hbar}\rho(E')| < k'|W|k > |^2\Big|_{E'=E_k+\hbar\omega} \end{aligned} \tag{1.2}$$

Our Eq. (1.2) is exactly the same as the Eq. (13.29) in Ref. [18]. That the $w_{k\to all\ states\ of\ k'}$ is independent of time is called Fermi golden rule.

Solution:

Let us give the method of this textbook to derive Fermi golden rule. We calculate the transition probability density (= differential transition probability) in first order approximation.

$$\begin{aligned} dP_{k\to(k')} &= dE'\rho(E')| < k'|U_I(t,t_\circ)|k > |^2 \\ &= dE'\rho(E')|S_{k'k}(t,t_\circ)|^2 \\ &= dE'\rho(E')\frac{1}{\hbar^2}\frac{| < k'|W|k > |^2}{[(\omega_{k'k} - \omega)/2]^2}sin^2[(\omega_{k'k} - \omega)(t - t_\circ)/2]. \end{aligned} \tag{1.3}$$

Under first order approximation, the differential transition rate into a final state energy interval $E' \to E' + dE'$ is

$$\begin{aligned} dw_{k\to k'} &= \frac{\partial}{\partial t} dP_{k\to k'}(t,t_\circ) = dE'\rho(E')\frac{\partial}{\partial t}| < k'|U_I(t,t_\circ)|k > |^2 \\ &= dE'\rho(E')\frac{\partial}{\partial t}|S_{k'k}(t,t_\circ)|^2 \\ &= dE'\rho(E')\frac{1}{\hbar^2}\frac{| < k'|W|k > |^2}{[(\omega_{k'k}-\omega)/2]^2}\frac{\partial}{\partial t}sin^2[(\omega_{k'k}-\omega)(t-t_\circ)/2]. \end{aligned} \quad (1.4)$$

Integration over the final state energy E' then yields an expression for the transition rate of initial state $|k>$

$$\begin{aligned} w_{k\to all\ k'} &= \int_0^\infty dw_{k\to k'} \\ &= \int_0^\infty dE'\rho(E')\frac{1}{\hbar^2}\frac{| < k'|W|k > |^2}{[(\omega_{k'k}-\omega)/2]^2} \\ &\times \frac{1}{t-t_\circ}sin^2[(\omega_{k'k}-\omega)(t-t_\circ)/2] \\ &\approx \frac{1}{\hbar^2}| < k'|W|k > |^2\rho(E')\frac{1}{t-t_\circ} \\ &\times \int_0^\infty dE'\frac{sin^2[(\omega_{k'k}-\omega)(t-t_\circ)\hbar/2\hbar]}{[(\omega_{k'k}-\omega)\hbar/2\hbar]^2} \\ &= \frac{1}{\hbar^2}| < k'|W|k > |^2\rho(E')\frac{1}{t-t_\circ}2\hbar(t-t_\circ) \\ &\times \int_{-(E_k-\hbar\omega)(t-t_\circ)/2\hbar}^\infty dx\frac{sin^2x}{x^2} \\ &\approx \frac{1}{\hbar^2}| < k'|W|k > |^2\rho(E')\frac{1}{t-t_\circ}2\hbar(t-t_\circ) \\ &\times \int_{-\infty}^\infty dx\frac{sin^2x}{x^2} \\ &= \frac{1}{\hbar^2}| < k'|W|k > |^2\rho(E')\frac{1}{t-t_\circ}2\hbar(t-t_\circ)\pi \\ &= \frac{2\pi}{\hbar}| < k'|W|k > |^2\rho(E'). \end{aligned} \quad (1.5)$$

The third and fourth equalities are an approximate expression, because we assume the $E'(k')$-independence of $| < k'|W|k > |^2\rho(E')$. Under this approximate condition, we derive the Fermi golden rule (= time independence of the transition rate of initial state $|k>$). From our derivation, we can see that Fermi golden is not an exactly rule. On the contrary, it is an quite approximate rule. This rule requires the independence of energy of final state of $| < k'|W|k > |^2\rho(E')$. In fact, generally, this condition cannot be satisfied. For example, always for a three dimensional system $\rho(E') \propto \sqrt{E'}$.

Let us discuss that why the two methods give the same equations in Eqs. (1.5) and (1.2)

(i). It is obviously that the duration time T of time-dependence perturbation has to have a magnitude order of relaxation time of the transition process. Generally, the relaxation time is from 10^{-7} to 10^{-15} second. Therefore, the limit $T \to \infty$ is not reasonable in physics. Therefore, when one uses the Fermi golden rule, one has to use the rule based the proof of this textbook, i. e., we should know that the Fermi golden rule is not exact, and thus one should: (a). consider the conditions in any particular cases, and make an estimations for the error; (b). abandon the energy conservation of transition process.

(ii). As is well known, when some references treat the problems in solid state, particle physics, nuclear physics, and so on, some references use the Fermi golden rule, and make further analyses based on the Fermi golden rule. By using this "cutting up with just one time knife", from 1930 up to 2013, some references make conclusions, the number of which will be more than 10^3. This textbook believes that in the 10^3 conclusions there are, at least, more than 500, which are of serious mistakes. If one person makes derivations once again, then one might have new discoveries.

(2). Some references, such as Refs. [18, 20], like to use partial wave method to treat scattering problem. The basic idea of this method is to expand the incident plane wave in a central symmetrical field $V(r) = W(r)$ as

$$\psi = \sum_{l=0}^{\infty} R_l(kr) Y_l^0(\theta).$$

After expansion some references can prove that the total scattering amplitudes equals the superposition of all scattering amplitudes of all the partial waves, i. e.,

$$f(\theta) = \sum_{l=0}^{\infty} f_l(\theta).$$

Please read this method, and give some comments. If you make a detail reading and comments, then is is might be that you can write a paper for publication.

(3). Prove that our complex total scattering amplitude is equal to the some references' scattering amplitude in condition of energy conservation.

(4). Please illustrate that in the time-dependent perturbation theory the reason of appearance of the density matrix in energy representation.

Chapter 5

Quantum Relaxation

AUNRTB's transition theory, including scattering theory, are based on the limit of *duration time* $\to \infty$, and therefore leads to the time independence of the transition rate. (AUNRTB=all until now references and textbooks.) AUNRTB define that the inverse of the transition rate is equal to the relaxation time τ. On the contrary, this textbook does not take the limit of *duration time* $\to \infty$ in our transition theory, and therefore leads to more or less time dependence of the transition rate and thus relaxation time. This chapter gives a new definition for the relaxation time, and proves that under the condition of time independence of transition rate our definition of relaxation time is equal to some references' definition. Ref. [5] is the first to address the quantum relaxation theory in quantum mechanics systematically.

Although the non-exponential relaxation phenomena and pink noise were observed more than 200 years ago, and are still the present research hot point, but unified microscopic mechanism and theory have not yet been proposed. This chapter will propose both unified mechanism and theory.

5.1 Introduction

Suppose that at $t = 0$ we have $N(0)$ particles in initial state. We denote the transition, including decay, probability by $P(t)$, and denote the number of particle keeping the initial state at time t as $N(t)$, then $-dN(t) = -[N(t+dt) - N(t)]$ represents the transited particle number during dt is

$$\begin{aligned} dN(t) &= N(t+dt) - N(t) = -\{-[N(t+dt) - N(t)]\} \\ &= -N(t)dP(t) = -N(t)\frac{dP(t)}{dt}dt. \end{aligned} \tag{5.1.1}$$

Eq. (5.1.1) can be rewritten as

$$\frac{dN(t)}{N(t)} = -dP(t). \tag{5.1.2}$$

Integrating the number of states from $N(0)$ to $N(t)$ as time varies from 0 to t, we obtain

$$\int_{N(0)}^{N(t)} \frac{dN(t)}{N(t)} = -\int_0^t dP(t). \tag{5.1.3}$$

From Eq. (5.1.3) we have

$$N(t) = N(0)e^{-[P(t)-P(0)]} = N(0)e^{-P(t)}, \tag{5.1.4}$$

where $P(0) = 0$ comes from theorem XXXXII in chapter 1.

To continue our derivation we need to know the possible time-dependence of $P(t)$. Although the time-dependence of $P(t)$ is very complex, but we still hope to know approximately a general analytical form of this time dependence as possible as we can. We assume $| < k'|W|k > |^2\rho(k') = | < k'|H'|k > |^2\rho(k') = |H'_{k'k}|^2\rho(k') = AE_{k'}^{\beta}$ (See chapter 4. For the convergence of the integral in transition probability, it is required that $-1 < \beta < 1$.)

$$
\begin{aligned}
P(t) &\equiv P_{all\leftarrow k}(t) \\
&= \int_{-\infty}^{+\infty} dE_{k'}\rho(k')\frac{|H'_{k'k}|^2}{\hbar^2}\left[\frac{sin[(E_{k'}-E_k-E)t/(2\hbar)]}{(E_{k'}-E_k-E)/(2\hbar)}\right]^2 \\
&= \frac{1}{\hbar^2}A\int_{-\infty}^{+\infty} dE_{k'}E_{k'}^{\beta}\left[\frac{sin[(E_{k'}-E_k-E)t/(2\hbar)]}{(E_{k'}-E_k-E)/(2\hbar)}\right]^2 \\
&= \frac{(2\hbar)^{1+\beta}A}{\hbar^2}\int_{-\infty}^{+\infty} d\frac{(E_{k'}-E_k-E)t}{2\hbar}\left(\frac{(E_{k'}-E_k-E+E_k+E)t}{2\hbar}\right)^{\beta} \\
&\times \frac{sin^2[(E_{k'}-E_k-E)t/(2\bar{)}]}{[(E_{k'}-E_k-E)t/(2\hbar)]^2}t^{1-\beta} \\
&= \frac{(2\hbar)^{1+\beta}A}{\hbar^2}\int_{-\infty}^{+\infty} dx\left(x+\frac{(E_k+E)t}{2\hbar}\right)^{\beta}\frac{sin^2x}{x^2}t^{1-\beta} \\
&\approx \frac{(2\hbar)^{1+\beta}A}{\hbar^2}\int_{-\infty}^{+\infty} dx|x|^{\beta}\frac{sin^2x}{x^2}t^{1-\beta} \\
&+ \frac{(2\hbar)^{1+\beta}A}{\hbar^2}\left|\frac{(E_k+E)t}{2\hbar}\right|^{\beta}\int_{-\infty}^{+\infty} dx\frac{sin^2x}{x^2}t^{1-\beta} \\
&= t^{1-\beta}\frac{(2\hbar)^{1+\beta}A}{\hbar^2} \\
&\times \left[\frac{\pi}{2^{\beta}\sin[(1-\beta)\pi/2]\Gamma(2-\beta)}\right. \\
&+ \left.\pi\left|\frac{(E_k+E)t}{2\hbar}\right|^{\beta}\right] \\
&= C_1(\beta)t^{1-\beta}+C_2[\beta,(E_k+E)]t \\
&\equiv P_1(t)+P_2(t). \qquad (5.1.5)
\end{aligned}
$$

The related integral formulas can be found in Refs. [28, 29] The subscript 1 denotes time-non-linear transition. The subscript 2 denotes time-linear transition.

If $\beta = 0$, then the first and the second terms in Eq. (5.1.5) are equal to each other, and both becomes time-linear transition.

If $\beta \neq 0$, let us estimate the magnitudes of the two terms.

(1). Boundary absorbtion. For examples: (i). Photoeffect under boundary absorbtion; (ii). The particle hopping from one position to a neighborhood position in solid by the boundary absorbtion of phonon (See section 5.3). In case of boundary absorption, the binding energy of a particle $E_k < 0$ is equal to the energy $E > 0$ of absorbing photon or phonon, i. e., $E_k + E = 0$. Then Eq. (5.1.5) gives

$$P(t) = P_1(t) = t^{1-\beta}\frac{(2\hbar)^{1+\beta}A}{\hbar^2}\frac{\pi}{2^{\beta}[\sin(1-\beta)\pi/2]\Gamma(2-\beta)}. \qquad (5.1.6)$$

(2). Non-boundary absorbtion $E_k + E > 0$. For examples: (i). Photoeffect under non-boundary absorbtion; (ii). The particle hopping from one position to a neighborhood position in solid by the non boundary absorbtion of phonon. Let us estimate the magnitudes of the two terms in Eq. (5.1.5). Assume $E_k + E \leq 0.1E$, where E denotes the energy of photon or phonon, and $E_k < 0$ denotes, for example, the binding energy of the electron in an atom or the binding energy of the hopping particle in a bounded position.

(2.1). Assume $\beta = 0.9$. Then we have

$$P_1(t) = \mathcal{Z}\frac{\pi}{2^\beta[\sin(1-\beta)\pi/2]\Gamma(2-\beta)} = 6.39, \tag{5.1.7}$$

and

$$P_2(t) = \mathcal{Z}\pi\left|\frac{(E_k+E)t}{2\hbar}\right|^\beta < \mathcal{Z}\pi\left|\frac{0.1\nu/(2\pi)t}{2}\right|^\beta \approx \mathcal{Z}\,|0.25|^{0.9} = 0.287. \tag{5.1.8}$$

where ν is the frequency, and T is the period of light or acoustic waves. (Assume that the duration time of an absorption process $t \approx 10T$.)

(2.2). Assume $\beta = 0$. Then we have

$$P_1(t) = \frac{\pi}{2^\beta \sin(1-\beta)\pi/2\Gamma(2-\beta)} = \pi, \tag{5.1.9}$$

and

$$P_2(t) = \pi\left|\frac{(E_k+E)t}{2\hbar}\right|^\beta = \pi. \tag{5.1.10}$$

From the comparison between Eqs. (5.1.7-10) we see that when $0 = \beta \leq 1$, the the time-linear and time-non-linear transitions are of the same contributions to the transition probability.

(2.3). Assume $\beta = -0.9$. Then we have

$$\frac{\pi}{2^{0.1}(1.9)\sin(1.9)\pi/2\Gamma(1.9)} = \frac{\pi}{(1.07)0.156\times 0.96} = 20.4, \tag{5.1.11}$$

and

$$\pi\left|\frac{(E_k+E)t}{2\hbar}\right|^\beta \approx (0.5\pi 0.1\omega t)^{-0.9} > (0.5\pi 0.1\nu/(2\pi)10T)^{-0.9} = 3.4. \tag{5.1.12}$$

(2.4). Assume $\beta = -0.1$. Then we have

$$\frac{\pi}{2^{0.9}(1.1)\sin(1.1)\pi/2\Gamma(1.1)} = \frac{\pi}{(1.87)0.988\times 0.95} = 1.79, \tag{5.1.13}$$

and

$$\pi\left|\frac{(E_k+E)t}{2\hbar}\right|^\beta \approx (0.5\pi 0.1\omega t)^{-0.1} > (0.5\pi 0.1\nu/(2\pi)10T)^{-0.1} = 0.87. \tag{5.1.14}$$

Eqs. (13.1.2.11-14) tell us that so long as $\beta \neq 0$, the time-linear transition probability in Eq. (5.1.5) is less or much less than the time-non-linear term. That is, Fermi golden rule does not hold except $\beta = 0$.

Next, we introduce a new physical quantity for relaxation processes.

Survival probability

In many practical problems, it is important to know the survival probability $Q(t)$, i. e., the fraction of the particle in initial state at $t = 0$ preserved after quantum transition. Considering Eq. (5.1.4-5), we obtain

$$\begin{aligned} Q(t) &= \frac{N(t)}{N(0)} = e^{-P(t)} = e^{-P_1(t)}e^{-P_2(t)} \\ &= e^{-C_1(\beta)t^{1-\beta}}e^{-C_2(\beta,E_k+E)t} = e^{-(t/\tau_1)^{1-\beta}}e^{-(t/\tau_2)} \\ &\equiv Q_1(t)Q_2(t), \end{aligned} \tag{5.1.15}$$

where

$$\tau_1 \equiv (C_1)^{-1/(1-\beta)}, \tag{5.1.16}$$

and

$$\tau_2 \equiv (C_2)^{-1}. \tag{5.1.17}$$

τ_1 is the time interval (lifetime or relaxation time) for the number of particles in initial state to fall to $1/e$ of the initial particle number in case of time-non-linear transition. τ_2 is the time interval (lifetime or relaxation time) for the number of particles in initial state to fall to $1/e$ of the initial particle number in the case of time-linear transition. $w_1 = 1/\tau_1$ and $w_2 = 1/\tau_2$ are transition rate or decay rate. If $\beta = 0$, then the transition rate here is the same as we defined in chapter 1. Note that the normalization of $Q(t)$ has not yet been made.

Because $Q(t)$ is a time function, the particles in initial state can have different lifetimes or relaxation times. So we define (mean) lifetime or (mean) relaxation time as

$$\tau_{mean} = -\int_0^\infty t dQ(t). \tag{5.1.18}$$

Let us make some illustrations for Eq. (5.1.18). $N(t) - N(t+dt) = N(0)[Q(t) - Q(t+dt)]$ represents the survival particle number at time $t \to t+dt$. $tN(0)[Q(t)-Q(t+dt)]$ represents the total survival time of the survival particles at time $t \to t + dt$.

$$\int_0^\infty tN(0)[Q(t) - Q(t+dt)]$$

represents the total survival time of the survival particles at time $t \to \infty$. Therefore, the average survival time, which is called (mean) lifetime, of one particle is given by Eq. (5.1.18).

Of course, we can substitute Eq. (5.1.15) into Eq. (5.1.18), and obtain lifetime. The expression will be quite complex. To get analytical expression, let us just discuss the factor $Q_1(t)$. Substituting $Q_1(t)$ in Eq. (5.1.15) into Eq. (5.1.18) yields

$$\tau_{1,mean} = \tau_1 \Gamma\left(\frac{2-\beta}{1-\beta}\right) \equiv \tau > \tau_1. \tag{5.1.19}$$

If $t = \tau_{1,mean}$, then

$$Q_1(\tau_{1,mean}) = e^{-(\tau_{1,mean}/\tau_1)^{1-\beta}} = e^{-\{\Gamma[(2-\beta)/(1-\beta)]\}^{1-\beta}}. \tag{5.1.20}$$

Eq. (5.1.5) gives

$$C_1 = \frac{A2\pi}{\hbar^{1-\beta}(1-\beta)\sin[(1-\beta)\pi/2]\Gamma(1-\beta)}. \tag{5.1.21}$$

From Eq. (5.1.18) we can introduce (mean) wavefunction Ψ.

$$\tau_{mean} = \int_0^\infty t\frac{-dQ(t)}{dt}dt \equiv \int_0^\infty t|\Psi(t)|^2 dt. \tag{5.1.22}$$

If in Eq. (5.1.15) we take $Q(t) = Q_1(t)$, then the (mean) wavefunction $\Psi(t)$ should have the following time dependence

$$\begin{aligned}\Psi(t) &= Ne^{-iE_k t/\hbar}|\Psi(t)| \\ &= Ne^{-iE_k t/\hbar}\sqrt{\frac{-dQ(t)}{dt}} \\ &= N\left(\frac{1}{\tau}\right)^{(1-\beta)/2}(1-\beta)^{1/2}e^{-iE_k t/\hbar}t^{-\beta/2}e^{-(t/\tau_1)^{1-\beta}/2}, \end{aligned} \quad (5.1.23)$$

where N is determined by normalization condition of the (mean) wavefunction. If $\beta = 0$, then the transition or decay is called exponential transition or decay. If $\beta \neq 0$, then the transition or decay is called non-exponential transition or decay or fractional-exponential transition or decay.

Next we study two cases: $\beta = 0$ and $-1 < \beta < 1,\ \beta \neq 0$.

(1). If $\beta = 0$, then the transition is time-linear. In this special case, we give the following formulas.

From Eq. (5.1.23) we get

$$\Psi(t) = N\frac{1}{\sqrt{\tau_1}}e^{-iE_c t/\hbar}, \qquad E_c \equiv E_k - \frac{i}{2}\Gamma. \quad (5.1.24)$$

$$Im(E_c) = \frac{\Gamma}{2}, \qquad \Gamma = \hbar w_1 = \frac{\hbar}{\tau_1}. \quad (5.1.25)$$

Since the time dependence of the wavefunction is not harmonic, the real energy of the state does not have a certain value. Its Fourier spectrum has many monochromatic components:

$$\Psi(t) = N\frac{1}{\sqrt{\tau_1}}e^{-iE_k t/\hbar}e^{-w_1 t/2} = \int_{-\infty}^{\infty} d\omega\Psi(\omega)e^{-i\omega t}. \quad (5.1.26)$$

From Eq. (5.1.26)

$$\begin{aligned}\Psi(\omega) &= \int_0^{\infty} dtN\frac{1}{\sqrt{\tau_1}}e^{-i\omega_k t}e^{-w_1 t/2}e^{i\omega t} \\ &= N\frac{1}{\sqrt{\tau_1}}\frac{i}{(\omega-\omega_k)+(iw_1/2)} = N\frac{1}{\sqrt{\tau_1}}\frac{i(\omega-\omega_k)+w_1/2}{(\omega-\omega_k)^2+(w_1/2)^2}. \end{aligned} \quad (5.1.27)$$

$\Psi(\omega)$ in Eq. (5.1.27) represents the probability amplitude of state $\Psi(t)$ at state $|\omega>$. Therefore, the probability of state $\Psi(t)$ at state $|\omega> \to |\omega + d\omega>$ is

$$\begin{aligned}\rho(\omega)d\omega &= |\Psi(\omega)|^2 d\omega = N^2\left(\frac{1}{\tau_1}\right)\left|\frac{i(\omega-\omega_k)+w_1/2}{(\omega-\omega_k)^2+(w_1/2)^2}\right|^2 d\omega \\ &= N^2\left(\frac{1}{\tau_1}\right)\frac{1}{(\omega-\omega_k)^2+(w_1/2)^2}d\omega. \end{aligned} \quad (5.1.28)$$

Note that N in Eq. (5.1.28) represents normalization constant other than particle number. From Eq. (5.1.28) and probability normalization condition of $\rho(\omega)$ or $\rho(E)$

$$\int_{-\infty}^{\infty} dE\rho(E) = 1, \quad (5.1.29)$$

we come to the Lorentzian distribution (or Breit-Wigner line shape or Breit-Wigner formula 1936 [30]) of probability density of energy or frequency

$$\rho(E) = \frac{1}{2\pi} \frac{\Gamma}{(E - E_k)^2 + \Gamma^2/4}. \tag{5.1.29$'$}$$

This Lorentzian distribution is similar to the resonance curve of a classical oscillator with frequency $E_k/\hbar$ and damping. From Eq. (5.1.29)$'$ we see that the maximum value of $\rho(E)$ is at $E = E_k$ and is equal to $2/(\pi\Gamma)$. The quantity Γ, equal to the full width at half-maximum, $FWHM$, is usual called level width or linewidth. Linewidth is a measure of the energy interval near the resonance value E_k (the centroid of the resonance curve) where the probability of finding the system is large [30].

Thus, for example, the nuclei or atoms in exited states and in perturbation field are unstable, and their lifetime τ_1 is related to the linewidth Γ by Eq. (5.1.25). If $\tau_1 \to \infty$, then the curve of $\rho(E)$ versus E is getting narrow around the centroid, $\Gamma \to 0$, and stretching up. The area under the curve stays equals to one so that

$$\Gamma \to 0: \qquad \rho(E) \to \delta(E - E_k), \tag{5.1.30}$$

and we come back to the initial stationary state with a certain energy E_k.

(2). The more general case is $\beta \neq 0$. If $\beta \neq 0$, then the transition is non-exponential transition (decay). We give the following formulas.

From Eq. (5.1.23) we get

$$\begin{aligned}
\Psi(t) &= N\sqrt{\left(\frac{1}{\tau_1}\right)^{1-\beta}(1-\beta)t^{-\beta/2}e^{-iE_kt/\hbar}e^{-(t/\tau_1)^{1-\beta}/2}} \\
&\equiv N\sqrt{\left(\frac{1}{\tau_1}\right)^{1-\beta}(1-\beta)t^{-\beta/2}e^{-iE_kt/\hbar}e^{-(w_1t)^{1-\beta}/2}} \\
&\equiv N\sqrt{\left(\frac{1}{\tau_1}\right)^{1-\beta}(1-\beta)t^{-\beta/2}e^{-iE_kt/\hbar}e^{-(\Gamma_1t/\hbar)^{1-\beta}/2}}.
\end{aligned} \tag{5.1.31}$$

Since the time dependence of the wavefunction is not harmonic, its Fourier spectrum has many monochromatic components:

$$\Psi(t) = N\sqrt{\left(\frac{1}{\tau_1}\right)^{1-\beta}(1-\beta)t^{-\beta/2}e^{-iE_kt/\hbar}e^{-(w_1t)^{1-\beta}/2}} = \int_{-\infty}^{\infty} d\omega \Psi(\omega) e^{-i\omega t}. \tag{5.1.32}$$

$$\begin{aligned}
\Psi(\omega) &= \int_0^{\infty} dt N\sqrt{\left(\frac{1}{\tau_1}\right)^{1-\beta}(1-\beta)t^{-\beta/2}e^{-i\omega_k t}e^{-(w_1t)^{1-\beta}/2}e^{i\omega t}} \\
&= \int_0^{\infty} dt N\sqrt{\left(\frac{1}{\tau_1}\right)^{1-\beta}(1-\beta)t^{-\beta/2}e^{-(w_1t)^{1-\beta}/2}}\cos(\omega-\omega_k)t \\
&+ i\int_0^{\infty} dt N\sqrt{\left(\frac{1}{\tau_1}\right)^{1-\beta}(1-\beta)t^{-\beta/2}e^{-(w_1t)^{1-\beta}/2}}\sin(\omega-\omega_k)t.
\end{aligned} \tag{5.1.33}$$

$|\Psi(\omega)|^2$ in Eq. (5.1.33) represents the probability amplitude of state $\Psi(t)$ at state $|\omega >$.

Therefore, the probability of state $\Psi(t)$ at states $|\omega>\to|\omega+d\omega>$ is

$$\begin{aligned}\rho(\omega) &= N^2\left(\frac{1}{\tau_1}\right)^{1-\beta}(1-\beta)\left[\left(\int_0^\infty dtt^{-\beta/2}e^{-(w_1t)^{1-\beta}/2}\cos(\omega-\omega_k)t\right)^2\right.\\ &+ \left.\left(\int_0^\infty dtt^{-\beta/2}e^{-(w_1t)^{1-\beta}/2}\sin(\omega-\omega_k)t\right)^2\right]. \quad (5.1.34)\end{aligned}$$

From Eq. (5.1.34) and probability normalization condition of $\rho(\omega)$ or $\rho(E)$

$$\int_{-\infty}^{\infty} dE\rho(E) = 1, \quad (5.1.35)$$

we can determine the normalization constant N. Eq. (5.1.34) is useful for seeking the product of linewidth-lifetime (See subsection 5.2.2).

While particle physicists use the lifetime to characterize the decay of unstable particles, nuclear physicists ordinarily refer to the half-life [31], namely the time for the number of nuclei to fall to one half of the initial number:

$$N(t_{1/2}) = \frac{1}{2}N(0), \quad (5.1.36)$$

or equivalently

$$\frac{1}{2} = e^{-t_{1/2}/\tau}. \quad (5.1.37)$$

Thus

$$t_{1/2} = \tau\ln 2 = 0.693\tau. \quad (5.1.38)$$

5.2 Linewidth-Lifetime Product

The linewidth-lifetime product is an application of the theory in section 5.1.

5.2.1 Existent Problems

This linewidth(ΔE)-lifetime(τ) product is a controversial issue since the advent of quantum theory [30], [32]-[36]. Here the linewidth is the FWHM (=full width at half-maximum), and lifetime is the (mean) lifetime. There are many derivations for the linewidth-lifetime product. If we write $\Delta E\tau = C\hbar$, then $C = 1$ [30], [37]-[40], $\pi/4$ or $\pi/2$ [41, 42], and $3\times 5^{-3/2}\pi$ [43]. Although Refs. [37]-[43], [9], [5] give different values of the product between linewidth and lifetime, but some references have the following common viewpoints and conclusions:

(1). Some references think that all systems should have unified value of the lowest non-zero bound for the product although different papers give different lowest bound value;
(2). Some references think that the linewidth-lifetime product is determined by Heisenberg energy-time uncertainty relation.

On the contrary, the two diagrams Figs. 5.1-2 in subsection 5.2.2 clearly show that:
(1)'. Any system has oneself value of the linewidth-lifetime product. For some systems the value of linewidth-lifetime product can be approach zero. One just can ask to what is the value of linewidth-lifetime product of some system equal? One can never ask what is the value of the linewidth-lifetime product of some system larger than? Because for linewidth-lifetime product there is no the so-called lowest bound.
(2)'. The linewidth-lifetime product does not have direct connection with Heisenberg energy-time uncertainty relation.

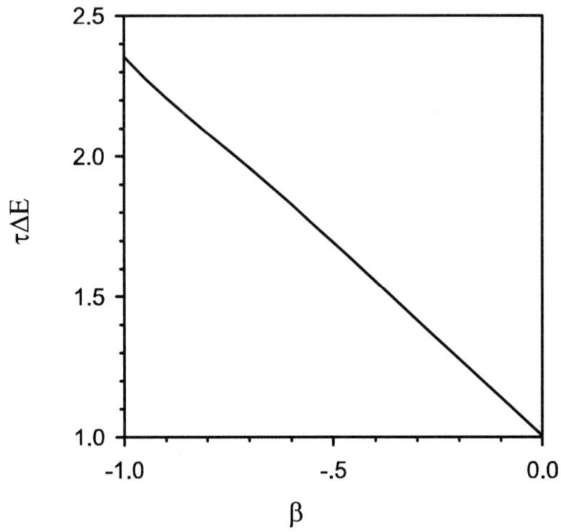

Figure 5.1: The curve of $\Delta E\tau$ vs β. The unit of $\Delta E\tau$ is $\hbar$. $-1 < \beta = 0$. Note that according to the definition in Eq. (5.1.19), $\tau = \tau_{1,mean} \neq \tau_1$.

5.2.2 Linewidth-Lifetime Product

For the $\rho(E)$ in Eq. (5.1.34) the probability density of energy is not the simple Lorentzian distribution (or Breit-Wigner line shape). We can make numerical calculations to verify this conclusion. The results of numerical calculations for the linewidth-lifetime product $\Delta E \cdot \tau$ are shown in Fig. 5.1 and Fig. 5.2.

From Fig. 5.1 and Fig. 5.2 we see that $2.348\hbar \geq \Delta E \cdot \tau \geq 4.595 \times 10^{-5}\hbar$ for $-0.999 \leq \beta \leq 0.9$. If $\beta \to 1$, then $\Delta E \cdot \tau \to 0$, which can not been shown in Fig. 5.2. For example, our numerical calculation gives $\Delta E \cdot \tau = 4.684 \times 10^{-10}$ for $\beta = 0.95$.

Therefore, the linewidth-lifetime product is strongly dependent on the value of β, which is determined by the energy-dependence of $|H'_{k'k}|^2\rho(k')$ (the so-called vacuum space energy support ability). If it depends strongly on energy (or frequency), then we should make numerical calculations or make estimation from Figs. 5.1-2 for the linewidth-lifetime product.

5.3 Non-Debye Relaxation in Condensed Matters

5.3.1 Introduction

In references non-Debye relaxation=non-exponential relaxation.. The species, including ions, electrons, electro-hole pairs, defects, polars, polaritons, dislocations, impurities, dipoles rotating at fixed site, and so on, can make migration, hopping, composition, diffusion, and so on. In relaxation theory one prefers to use relaxation time instead of duration time or lifetime, relaxation function instead of survival probability, and activation energy instead of binding energy or excitation energy. The relaxations in condensed matter are related to diffusion and viscosity in condensed matter through simple transformations. For example, the viscosity $\eta = G_N\tau$, where G_N being the plateaumodulus [44]. Although there are accumulated many experimental data on the relaxation, diffusion and viscosity since 200 years ago, but the microscopic mechanism of relaxation is still not clear [45]-[53]. It is well known that the survival probability of many relaxation processes follows empirical

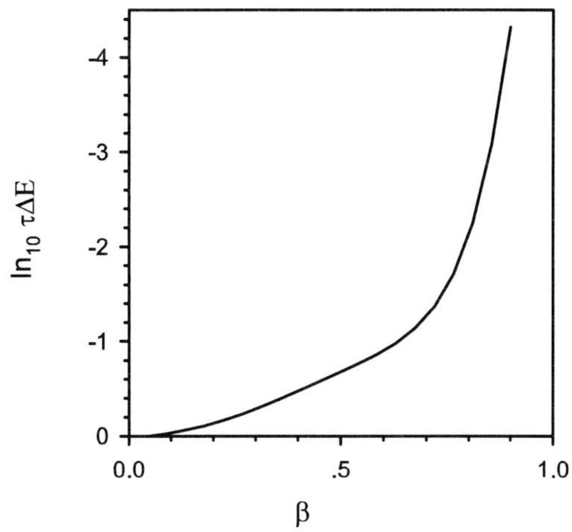

Figure 5.2: The curve of $\ln_{10}\Delta E\tau$ vs β. The unit of $\Delta E\tau$ is $\hbar$. $0 = \beta < 1$. If $\beta = 0.95$, then $\Delta E\tau = 4.684 \times 10^{-10}$.

Kohlrausch-Williams-Watts (KWW) rule [45]

$$Q(t) = e^{-(t/\tau)^{\beta_{KWW}}}, \tag{5.3.1.1}$$

where $0 < \beta_{KWW} \leq 1$, τ is the lifetime, which is called relaxation time in condensed matter physics. If $\beta = 1$, it is called Debye relaxation. If $\beta < 1$, it is called non-Debye relaxation.

5.3.2 Relaxation Caused by Species-Phonon Interaction

The interaction $H'(t)$ between the species at $\mathbf{r}$ and all the ions of the condensed matter is given by Eq. (3.5.1). Suppose the position space is one dimensional. The ions number is N. Considering Eq. (3.4.19.b) yields

$$\begin{aligned} H' &= \sum_{l}^{N} \frac{\partial U(r-R_l)}{\partial R_l} x_l \\ &= \sum_{l} \frac{\partial U(r-R_l)}{\partial R_l} \sum_{k} \sqrt{\frac{\hbar}{2NM\omega(k)}} \\ &\times \left(e^{ikal-i\omega(k)t} b_k + e^{-ikal+i\omega(k)t} b_k^{+}\right), \end{aligned} \tag{5.3.2.1}$$

where the summation over l is only for the unit cells around r, because $U(r - R_l)$ as a distance function decays very fast, and thus for an one dimensional atom chain $l = 0, \pm 1, \pm 2$ is enough. The stationary states of the species at n and n' are

$$e^{-iEt/\hbar} a_n |n>, \qquad e^{-iEt/\hbar} a_{n'} |n'>, \tag{5.3.2.2}$$

where we suppose that the energies at site n and n' are the same, and $R_{n'} = R_n + a'$.

The transition probability from n into n' in first order perturbation approximation of absorption and emission of photon is

$$\begin{aligned}
P_{n'n}(t) &\overset{(1)}{=} \sum_{n_{k'}=0}^{\infty} \sum_{\{n_k\}k'} \left| \frac{1}{i\hbar} \int_0^t dt' < n'|a_{n'}^{+} e^{iEt'/\hbar} \right. \\
&\left. < n_{k'}|e^{iE_{k'}n_{k'}t'/\hbar} H' e^{-iE_k t'/\hbar}|n_k > e^{-iEt'/\hbar} a_n|n > \right|^2 \\
&\overset{(2)}{=} \sum_{n_{k'}\{n_k\}k'} \left| \frac{1}{i\hbar} \int_0^t dt' < n'|a_{n'}^{+} < n_{k'}|e^{iE_{k'}t'/\hbar} \sum_{l=1}^{N} \frac{\partial U(r-R_l)}{\partial R_l} \right. \\
&\times \left. \sum_{k''} \sqrt{\frac{\hbar}{2NM\omega(k'')}} e^{ik''al} b_{k''} e^{-iE_k t'/\hbar}|n_k > a_n|n > \right|^2 \\
&+ \sum_{\{n_{k'}\}\{n_k\}k'k} \left| \frac{1}{i\hbar} \int_0^t dt' < n'|a_{n'}^{+} < n_{k'}|e^{iE_{k'}t'/\hbar} \sum_{l=1}^{N} \frac{\partial U(r-R_l)}{\partial R_l} \right. \\
&\times \left. \sum_{k''} \sqrt{\frac{\hbar}{2NM\omega(k'')}} e^{-ik''al+i\omega(k'')t'} b_{k''}^{+} e^{-iE_k t'/\hbar}|n_k > a_n|n > \right|^2 \\
&\overset{(3)}{=} \sum_{n_{k'}\{n_k\}k'} \left| < n_{k'}| \sum_{k''} \sqrt{\frac{\hbar}{2NM\omega(k'')}} \sum_{l=1}^{N} e^{ik''al} b_{k''}|n_k > \right. \\
&\times \left. < n'|\frac{\partial U(r-R_l)}{\partial R_l}|n > \right|^2 \frac{\sin^2[(E_{k'}-E_k)t/(2\hbar)]}{[(E_{k'}-E_k)/2]^2} \\
&+ \sum_{n_{k'}\{n_k\}k'} \left| < n_{k'}| \sum_{k''=1} \sqrt{\frac{\hbar}{2NM\omega(k'')}} \sum_{l=1}^{N} e^{-ik''al} b_{k''}^{+}|n_k > |n > \right. \\
&\times \left. < n'|\frac{\partial U(r-R_l)}{\partial R_l}|n > \right|^2 \frac{\sin^2[(E_{k'}-E_k)t/(2\hbar)]}{[(E_{k'}-E_k)/2]^2} \\
&\overset{(4)}{=} \sum_{n_{k'}\{n_k\}k'} \left| \sqrt{\frac{\hbar}{2NM\omega(k'-k)}} \sum_{l=1}^{N} e^{i(k'-k)al} \sqrt{n_{k'-k}} \delta_{n_k,n_{k'}+1} \right. \\
&\times \left. < n'|\frac{\partial U(r-R_l)}{\partial R_l}|n > \right|^2 \frac{\sin^2[(E_{k'}-E_k)t/(2\hbar)]}{[(E_{k'}-E_k)/2]^2} \\
&+ \sum_{n_{k'}\{n_k\}k'} \left| \sqrt{\frac{\hbar}{2NM\omega(k-k')}} \sum_{l=1}^{N} e^{-i(k-k')al} \sqrt{(n_{k-k'}+1)} \delta_{n_k,n_{k'}-1} \right. \\
&\times \left. < n'|\frac{\partial U(r-R_l)}{\partial R_l}|n > \right|^2 \frac{\sin^2[(E_{k'}-E_k)t/(2\hbar)]}{[(E_{k'}-E_k)/2]^2} \\
&\overset{(5)}{=} \sum_{\{n_k\}k'} \frac{\hbar}{2NM\omega(k'-k)} |\mathfrak{N}|^2 (2n_{k-k'}+1) \frac{\sin^2[(E_{k'}-E_k)t/(2\hbar)]}{[(E_{k'}-E_k)/2]^2} \\
&\overset{(6)}{=} \sum_{\{n_k\}(k-k')} \frac{\hbar}{2NM\omega(k'-k)} |\mathfrak{N}|^2 (2n_{k-k'}+1) \frac{\sin^2[(E_{k'}-E_k)t/(2\hbar)]}{[(E_{k'}-E_k)/2]^2} \\
&\overset{(7)}{=} \sum_{k-k')} \frac{\hbar}{2NM\omega(k'-k)} |\mathfrak{N}|^2 (2\overline{n_{k-k'}}+1) \frac{\sin^2[(E_{k'}-E_k)t/(2\hbar)]}{[(E_{k'}-E_k)/2]^2}
\end{aligned}$$

$$
\begin{aligned}
&\overset{(8)}{=} \sum_k \frac{\hbar}{2NM\omega(k)} |\mathfrak{N}|^2 (2\overline{n_k}+1) \frac{\sin^2[E_k t/(2\hbar)]}{(E_k/2)^2} \\
&\overset{(9)}{=} \int_{-\infty}^{\infty} dE_k \rho(E_k) \frac{\hbar}{2NM\omega(k)} |\mathfrak{N}(k)|^2 (2\overline{n}_k+1) \frac{\sin^2[E_k t/(2\hbar)]}{(E_k/2)^2} \\
&\overset{(10)}{=} \int_{-\infty}^{\infty} dE_k V \frac{3}{2\pi^2} \frac{1}{\hbar^3 v_{ph}^3} \frac{2+\beta}{3} E_D^{1-\beta} E_k^{1+\beta} \frac{1}{2NM} \frac{\hbar^2}{\hbar\omega(k)} \\
&\quad \overline{(\mathfrak{N})^2} (2\overline{n}_k+1) \frac{\sin^2[E_k t/(2\hbar)]}{(E_k/2)^2} \\
&\overset{(11)}{\equiv} A' \int_{-\infty}^{\infty} dE_k E_k^{\beta} (2\overline{n_k}+1) \frac{\sin^2[E_k t/(2\hbar)]}{(E_k/2)^2} \\
&\overset{(12)}{\approx} A' \int_{-\infty}^{\infty} dE_k E_k^{\beta} (2\overline{n}_k) \frac{\sin^2[E_k t/(2\hbar)]}{(E_k/2)^2} \\
&+ A' \int_{0}^{\infty} dE_k E_k^{\beta} (1) \frac{\sin^2[E_k t/(2\hbar)]}{(E_k/2)^2} \\
&\overset{(13)}{=} A'\hbar^{\beta-1} \int_{-\infty}^{\infty} d\omega_k \omega_k^{\beta} (2\overline{n}_k) \frac{\sin^2(\omega_k t/2)}{\omega_k^2} \\
&+ A'\hbar^{\beta-1} \int_{-\infty}^{\infty} d\omega_k \omega_k^{\beta} \frac{\sin^2(\omega_k t/2)}{\omega_k^2} \\
&\overset{(14)}{=} A \int_{-\infty}^{\infty} d\omega \omega^{\beta-2} (2\overline{n}_k) \sin^2(\omega t/2) + A \int_{-\infty}^{\infty} d\omega \omega^{\beta-2} \sin^2(\omega t/2) \\
&\overset{(15)}{=} A \int_{-\infty}^{\infty} d\omega \omega^{\beta-2} (2\overline{n}_k) \sin^2(\omega t/2) + A t^{1-\beta} \int_{-\infty}^{\infty} dx x^{\beta-2} \sin^2(x/2) \\
&\overset{(16)}{=} A \int_{-\infty}^{\infty} d\omega \omega^{\beta-2} (2\overline{n}_k) \sin^2(\omega t/2) \\
&+ A \frac{\pi}{2\Gamma(2-\beta)\sin[\pi(1-\beta)/2]} t^{1-\beta} \\
&\overset{(17)}{\equiv} A \int_{-\infty}^{\infty} d\omega \omega^{\beta-2} (2\overline{n}_k) \sin^2(\omega t/2) + A'' t^{1-\beta} \\
&\overset{(18)}{\equiv} P_1(t) + P_2(t). \qquad (5.3.2.3)
\end{aligned}
$$

In the first equality, there is the summation over $\{n_k\}$, which means the average over initial states of phonon system. In the first equality, there is the summation over $n_{k'}$, which means the summation over final states of phonon system. In the seventh equality, we accomplish the average over initial states of phonon system, and obtain Bose distribution $\overline{n_{k-k'}}$. In the ninth equality, $\rho(E_k)$ represents the energy state density of phonon in volume V, and $E_k \lesseqgtr 0$. The second term in the twelfth equality is the contribution of zero point oscillation. In the sixteenth equality, we uses the mathematical formula in Ref. [54].

For perfect $\rho(E_k) \propto E_k^2$ [55]. Actually, for real solid any textbooks on solid state show that $\rho(E_k)$ is a very complex function of E_k. Considering that in cases of disorder systems there should be some average effect, we assume $\rho(E_k) \propto E_k^1$. Therefore, we take for 3-

dimensional system

$$\begin{aligned}\rho(E_k) &= V\left(\frac{3}{2\pi^2}\frac{1}{\hbar v_{ph}^3}\right)\frac{2+\beta}{3}\omega_D^{1-\beta}\omega_k^{1+\beta} \\ &= V\frac{3}{2\pi^2}\frac{1}{\hbar^3 v_{ph}^3}\frac{2+\beta}{3}E_D^{1-\beta}E_k^{1+\beta},\end{aligned} \quad (5.3.2.4)$$

where ω_D represents Debye frequency, ρ satisfies

$$\int_0^{E_D} dE_k\rho(E_k) = N, \quad (5.3.2.5)$$

and

$$\beta = \begin{cases} 0, & for \quad complete \quad disorder \quad system \\ 1, & for \quad perfect \quad order \quad system \end{cases} \quad (5.3.2.6)$$

$$A' = V\frac{3}{2\pi^2}\frac{1}{\hbar^3 v_{ph}^3}\frac{2+\beta}{3}E_D^{1-\beta}\frac{\hbar^2}{2NM}\overline{(\mathfrak{N})^2}. \quad (5.3.2.7)$$

Next, at first, let us discuss the second term of the thirteenth equality in Eq. (5.3.2.3).

$$P_2(t) = \frac{A\pi}{2^{1-\beta}(1-\beta)\Gamma(1-\beta)\sin[\pi(1-\beta)/2]}t^{1-\beta}, (-1<\beta<1). \quad (5.3.2.8)$$

Actually, we are interesting just in $0<\beta<1$ (See Eq. (5.3.2.6).).

Let us give the formulas for the survival probability in condensed matters. In this case we should consider that when a species moves from one site to another site or from one direction to another direction, the species absorbs and emits phonon, and has to overcomes the potential barrier $E>0$ at any site or direction. E is called primitive activation energy [56]. The probability to overcome the potential barrier to move is given by Ref. [53, 56]

$$G(T) = e^{-E/T} = e^{-E_\circ/(T-T_\circ)}, \qquad E = \frac{E_\circ T}{T-T_\circ}. \quad (5.3.2.9)$$

Therefore, the actual quantum transition probability in condensed matter is $P(t)G(T)$. From Eqs. (5.3.2.3-9) we obtain the survival probability:

$$Q_2(t) = e^{-P_2(t)G(T)}, \quad (5.3.2.10)$$

$$Q_2(t,T) = e^{-A''t^{1-\beta}G(T)}, \ (0<\beta<1). \quad (5.3.2.11)$$

Let us make comparison of Eq. (5.3.2.11) with experiments.

Set

$$Q_2(t) = e^{-(t/\tau_2)^{1-\beta}}, \quad (5.3.2.12)$$

where the relaxation time (i. e., lifetime or the duration time of one jumping of species)

$$\begin{aligned}\tau_2 &= \left[\frac{1}{A''G(T)}\right]^{1/(1-\beta)} \\ &= \left[\frac{1}{A''}\right]^{1/(1-\beta)} e^{-E_\circ/[(1-\beta)(T-T_\circ)]} \\ &= \tau_{20}e^{E_\circ/[(1-\beta)(T-T_\circ)]} = \tau_{20}e^{-E/[(1-\beta)T]}.\end{aligned} \quad (5.3.2.13)$$

We use Eq. (5.3.2.12) to study practical cases $0\le\beta\le 1$, and therefore $0\le 1-\beta=\beta_{KWW}\le 1$. Thus, $Q_2(t)$ is the empirical KWW non-Debye (or called as non-exponential) relaxation rule. This indicates that, at least, the interaction between species and zero point energy phonons is a possible mechanism of KWW relaxation. If we consider that our $Q_2(t)$ can also explain various experimental data, then we can conclude that the interaction between species and zero point energy phonons is certainly the mechanism of KWW relaxation. The various experimental verifications for our relaxation formula, Eq. (5.3.2.12-13), is as follows:

(1.1). To describe the temperature-dependence of the relaxation time τ_1, some references accept the pure empirical super-Arrhenius Vogel-Fulcher-Tammann (VFT) law, $\tau_1 = \tau_{10}\exp[E'/(T - T_\circ)]$, proposed in 1921 Refs. [57, 45]. Our Eq. (5.3.2.10) can explain VFT law. Moreover, Ref. [45] lists many experimental data from many different materials to show that the VFT law has to be modified as

$$\tau_1 = \tau_{10}e^{-E_\circ/[\beta_{KWW}(T-T_\circ)]} = \tau_{10}e^{-E/(\beta_{KWW}T)}, \tag{5.3.2.14}$$

and calls the E/β_{KWW} the apparent activation energy. The E is called primitive activation energy. Some references think that that the E in condensed matter medium becomes larger $E/\beta_{KWW} > E$ is due to the interaction between species and medium. However, some references do not know what interaction it is. Our Eq. (5.3.2.12-13) can explain Eq. (5.3.2.14) very good. We conclude that the interaction, proposed in many references, is just the interaction between species and zero point energy phonons. We call the mechanism of KWW relaxation rule the zero point energy phonon mechanism.

(1.2). According to our zero point energy phonon mechanism, the value of $\beta_{KWW} = 1 - \beta$ should bo temperature-dependent. As is well known, when temperature goes up, the randomness of the system goes up as well. Therefore, when temperature goes up, the value of β_{KWW} goes up as well. This temperature dependence of β_{KWW} has been verified by experimental data of Ref. [46, 47] (See Figs. 5.3 and 5.4).

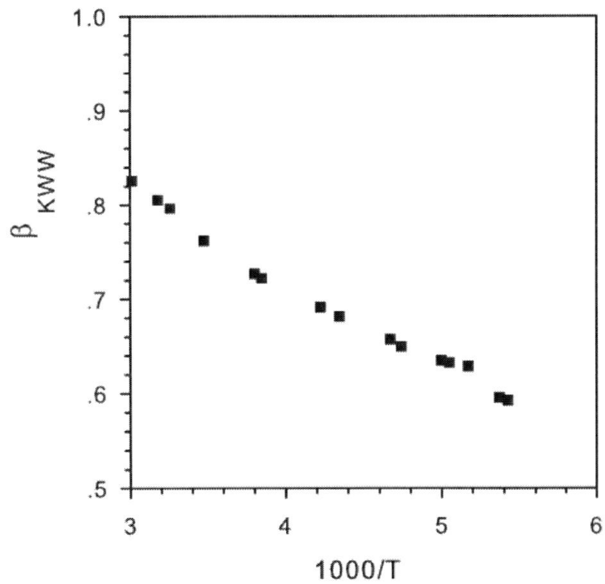

Figure 5.3: The data of relations between β_{KWW} and 1000/T for glycerol.

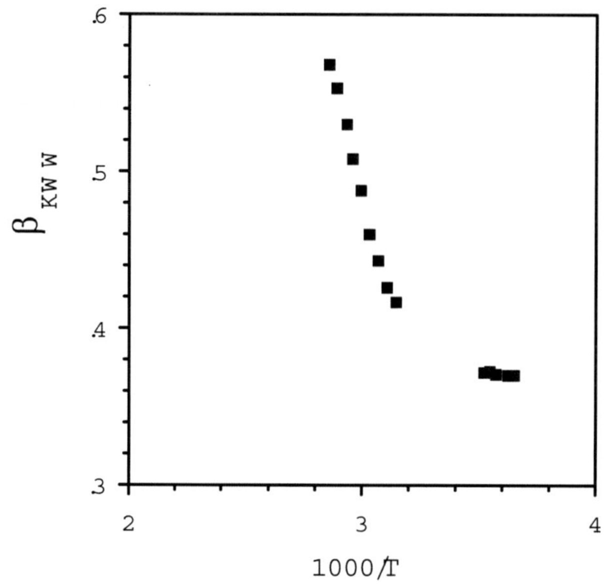

Figure 5.4: The data of relations between β_{KWW} and 1000/T for sorbitol.

(1.3). Let us study isotope effect of relaxation time. Historically, from the observed isotope effect physicists inferred that the conventional superconductivity is caused by electron-phonon interaction and got success. Historical experience is worthy to use for reference. Chapter 3 gives us the following formulas:

Using the phonon language, the displacement of l-th ion from its equilibrium, x_l, can be expressed as a Hermitian superposition of normal modes (Refer to chapter 3.).

$$x_l = \sum_{k=1} \sqrt{\frac{\hbar}{2NM\omega(k)}} \left(e^{ikal} b_k + e^{-ikal} b_k^+\right). \tag{5.3.2.15}$$

For a three dimensional system the denominator has $\sqrt{M}$ as well. Chapter 3 tells us that the interaction of an electron with the ion at site l at origin of coordinate is the simplest case, and is

$$H'_{i-l} = -\frac{\partial U(\mathbf{r}_i - \mathbf{R}_l)}{\partial \mathbf{R}_l} \mathbf{x}_l = -\frac{\partial U(\mathbf{r}_i - \mathbf{R}_l)}{\partial \mathbf{R}_l} \sum_{\mathbf{k}} \sqrt{\frac{\hbar}{2NM\omega(\mathbf{k})}} \left(b_{\mathbf{k}} e^{i\mathbf{k}\cdot\mathbf{R}_l} + b_{\mathbf{k}}^+ e^{-i\mathbf{k}\cdot\mathbf{R}_l}\right), \tag{5.3.2.16}$$

where $U(\mathbf{r}_i - \mathbf{R}_l)$ represents the interaction potential energy between species at $\mathbf{r}_i$ and the ion at lattice site l. If we make harmonic oscillator approximation, then the potential should be dependent on the classical oscillator frequency. Therefore, $U(\mathbf{r}_i - \mathbf{R}_l) \propto \hbar\omega_\circ \propto 1/\sqrt{m}$, where m is the mass of one species. For example, the species can be an interstitial solute atom of Li in solid. The atom of Li makes diffusion motion. From Eq. (5.3.2.12) and then from Eq. (5.3.2.10) we have $\tau_1 \propto m^{2/(1-\beta)}$.

Let us study the isotope effect of the relaxing species Li^7 and Li^6 in lithium borate glasses $({}^6Li, {}^7Li)_2 : 2.88B_2O_3$ [58]. Ref. [58] gives the following formula to study the isotope effect. The relaxation time is anti-proportional to the conductivity σ. The conductivity is easy to be measured in experiments. Eqs. (5.3.2.18-21) can give us the expression of $\tau_{1,7}/\tau_{1,6}$. At last we have

$$\frac{\sigma_6}{\sigma^7} = \frac{\tau_{1,7}}{\tau_{1,6}} = \left(\frac{m_{^7Li}}{m_{^6Li}}\right)^{1/(2-\beta)} = \left(\frac{7}{6}\right)^{1/\beta_{KWW}}. \tag{5.3.2.25}$$

Considering the temperature dependence of β_{KWW}, the observed temperature-dependence of σ^6/σ^7 (See Fig. 6 in Ref. [58]) can be explained very well.

By the way, Ref. [58] also proposed a theory to explain the observed temperature-dependence of σ^6/σ^7, and their theory fits experimental data quite well. However, according to the theory β in Ref. [58] is independent on temperature. Our viewpoint is that many experimental facts have shown that β is sensitive to temperature, and thus the theory in Ref. [58] cannot be correct. In mechanism, we think that the non-Debye (i. e., non-exponential) relaxation process in condensed matter is determined by the interaction between the species and zero point energy phonons and is randomness-dependent.

(1.4). Let us explain the celebrated $M^{3.4}$ molecular-weight-dependence of sher viscosity in entangled polymer melts, where M is the molecule weight [44]. Ref. [44] found that the A in Eq. (5.3.2.12) is proportional to M^2. Hence, $\tau_2 \propto M^{2/(1-\beta)} = M^{2/\beta_{KWW}}$. Ref. [44] also gave: the viscosity $\eta = G_N\tau$; $0.54 \leq \beta_{KWW} \leq 0.6$, where G_N is the plateau modulus. From the data in Ref. [44] we obtain that η has molecular-weight-dependence $M^{3.33}$ – $M^{3.70}$. Therefore, our theory can explain the celebrated $M^{3.4}$ molecular-weight-dependence of sher viscosity in entangled polymer melts.

(2). The above derivations do not confine in the interaction between species and phonons. Actually, $|\mathfrak{N}|_k^2$ and b^+, b can represent any interactions.

From the above two points, we conclude that the relaxation processes of KWW in condensed matter (or complex system) is determined by:

(a). The interaction between species and zero point energy phonons (or more generally, any zero point energy excitations);

(b). The randomness of the medium; (the temperature dependence of β and disorder)

(c). The relaxation phenomena of KWW might occur even in social science, if the variations of any species (here, any species=any social event) are caused by some stochastic influences.

As an interesting application of the formulas in this subsection, we study the pink noise, i. e., $1/f$ noise, in next subsection, which now is still an open hot problem lasting more than 200 years..

5.3.3 Pink Noise

Pink noise or $1/f$ noise (sometimes also called flicker noise) is a signal or process with a frequency spectrum such that the power spectral density (energy or power per Hz) is inversely proportional to the frequency. In pink noise, each octave carries an equal amount of noise power. The name arises from the pink appearance of visible light with this power spectrum [59]. Within the scientific literature the term $1/f$ noise is sometimes used a little more loosely to refer to any noise with a power spectral density of the form

$$S(f) \propto \frac{1}{f^\alpha}, \tag{5.3.3.1}$$

where f is frequency and $0.5 < \alpha < 2$, with exponent α usually close to 1. These $1/f$-like noises occur widely in nature and are a source of considerable interest in many fields. The term flicker noise is sometimes used to refer to $1/f$ noise, although this is more properly applied only to its occurrence in electronic devices due to a direct current. Mandelbrot and Van Ness proposed the name fractional noise (sometimes since called fractal noise) to emphasize that the exponent of the spectrum could take non-integer values and be closely related to fractional Brownian motion, but the term is very rarely used.

$1/f$ noise occurs in many physical, biological and economic systems (Stock market and GNP). Some researchers describe it as being ubiquitous [59]. In physical systems, it is present in some meteorological data series, the electromagnetic radiation output of some astronomical bodies, and in almost all electronic devices (referred to as flicker noise). In biological systems, it is present in, for example, heart beat rhythms, neural activity, and the statistics of DNA sequences, as a generalized pattern [59]. In financial systems, it is often referred to as a long-term memory effect. Also, it describes the statistical structure of many natural images (images from the natural environment). Recently, $1/f$ noise has also been successfully applied to the modeling of mental states in psychology, and used to explain stylistic variations in music from different cultures and historic periods. Voss and Clarke claim that almost all musical melodies, when each successive note is plotted on a scale of pitches, will tend towards a pink noise spectrum [59]. Similarly, a general $1/f$ distribution pattern has been observed in film shot length by James, in the study of 150 popular movies released from 1935 to 2005 [59].

There are many theories of the origin of $1/f$ noise. Some theories attempt to be universal, while others are applicable to only a certain type of material, such as semiconductors. Universal theories of $1/f$ noise remain a matter of current research interest [59]. A hypothesis has been proposed to explain the genesis of $1/f$ noise on the basis of a mathematical convergence theorem related to the central limit theorem of statistics.

To explain the KWW relaxation rule, Ngai gave phenomenologically the correct form of relaxation function, which is exactly same as that in subsection 5.3.2. After Ngai's works, Yu and Liu proved that $1/f$ noise as well as the KWW relaxation comes from the same relaxation function [60]. Next I introduce briefly the Yu-Liu phenomenological theory of pink noise. If the theory in Ref. [60] is correct, then we can conclude that the origin of pink noise is the interaction between species and zero point energy excitation, and is temperature-dependent.

For convenience of following statement, let me introduce briefly Markov process. A markov process is stochastic process $X(t)$ with a short memory, that is a process which rapidly forget its past history [52]. On the contrary, a non-Markov process is also stochastic process $X(t)$ with a long memory, for which a memory function has to be introduced.

The work of Ref. [60] is as follows:

$$\frac{\partial n(\mathbf{r},t)}{\partial t} = \frac{a_\circ^2}{6}\int_0^t \phi(t-t')\nabla^2 n(\mathbf{r},t')dt', \tag{5.3.3.2}$$

where $n(\mathbf{r},t)$ is the number density of species, a_0 is the free path of species, ϕ is a memory function. Green function equation of Eq. (5.3.3.2) is

$$\frac{\partial G(\mathbf{r},t;\mathbf{r}')}{\partial t} = \frac{a_\circ^2}{6}\int_0^t \phi(t-t')\nabla^2 G(\mathbf{r},t';\mathbf{r}')dt' + \delta(\mathbf{r}-\mathbf{r}')\delta(t). \tag{5.3.3.3}$$

Making Laplace transformation for Eq. (5.3.3.3) yields

$$ifG_{Laplace}(\mathbf{r},if;\mathbf{r}') = \frac{a_\circ^2}{6}\phi_{Laplace}(if)\nabla^2 G_{Laplace}(\mathbf{r},if;\mathbf{r}')dt' + \delta(\mathbf{r}-\mathbf{r}'), \tag{5.3.3.4}$$

where f represents frequency. The spectra of correlation function of fluctuation of species number is

$$S_N(f) = \frac{4<(\Delta N)^2>}{V_s}Re\int_{V_s}\int_{V_s} G_{Laplace}(\mathbf{r},if;\mathbf{r}')d\mathbf{r}d\mathbf{r}', \tag{5.3.3.5}$$

where N is the total number of species in volume V_s. The noise spectra of current and voltage are

$$S_I(f) = \frac{<I>^2}{N^2} S_N(f) \quad and \quad S_V(f) = \frac{<V>^2}{N^2} S_N(f), \tag{5.3.3.5}$$

respectively.

There is a relation between the Laplace transformations of memory function ϕ and mean wavefunction, defined by Eq. (5.1.22), Ψ:

$$\phi_{Laplace}(if) = \frac{if\Psi_{Laplace}(if)}{1 - \Psi_{Laplace}(if)}. \tag{5.3.3.6}$$

In terms of Eqs. (5.3.3.6-2), Yu and Liu took

$$\Psi(t) = -\frac{dQ(t)}{dt} \propto t^{n-1} e^{-(t/\tau')^n}, \quad 0 < n < 1, \quad n = 1 - \beta, \tag{5.3.3.7}$$

and proved Eq. (5.3.3.1).

From the analysis at the end of this subsection, it can be concluded that the origin of all pink noises in many different systems is the influence of zero point energy excitation on the species motion.

5.4 General Study for Survival Probability

Theorem XXXXIX: Non-exponential survival probability: If a system makes transition from initial state at $t = 0$ under the action of a time independent Hamiltonian H, and the energy spectrum of H has the lowest limit E_{min}, then survival probability

$$Q(t) \quad \stackrel{t\to\infty}{\longrightarrow} e^{-At^a}, \quad (A > 0, 0 < a < 1). \tag{5.4.1}$$

This theorem means that the existence of ground state of a quantum system leads to require that: in long time transition process, the survival probability can deviate from exponential reduction [61, 10].

Proof:

If we do not use the time-dependent perturbation theory, then the definition of $Q(t)$ should be

$$Q(t) = | < \Phi(0)|\Phi(t) > |^2 \equiv |B(t)|^2, \tag{5.4.2}$$

where

$$|\Phi(t) >= e^{-iHt/\hbar}|\Phi(0) > . \tag{5.4.3}$$

Here, we should emphasize that $\Phi(t) \neq \Psi(t)$. The time dependence of Φ comes from time evolution operator, and thus comes directly from Schrödinger equation. However, $\Psi(t)$ in relaxation process comes from quantum transition. At the end of this section, we will make discussions for this point.

Suppose the eigenfunction set is $|\phi_{E,\alpha}>$, and (E, α) represent all eigenvalues of eigenstates. Then we have

$$\begin{aligned} B(t) &= < \Phi(0)|\Phi(t) >=< \Phi(0)|e^{-iHt/\hbar}|\Phi(0) > \\ &= \int dE \int d\alpha < \Phi(0)|\phi_{E,\alpha} >< \phi_{E,\alpha}|e^{-iHt/\hbar}|\Phi(0) > \\ &= \int_{E_{min}}^{\infty} dE \int d\alpha| < \phi_{E,\alpha}|\Phi(0) > |^2 e^{-iEt/\hbar} \\ &\equiv \int_{E_{min}}^{\infty} dE\omega(E) e^{-iEt/\hbar}, \end{aligned} \tag{5.4.4}$$

where $\omega(E)$ is the Fourier transform of $B(t)$, i. e.,

$$\omega(E) = \int d\alpha | < \phi_{E,\alpha} | \Phi(0) > |^2.$$

From Eq. (5.4.4) we have

$$\int_{E_{min}}^{\infty} dE\omega(E) =< \Phi(0)|\Phi(0) >= 1. \quad (5.4.5)$$

Due to the integrable property of $\omega(E)$, Riemann-Lebesque theorem in Fourier analysis [28] concludes

$$\lim_{t \to \infty} B(t) = 0. \quad (5.4.6)$$

Due to the Fourier transform of $B(t)$ $\omega(E) = 0 \ \ for \ \ E \leq E_{min}$, Payley-Wiener theorem in Fourier analysis [28] concludes

$$\int_{-\infty}^{\infty} dt \frac{|\ln |B(t)||}{1+t^2} < +\infty. \quad (5.4.7)$$

The convergence of integral at $t = \infty$ requires

$$|\ln |B(t)|| \stackrel{t \to \infty}{\longrightarrow} A' t^a, \ \ (0 < a < 1). \quad (5.4.8)$$

Considering Eq. (5.4.6), we have $B(t) \stackrel{t \to \infty}{\longrightarrow} 0$, $\ln |B(t)| \stackrel{t \to \infty}{<} 0$, and thus $A' < 0$. QED.

Comments:

(1). The condition of this theorem that the energy spectrum of H has the lowest limit E_{min} is always be satisfied. Because considering the existence of spontaneous emission, if there is no E_{min}, then the state of system is not stable.

(2). Although the duration time $t \to \infty$ does not correspond to any actual case, it has theoretical significance. Next, we prove one corollary.

Corollary XVII: General form of time dependence of transition probability: If

$$P_{k \to all \ final \ srates} = (t/\tau)^a, \ and \ \tau \ is \ independent \ of \ time, \ then \ 0 < a < 1. \quad (5.4.9)$$

Proof:

In physics,

$$Q(t) = e^{-P_{k \to all \ final \ states}}. \quad (5.4.10)$$

Therefore, Eq. (5.4.1) can prove Eq. (5.4.9). QED.

As one end of chapters 4 and 5, we would like to mention Refs. [48, 50]. Ref. [48] studies open system by using quantum mechanics, and derives a series of formulas. The problem is in that: Ref. [48] uses the quantum transition theory with energy conservation. Therefore, according to our big theorem of energy, Ref. [48] needs to make systematic modifications. One should use quantum mechanics to study the contents in Ref. [50].

5.5 Exercises and Solutions

(1). Some references, such as Res. [31, 32], think that

$$Q(t) = e^{-t/\tau} = e^{-wt}, \tag{1}$$

and think τ represents lifetime, w represents transition (or decay) rate.

(1.1) Prove that the definition of lifetime is the same as from our formula

$$\tau = -\int_0^\infty t dQ(t).$$

(1.2). Make some discussions.

Solution:

(1.1).

$$\tau = -\int_0^t t dQ(t) = -\int_0^\infty t de^{-t/tau'} = \tau = \frac{1}{w}. \tag{2}$$

(1.2). The $Q(t)$ in Eq. (1) is only a special case ($\beta = 1$) of our general formula. In more general cases Eq. (1) can not give the definition of τ.

(2). Using Eq. (5.1.20) to estimate the magnitudes of survival probability $Q_1(t)$ at $t = \tau_1$.

Solution:

If $\beta = 0.9$, then $Q_1(\tau_1) = \exp(-4.528)$. If $\beta = -0.9$, then $Q_1(\tau_1) = \exp(-0.896)$. If $\beta = 0$, then $Q_1(\tau_1) = \exp(-1)$.

(3). After Eq. (5.1.17) we defined that: "τ_1 is the time duration (interval, lifetime) for the number of particles in initial state to fall to $1/e$ of the initial particle number in case of non-exponential transition. τ_2 is the time interval (lifetime) for the number of particles in initial state to fall to $1/e$ of the initial particle number in the case of exponential transition. Please use probability language instead of the number to define τ_1 and τ_2.

(4). If the initial state is an excited state of nucleus, then the transition process is called decay of nucleus. Give the definitions of τ_1 and τ_2 in case of decay.

Solution:

For example, τ_2 is a time interval (lifetime) for the number of nuclei to fall to $1/e$ of the initial number. τ_2 is referred to as the lifetime of the unstable nuclei.

(5). For the 2P to 1S transition in hydrogen atom, we have calculate the transition rate of $w = 0.6 \times 10^9$ per second. Calculate the FWHM Γ of the photon line, and make comparison with the energy of photon.

Solution:

$$\Gamma = \hbar w = \frac{(1.05 \times 10^{-27}\ erg\ sec)(0.6 \times 10^9\ sec^{-1})}{1.602 \times 10^{-12}\ erg/eV} \approx 0.4 \times 10^{-6}\ eV.$$

Since the energy of photon is about 10 eV, the width is about 10^{-7} of the photon energy.

(6). Using Eq. (5.1.29) draws the diagram of $\rho(E)$ versus E to see what are the FWHM and Lorenzian distributioin. Put the $E = E_k$ as the centroid.

(7). Do you think that the width Γ in the diagram of exercise (5) characterizes the Heisenberg energy uncertainty of a decaying or transiting systems? This viewpoint can be found in Re. [32].

Solution:

No. Heisenberg energy-time uncertainty relation is a general property of a time-dependent function. However, here the Γ comes from transition (or decay) processes, it is a purely physical effect, and it is connected with the energy-non-conservation of transition processes. We can say that Heisenberg energy uncertainty and γ are totally unrelated, or in other words, they have absolutely nothing to do with each other.

(8). From your diagram drawn in exercise (6), you can see that the Lorenzian curve has a long tail as $|E - E_k| >> \Gamma$. With a slow decrease of $\rho(E)$, the mean square deviation of energy from centroid,

$$(\Delta E)^2 = \int_{-\infty}^{\infty} dE \rho(E)(E - E_k)^2, \tag{1}$$

diverges. For this divergence, some references, such as Ref. [32, 11], have the following viewpoint:

(8.1). This infinity is an unphysical feature of the Lorentzian distribution.

(8.2). In any real system, the energy spectrum does not include all energies from $-\infty$ to ∞; it cuts off from below at the ground state energy;

(8.3). Therefore, the energy distribution of an decaying or transiting system cannot be exactly Lorentzian;

(8.4). The remote energy tails correspond, in spirit of the Heisenberg energy-time uncertainty relation, to very short time. Therefore, the initial stage (short time) of the decay or transition should deviate from the exponential [i. e., $\exp(-t/\tau)$]. Thus, the deviations are expected at short time;

(8.5). The deviations are expected at long time ($t >> \tau$) as well;

(8.6). Just within the physical important time interval, the decay or the transition are close to the exponential, and Lorentzian is a good approximation to reality not very far from the centroid (E_k).

Solution:

(8.1)'. No. This infinity of the Lorentzian distribution in Eq. (1) is a physical feature of the Lorentzan distribution;

(8.2)'. No. There exists a very big confusion. The energy spectrum has to include all energies from $-\infty$ to ∞; it cannot be cut off from below at the ground state energy. Our reason for this is simple. When we make the Fourier transformation, we have to remember that the energy spectrum appearing in Fourier transformation does not have any connection with the atomic discrete spectrum;

(8.3)'. No. Therefore, the energy distribution of a decaying or transiting system is exactly Lorentzian because $\Psi(t) \propto \exp(-t/\tau)$.;

(8.4)'. No. Although the remote energy tails of the Lorentzian distribution correspond, in spirit of the Heisenberg energy-time uncertainty relation, to very short time, but due to that the Heisenberg energy-time uncertainty relation just gives a lowest bound, we do not have any reason to think that the initial stage (short time) of the decay or transition should deviate from the exponential [i. e., $\exp(-t/\tau)$]. In one word, Heisenberg energy-time uncertainty relation just says $\Delta E \Delta t \geq \hbar/2$. If $\Delta E = \infty$, then this does not completely violate Heisenberg energy-time uncertainty relation;

(8.5)'. No. The deviations do not be expected at long time ($t >> \tau$) as well;

(8.6)'. No. Within all the time interval, the decay or the transition are the exponential time-dependence, and Lorentzian is a good for any value of E.

(8.7)'. Of course, we can improve transition and decay theory, but the reasons are not the (9.1)-(9.6).

(9). The relative natural abundance on Earth of the two isotopes ^{235}U and ^{238}U is roughly 1/138. If we assume that the two elements were initially formed with the same abundance, then how much time has elapsed since these elements were created? The half-life of ^{235}U is 7.04×10^8 year and the half-life of ^{238}U is 4.47×10^9 year.

Solution:

From the half-lives we know that $\tau(235_U) = 1.02 \times 10^9$ year and $\tau(238_U) = 6.45 \times 10^9$ year. Since

$$N(t) = N(0)e^{-t/\tau}$$

therefore, $N_{235_U} = N(0)\exp(-t/\tau 235_U)$ and $N_{238_U} = N(o)\exp(-t/\tau 238_U)$. Hence

$$\frac{1}{138} = \frac{\exp[-t/\tau(235_U)]}{\exp[-t/\tau(238_U)]}$$

Substituting in the lifetimes, we find t=6.0×10^9 year.

(10). In the atmosphere, cosmic ray protons collide with the nuclei of air molecules and produce neutrons. These neutrons are absorbed by ^{14}N nuclei in the reaction

$$n +^{14} N \rightarrow \ ^{14}C + p.$$

In the steady state, the ratio $^{14}C/all\ Carbon \ = \ 1.3 \times 10^{-12}$. ^{14}C then beta decays, i. e., ^{14}C emits electron, to ^{14}N in the reaction

$$^{14}C \ \rightarrow \ ^{14}N + e^- + \nu_e$$

with $t_{1/2} \ = \ 5730$ year. Consequently, all planets are radioactive, but the radioactivity diminishes after the planet dies since the ^{14}C is no longer replenished by photosynthesis. If the activity of the living wood is 15 decays/minute per gram of C, how old in the building?

Solution:

$$\frac{dN(t)}{dt} = \frac{d}{dt}\left(N(0)e^{-t/\tau}\right) = -\frac{N(0)}{\tau}e^{-t/\tau}.$$

Thus,

$$\ln\frac{-dN(t)/dt}{N(0)/\tau} = -\frac{t}{\tau}$$

The initial activity is $N(0)/\tau$ and the activity at time t is $-dN(t)/dt$. Therefore,

$$t = \tau \ln\frac{N(0)/\tau}{-dN(t)/dt} = (8270\ year)\ln(15/12) = 1845\ year,$$

where $\tau = t_{1/2}/0.693 = 8270$ year.

(11). In chapter 4 we addresses the Rutherford scattering. Until now there is still no body to calculate the scattering duration time, or even no body has the concept of scattering duration time. In physics, this duration time is equal to the relaxation time in this chapter 5. Estimate the duration time of the collision between the electron and nuclear in the Rutherford scattering.

Solution:
The relaxation time formula is

$$\tau = -\int tde^{-P(t)}. \tag{11.1}$$

Chapter 4 gives two methods to calculate the total transition probability: (i). This textbook's method. In the formula of total transition probability from initial state of free electron with momentum **k** to all possible final free states **k**$'$, this textbook considers that the density of final states $\rho(k') = A\sqrt{E'}$, and makes exact energy integration. The detail result con be found in section on Rutherford scattering. (ii). Some references' method. In the formula of total transition probability from initial state of free electron with momentum **k** to all possible final free states **k**$'$, although somereferences considers that the density of final states $\rho(k') = A\sqrt{E'}$, but thinks that it is a constant and tales it out of the integration for E'. Further, some references takes limit duration time $(t - t_\circ) \to \infty$.

Hope the readers to use the two methods to calculate the relaxation times, to make comparisons and more discussions. If your discussions are good enough, then it is might be that you finish a nice paper.

For the reader's convenience, we give some necessary data. The mass density of nucleus is $\rho_p = 2.7 \times 10^{17}\ kg/m^3$. The radius of atomic nuclear is $R_0 = 1.2 \times 10^{-15} A^{1/3}$, where A is the mass number, and is equal to the total number of neutrons plus protons.

In physics, if we do not make any numerical calculations, we can judge that $\tau_{this\ textbook} << \tau_{some\ references}$. Would the readers please address the reason.

(12). As is well known, the molecular relaxation in liquids is one of the subjects in science frontier [62]. However, all the present relaxation theories in liquids are based on the Fermi golden rule without even one exception [62].

Please take one example from Ref. [62], and make a new numerical calculations. Generally, you will find that your theory can explain the experimental data much better than the now available theory. Please write a paper, and try to submit for publication.

Chapter 6

Applications in Energy Sources

As an application of chapters 1-5, this chapter calculate the energies in cold fusion and internal conversion of nuclear physics. In these processes, the energies are not conservative. The author of this textbook hopes that the readers design many new processes of producing energy in terms of the energy nonconservation property in transition processes.

6.1 Cold Fusion

In 1989 there were two papers had claimed evidence for a surprisingly high rate of the nuclear fusion reaction $d+d \rightarrow^3 He+n$ in deuterium trapped in palladium and titanium at room temperature [63, 64]. This reaction is called cold fusion. Since 1989 up to now the flurry of interest produced by the announcement of cold fusion polarizes the scientific community into two groups: (a). Minority-diehard enthusiasts; (b). Majority-extreme skeptics. The latter call cold fusion pseudoscience, because, generally speaking, the reaction $d+d \rightarrow^3 He+n$ needs temperature $T > 10^7 >> 300$ K. Refs. [65, 66] made exact demonstrations to prove that the cold fusion are not possible absolutely. However, in 2004 the acoustic cavitation experiment further supports the cold fusion Ref. [67].

This textbook would like to point out that the reasons to negative cold fusion in Refs. [65, 66] are of big one-sidedness. Although all the derivations in Refs. [65, 66] are correct, these derivations are confined only in consideration of electron screening effect for deuteron. It is obvious that in condensed matter systems we have to try to study the influence of phonons and to use transition theory for the jumping between different positions of deuterons. Based on our theory in chapter 5 and a few solid state physical knowledge, the author of this textbook tried to use the interaction between deuteron and phonons to prove that the cold fusion is possible [68, 69, 70, 71]. However, the demonstrations in Refs. [68, 69, 70, 71] are not exact. The following will prove that the interaction between deuteron and zero point energy phonons is the origin of the cold fusion.

Using Eq. (5.3.2.12) and the formula of relaxation time in chapter 5, the relaxation time, caused by the interaction between deuteron and zero point energy phonons, is

$$\tau^{(2)} = -\int_0^\infty t dQ_2(t),\ Q_2(t) = e^{-(t/\tau_2)^{1-\beta}},\ \tau^{(2)} = \tau_2 \Gamma\left(\frac{1}{1-\beta}\right). \tag{6.1.1}$$

Let us estimate the value of $\overline{(\mathfrak{N})}^2$ in Eq. (5.3.2.7). The definition of $\overline{(\mathfrak{N})}^2$ is given by the fifth equality of Eq. (5.3.2.3). Therefore,

$$\begin{aligned}\overline{(\mathfrak{N})^2} &\approx \overline{\left(\frac{\partial U(r')}{\partial r'}\right)^2\Bigg|_{over\ cell}} \\ &\approx \left(\frac{\partial U(r')}{\partial r'}\Bigg|_{r'=0.3nm}\right)^2 \\ &\approx \left(\frac{\partial}{\partial r'}\frac{e^2}{r'}\Bigg|_{r'=0.3nm}\right)^2 = 6.55\times 10^{-8}\ (erg/cm)^2, \end{aligned} \qquad (6.1.2)$$

We consider the solid Pd, and use the following data to calculate the relaxation time of deuteron jumping from any site to its nearest neighbour site, which is equal to deuteron-deuteron collision time. $v_{ph} = 1.68\times 10^5$ cm/second, $a = 4$ Å, $M = 46\times 1.67\times 10^{-24}$ g, $\beta = 0.8$, then $1-\beta = 0.2$. Suppose that at room temperature the activation energy is zero. Eq. (5.3.2.13) becomes

$$\begin{aligned}\tau_2 &= \left[\frac{1}{A''G(T)}\right]^{-1/(1-\beta)} \\ &= \left[\frac{1}{A''}\right]^{-1/(1-\beta)} e^{E_\circ/[(1-\beta)(T-T_\circ)]} \\ &\approx \left[A'\hbar^{\beta-1}\frac{\pi}{2\Gamma(2-\beta)\sin[\pi(1-\beta)/2]}\right]^{-1/(1-\beta)}, \end{aligned} \qquad (6.1.3)$$

where

$$A' = V\frac{3}{2\pi^2}\frac{1}{\hbar^3 v_{ph}^3}\frac{2+\beta}{3}E_D^{1-\beta}\frac{\hbar^2}{2NM}\overline{(\mathfrak{N})^2}. \qquad (5.3.2.7)'$$

$$\begin{aligned}\tau_{2,\beta=0.8} &= \left\{A'\hbar^{(\beta-1)}\times\frac{\pi}{2\Gamma(2-\beta)\sin[(1-\beta)\pi/2]}\right\}^{-\frac{1}{1-\beta}} \\ &= \left\{12.6\times 10^2\times 2.5\times 10^5\times 5.5\right\}^{-\frac{1}{0.2}} \\ &= \left\{1.73\times 10^9\right\}^{-5} = 6.45\times 10^{-43} second, \end{aligned} \qquad (6.1.4)$$

which is a very small quantity. The readers might ask that why do we never hear so small relaxation time? Let me show you the origin. From section 5.3.2 of chapter 5 we know that when all until now references and textbooks calculate this τ, they always use $1-\beta = 1-0 = 1$. If we use $\beta = 0$, then

$$\begin{aligned}\tau_{2,\beta=0} &= \left\{A'\hbar^{(-1)}\times\frac{\pi}{2\Gamma(2)\sin[(1-\beta)\pi/2]}\right\}^{-\frac{1}{1-\beta}}\Bigg|_{\beta=0} \\ &= \left\{1.5\times 10^{-12}\times 10^{27}\times\frac{\pi}{2}\right\}^{-1} \\ &= 2.35\times 10^{-14}\ second, \end{aligned} \qquad (6.1.5)$$

which clearly shows that in our derivations, if we take $\beta = 0$, then we will obtain the same result as that in other references, and confirms that the so-called cold fusion is not possible absolutely. However, from chapter 5 we know that $\beta \neq 0$.

Substituting Eqs. (6.1.4) into Eq. (6.1.1) yields

$$\tau^{(2)}_{\beta=0.8} = \tau_{2,\beta=0.8}\Gamma(6) = 7.74\times 10^{-41}\ second. \qquad (6.1.6)$$

Let us use our numerical estimations to analyze the true or the false of the cold fusion. If the cold fusion happens truly, then it requires that the fusion rate $R_{cold,exp} \geqq 10^{-23}/second$. However, the theoretical value is only $R_{cold,theo55} \leqq 10^{-41}/second$ [65]. Ref. [66] considered screening effect of electron in palladium for the two deuterons, and gave $R_{cold,theo56} \leqq 10^{-31}/second$. According to this reaction rate, Ref. [66] said that if the number density of deuterons in palladium is $10^{22}/cm^3$, then this rate implies approximately one or two deuteron-deuteron fusion reaction per year per unit cm^3, and thus the deuteron fusion at room temperature is actually still false.

The problem in Refs. [65, 66] is not to consider the phonon-mediated deuteron interaction. In our calculations, $\beta = 0$ is equivalent to consider phonon-mediated deuteron interaction in a complete disorder palladium, and $\beta \to 1$ is equivalent to consider phonon-mediated deuteron interaction in a perfect palladium crystal. Actually, any metals cannot be perfect. Therefore, we consider the case $\beta = 0.8$. The $\tau_{2,\beta=0.8} = 7.74 \times 10^{-41}$ $second$ in Eq. (6.1.6) means that the collision rate is $W_{\beta=0.8} = 1.3 \times 10^{40}$. This collision rate is a rate that a deuteron collides with any deuteron around it. The rate that a deuteron collides with an assigned deuteron is $W_{\beta=0.8} \times \pi r^2/(4\pi R^2) \approx 1.3 \times 10^{30}$, where $r = 10^{-13}$ cm is the radius of deuteron, and $R = 4 \times 10^{-8}$ cm is the average distance between two nearest neighbor deuterons. Therefore, if we consider the phonon-mediated interaction, then $R_{cold,theo} = R_{cold,theo56} \times W_{\beta=0.8} \times second = 1.2 \times 10^{-1}/second$. Therefore, if we consider the phono-mediated interaction, then the cold fusion can happen in principle. In the cold fusion experiments people found that the more perfect the palladium crystal is, the easier the occurrence of the cold fusion is. This experimental fact can be explained in principle by the phonon-mediated interaction, because the more perfect the palladium crystal is, the larger the occurrence of the cold fusion is.. The cold fusion experiment also found that the deuteron can also collide with the atom Pd. Owing to the very high collision rate, caused by the phono-mediated interaction, a few deuterons can penetrate the high potential between the deuteron and atom Pd, and the deuteron-Pd fusion is produced.

Another example of cold fusion is the deuteron cold fusion during acoustic cavitation in Ca_3D_6O liquid [67, 72]. Taleyarkhan *et al.* measured time spectra of neutron in acoustic cavitation experiments with chilled deuterated acetone. Statistically significant neutron emission was measured. The neutron emission rate was up to $\sim 4 \times 10^5$ n/sec, where n indicates neutron. Taleyarkhan *et al.* thought that the neutron emission is due to D-D fusion inside the cavitation vapor bubbles nucleated in highly tensioned deuterated acetone by means of fast neutrons. Here, D indicates deuteron. The D-D fusion mechanism imagined by Taleyarkhan *et al.* is as follows. The intense implosive collapse of bubbles, including acoustic cavitation bubbles, can lead to extremely high compression, temperature (about 10^9 K), and density of D. Their duration is large enough. These conditions yield the D-D fusion inside the cavitation vapor bubbles.

The difficulty of Taleyarkhan *et al*'s mechanism of D-D fusion is that there is no direct experimental evidence to show the high temperature of 10^9 K in the cavitation vapor bubbles. As far as we know, there is still no exact microscopic theory for the D-D fusion during acoustic cavitation. Basing on the exact calculation of the transition probability of deuteron under phonon-deuteron interaction, this section proposes a microscopic theory for the D-D fusion. Our mechanism is as follows. During the acoustic cavitation a large number of D inside the bubbles enter the spherical surface layer of the chilled deuterated acetone by means of high compressions and high density of D. Note that the spherical surface layer surrounds the cavitation vapor babbles. We call the D entered the spherical surface layer the entered D, and is expressed as D_e or D(entered). The state of D_e in the spherical surface layer is of non-equilibrium and of high density. We call the spherical surface layer D_e-C_3D_6O the system. There is D_e-phonon interaction in the spherical surface layer. In the following, we will prove that the D_e-phonon interaction leads to the D_e-D_e fusion in the spherical surface layer.

Let us at first make physical analyses. The intense implosive collapse of cavitation vapor bubbles can lead to extremely high compressions and high density of D [67, 72]. During this process, a large number of D inside the bubbles enter the spherical surface layer, surrounding the cavitation vapor bubble, of the chilled deuterated acetone, and form a D_e-C_3D_6O system. The C_3D_6O is a liquid. The model of transitional motion in liquid, proposed by Kruus [73] and Larsson [74], is as follows. Suppose that a molecule to undergo vibration in its cell on the average for a time τ_o. It then is displaced into another cell through some distance l, where it again oscillates for a time τ_o, etc. We can describe such collective and translational motion by longitudinal phonons [73, 74]. Simply speaking, the density oscillations can be described by the longitudinal phonons. Neutron scattering data for liquid had been interpreted in terms of the presence of longitudinal phonon successfully [73, 74].

Owing to the existence of phonons, there is the phonon-mediated interaction between the two nearest neighbor deuterons, and thus the D_e-D_e cold fusion can occur in principle.

The phonon-mediated interaction in this section supports the D_e-D_e fusion in Refs. [67, 72] and predicts that there should be excess heat. From ${}^2_1D + {}^2_1D \longrightarrow {}^3_2He + {}^1_0n + 3.3$ MeV and the observed neutron emission rate ($\approx 4 \times 10^5$n/sec), we know that the excess heat rate is $\approx 1.3 \times 10^6$ MeV/sec. The reason that the excess heat is difficult to be observed in Ref. [72] is that the volume of the total spherical surface layer is too small (only 9.7×10^{-15} cm^3), and thus the number of moving D_es are too small.

The fusion imagined by Refs. [67, 72] occurs in the cavitation bubbles. This section argues that the fusion occurs in liquid.

In the above theory for the D_e-D_e fusion during acoustic cavitation, the sufficient conditions for occurrence of the D_e-D_e fusion are the existences of D_es in high density non-equilibrium state, i. e., there are many deuterons in moving states, and the existences of longitudinal phonons in a condensed matter system. If the compression and the density of moving H in the acoustic cavitation vapor bubbles in water are high enough, then a number of moving H can enter the spherical surface layer surrounding the cavitation bubbles and form non-equilibrium state of the entered H, at least, in a layer with thickness of one molecule of water. The H(entered)-H(entered) fusion can also occur in the spherical surface layer outside the bubbles, i. e., in water. However, there was no statistically significant neutron emission, when one used C_3H_6O liquid instead of C_3D_6O liquid [72]. We think that the H(entered)-H(entered) fusion already occurred in Ref. [72], when one used C_3H_6O liquid. The reason is simple. The last products of H(entered)-H(entered) fusion are ${}^4_2$He and excess heat. Therefore, that there is no neutron emission does not imply that there is no H(entered)-H(entered) fusion in C_3H_6O liquid. If the H(entered)-H(entered) fusion in water can be realized in suitable acoustic cavitation conditions, then the water can provide nearly good energy source for mankind. The cavitation vapor bubbles in water can be nucleated by fast neutron or by laser.

Owing to the high collision rate, the moving D_e (or H(entered)) can also overcome the high potential barriers between the moving D_e (or H(entered)) and any atom in C_3D_6O (or C_3H_6O and H_2O). Therefore, we predict that there are fusions between D_e (or H(entered)) and any atom of C_3D_6O (or C_3H_6O and H_2O) as well.

After the finding of neutron emission during acoustic cavitation in Refs. [67, 72], most people doubt the correctness of quantum mechanics. Our theory clearly shows that the observed fusion in Refs. [67, 72] is true, and the quantum mechanics is correct as well.

6.2 Internal Conversion

Let us consider quantum transition under the influence of time independent interactions. Examples of such transitions are: (1). Internal conversion, that is, the process in which an exited nucleus transfers its energy to the atomic electrons; (2). The Auger effect, that is, the readjustments of the electron shells of atoms with several electrons, accompanied by the ejection of one electron from atom.

In the present section, we shell consider internal conversion. This term reflects the original, incorrect viewpoint according to which the transfer of the excitation energy of the nucleus to the electrons in the atom wads considered to be an intranuclear photoeffect, caused by the photons emitted by the nucleus. Afterwards it becomes clear that the excited nucleus can transfer its energy to the electron also when the emission of a single photon is absolutely forbidden, that is, when the transition is between states with zero total angular momentum ($0 \to 0$ transitions; see the photoeffect in many textbooks on quantum mechanics such as Ref. [20]. Internal conversion and emission of photons by the nucleus must be considered as two alternative possibilities, which can realized when an atomic nucleus makes a transition from an excited state to the ground state. Many authors have considered the problem of calculating the transition probability for internal conversion; their papers differ from one another in the various approximations made for the wavefunctions of the atomic electrons and for the operator determining the transition. We shell consider here an elementary theory of internal conversion in which we choose the wavefunctions of the ejected electrons in the form of plane waves and use a non-relativistic approximation.

We shell thus use for the initial state of the electron

$$\Psi_k(\mathbf{r}) = \frac{1}{\sqrt{\pi a^3}} e^{-r/a}, \qquad a = \frac{\hbar^2}{\mu e^2 Z}, \tag{6.2.1}$$

where Ψ_k can be a wavefunction of hydrogen-like K-electron, μ is the mass of electron, Z is the proton's number in a nucleus. For the final state

$$\Psi_{k'}(\mathbf{r}) = \frac{1}{\sqrt{V}} e^{i\mathbf{k}'\cdot\mathbf{r}}, \tag{6.2.2}$$

which is a plane wave of the ejected electron. If we denote the wavefunctions of the initial and final states of the nucleus by $\phi_i(q)$ and $\phi_f(q)$, respectively, we have for the wavefunctions of the initial and final states of whole system

$$|ki >= \Psi_k(\mathbf{r})\phi_i(q) \quad and \quad |k'f >= \Psi_{k'}(\mathbf{r})\phi_f(q). \tag{6.2.3}$$

The transition probability $P_{all\leftarrow k}$ of internal conversation of a nucleus from initial excited state $\phi_i(q)$ to final state $\phi_f(q)$ and one of the atomic K-shell electrons going from initial state $\Psi_k(\mathbf{r})$ to final ejected state $\Psi_{k'}(\mathbf{r})$ is

$$\begin{aligned}
P_{all\leftarrow k}(t \geq t_\circ) &= \int d\Omega P_{all\leftarrow k,d\Omega} \\
&= \int d\Omega \frac{1}{\hbar^2} \int_0^{+\infty} |H'_{k'k}|^2 \rho(\omega_{k'}) \frac{\sin^2 \frac{(\omega_{k'}-\omega_\Delta+\omega_b)t_\circ}{2}}{\left[\frac{\omega_{k'}-\omega_\Delta+\omega_b}{2}\right]^2} d\omega_{k'} \\
&= \int d\Omega \frac{1}{\hbar^2} \int_0^{+\infty} |H'_{k'k}|^2 \rho(E_{k'}) \\
&\times \frac{\sin^2 \frac{(E_{k'}-E_\Delta+E_b)t_\circ}{2\hbar}}{\left[\frac{E_{k'}-E_\Delta+E_b}{2\hbar}\right]^2} dE_{k'},
\end{aligned} \tag{6.2.4}$$

where $\omega_{k'}$, ω_Δ, and ω_b are the angular frequency of the ejected electron, the excitation of nucleus, and the atomic K-electron binding, respectively, $\omega_{k'k} = \omega_{k'} - \omega_k \equiv \omega_{k'} - (\omega_\Delta - \omega_b)$, $\rho(\omega_{k'})$ is the density of states per unit angular frequency interval of ejected electron within a solid angle $d\Omega$, $E_{k'}$, E_Δ, and E_b are the energies of the ejected electron, excitation of nucleus, and atomic K-electron binding, respectively, $E_{k'k} = E_{k'} - E_k \equiv E_{k'} - (E_\Delta - E_b)$, $\rho(k')$ is the density of states per unit energy interval of ejected electron within a solid angle $d\Omega$, and $t_\circ$ is the duration time of the perturbation Hamiltonian.

$$\rho(k') = V\frac{\mu\hbar k'}{(2\pi\hbar)^3}d\Omega. \tag{6.2.5}$$

$H'_{k'k}$ is the matrix element of perturbation Hamiltonian.

$$H'_{k'k} =< k'f|H'|ki >\equiv< k'f|\sum_{j=1}^{Z}\frac{e^2}{|\mathbf{r}-\mathbf{q}_j|}|ki >, \tag{6.2.6}$$

where H' represents the Coulomb interactions between all Z protons in a nucleus at $\mathbf{q}_j$ and one electron at $\mathbf{r}$ (take the center of the nucleus as origin), and the summation is over all the protons in the nucleus. Because the perturbation operator H' in Eq. (6.2.6) does not contain spin variables, so that we cannot describe nuclear transition corresponding to magnetic multiple radiations.

If $|\mathbf{r}| >> |\mathbf{q_j}|$, we can expand the perturbation operator H' in spherical harmonics,

$$H' = \sum_{j=0}^{Z}\sum_{l=0}^{\infty}\sum_{m}\frac{4\pi e^2}{(2l+1)r}\left(\frac{q_j}{r}\right)^l Y_l^m(\Theta,\Phi)Y_l^{m*}(\theta_j,\phi_j), \tag{6.2.7}$$

where (Θ, Φ) are the angles defining the direction of $\mathbf{r}$ and (θ_j, ϕ_j) are the angles defining the direction of the $\mathbf{q}_j$

Using Eqs. (6.2.7) and (6.2.3), we can write

$$H'_{k'k} =< k'f|H'|ki >= \sum_{l,m}\frac{4\pi e^2}{2l+1} < f|\sum_j q_j^l Y_l^{m*}(\theta_j\phi_j)|i >< k'|\frac{Y_l^m(\Theta\Phi)}{r^{l+1}}|k >, \tag{6.2.8}$$

After substituting Eqs. (6.2.1-2) into the matrix element containing the integration over the electronic coordinates, we get

$$< k'|Y_l^m(\Theta\Phi)/r^{l+1}|k >= \frac{1}{\sqrt{\pi a^3 V}}\int\frac{Y_l^m(\Theta\Phi)}{r^{l+1}}e^{-r/a}e^{-i\mathbf{k}'\cdot\mathbf{r}}d\mathbf{r}. \tag{6.2.9}$$

Expanding $exp(-i\mathbf{k}'\cdot\mathbf{r})$ in spherical harmonics,

$$e^{-i\mathbf{k}'\cdot\mathbf{r}} = 4\pi\sum_{lm}(-i)^l j_l(k'r)Y_l^m(\theta\phi)Y_l^{m*}(\Theta\Phi), \tag{6.2.10}$$

where (θ, ϕ) are angles defining the direction of the wavevector $\mathbf{k}'$, and integrating over angles (Θ, Φ), we get

$$< k'|Y_l^m(\Theta\Phi)/r^{l+1}|k >= 4\sqrt{\frac{\pi}{a^3V}}(-i)^l Y_l^m(\theta\phi)\int\frac{1}{r^{l+1}}j(k'r)e^{-r/a}dr. \tag{6.2.11}$$

When evaluating the the integral in Eq. (6.2.11), we must bear in mind that in our approximation, $k'a >> 1$, only small values of r are important in the integral because of the fast oscillation of the spherical Bessel function; therefore

$$\int\frac{1}{r^{l+1}}j(k'r)e^{-r/a}dr \approx \frac{k'^{l-2}}{1\cdot2\cdot3\cdots(2l-1)}. \tag{6.2.12}$$

Substituting Eq. (6.2.12) into Eq. (6.2.11), and then Eqs. (6.2.8) and (6.2.5) into Eq. (6.2.4), integrating over the angles defining the direction of the emission of electron, i. e., integrating over the solid angle in Eq. (6.2.4) using the orthogonality of the spherical harmonics:

$$\int d\Omega Y_l^{m*}(\theta,\phi)Y_{l'}^{m'}(\theta,\phi) = \delta_{ll'}\delta_{mm'}, \tag{6.2.13}$$

we find the transition probability per ejected electron:

$$\begin{aligned}
P \equiv P_{all\leftarrow k}(t \geq t_\circ) &= \frac{128\pi e^4\mu}{\hbar(a\hbar)^3}\sum_{lm}\frac{|<f|\sum_{j=1}^{Z} q_j^l Y_l^m(\theta_j\phi_j)|^2}{((2l+1)!!)^2}\int_0^{+\infty} dE_{k'}k'^{2l-3} \\
&\times \frac{\sin^2\frac{(E_{k'}-E_\Delta+E_b)t_\circ}{2\hbar}}{\left[\frac{E_{k'}-E_\Delta+E_b}{2\hbar}\right]^2} \\
&\approx \frac{128\pi e^4\mu}{\hbar(a\hbar)^3}\sum_{lm}\frac{Z^2d^{2l}}{((2l+1)!!)^2}\int_0^{+\infty} dE_{k'}k'^{2l-3}\frac{\sin^2\frac{(E_{k'}-E_\Delta+E_b)t_\circ}{2\hbar}}{\left[\frac{E_{k'}-E_\Delta+E_b}{2\hbar}\right]^2} \\
&= \frac{128\pi e^4\mu}{\hbar(a\hbar)^3}\sum_{lm}\frac{Z^2d^{2l}}{((2l+1)!!)^2}\left(\frac{2\mu}{\hbar^2}\right)^{(2l-3)/2} \\
&\times \int_0^{+\infty} dE_{k'}E_{k'}^{(2l-3)/2}\frac{\sin^2\frac{(E_{k'}-E_\Delta+E_b)t_\circ}{2\hbar}}{\left[\frac{E_{k'}-E_\Delta+E_b}{2\hbar}\right]^2},
\end{aligned} \tag{6.2.14}$$

where the matrix element in the first equality of Eq. (6.2.14) is approximately taken as

$$\left|<f|\sum_{j=1}^{Z} q_j^l Y_l^m(\theta_j\phi_j)|i>\right|^2 \approx Z^2d^{2l}, \tag{6.2.15}$$

where d is the linear magnitude of the nucleus. The square of the matrix element occurring in Eq. (6.2.14) is proportional to the probability of the nuclear transition corresponding to an El multiple radiation. We must remind the reader that Eq. (6.2.14) was derived under the conditions $v << c$, and $Ze^2/(\hbar v) << 1$.

To seek the expression of Eq. (6.2.14), All until now references and textbooks (AUNRTB) calculate the integral in Eq. (6.2.14) by taking the following two approximations. First approximation is that the factor in the integrand $E_{k'}^{(2l-3)/2}$ is taken as a constant independent of integration variable $E_{k'}$. Second approximation is to assume that $t_\circ$ approaches to infinity. Then AUNRTB find from Eq. (6.2.4) that

$$P_{all\leftarrow k}(t_\circ \to \infty) = \lim_{t_\circ\to\infty} t_\circ\frac{128Z^2\pi e^4\mu}{(\hbar a)^3}\sum_{lm}\frac{k'^{2l-3}d^{2l}}{(2l+1)!!}. \tag{6.2.16}$$

Eq. (6.2.16) is AUNRTB's formula to calculate the internal conversion process. The transition rate is

$$W_{all\leftarrow k}(t_\circ \to \infty) \equiv \frac{P_{all\leftarrow k}(t_\circ \to \infty)}{t_\circ} = \frac{64Z^2\pi e^4\mu}{(\hbar a)^3}\sum_{lm}\frac{k'^{2l-3}d^{2l}}{(2l+1)!!}. \tag{6.2.17}$$

According to the two approximations made by AUMRTB the energy in internal conversion is conservative.

Comments for the above AUNRTB's treatment of the internal conversion. Generally speaking, the experimental value for time needed by one nucleus to emit one electron through internal conversion is only, for example 10^{-7} second. Therefore, the $t_\circ \rightarrow \infty$ should be bad. The approximation that neglects the energy dependence of $E_{k'}^{(2l-3)/2}$ in Eq. (6.2.14) should be bad as well because the integration up limit is ∞. From the integrand of Eq. (6.2.14), it is possible that the larger the $E_{k'}$ is, the larger the transition probability is. Therefore, let us try to establish a more exact internal conversion theory by using exact calculations.

The treatment of this textbook for the internal conversion. For simplicity, we consider parity-favored electric fourth multipole transition $E4$ ($l = 4$) in Eq. (6.2.14). The factor $E_k^{(2l-3)/2} = E_k^{5/2}$ in the integrand of Eq. (6.2.14) is strongly E_k-dependent. For element Bi, Z=83, $E_b = 95.7\times10^3$ eV, $E_\Delta = 5$ MeV, and the lifetime of excited nucleus $t_\circ = 10^{-7} << \infty$ (second). It is obvious that if we abandon the two approximations taking by AUNRTB, then the theoretical results should be better. Therefore, let us continue our calculations.

The exact formula for internal conversion is Eq. (6.2.14). To keep that perturbation theory of D-C transition process holds water, we should take, for example, $P = 0.05 << 1$. In this case the up limit of integral in Eq. (6.2.14) should be finite. If we take that the down limit of integral is zero, then the integral should be timed by a factor 2. Numerical calculations give the up limit of integral is $E_{k,up} = 6.4 \times 10^9$ erg $= 4 \times 10^{15}$ Mev, which is much larger than the excitation energy of nucleus $E_\Delta = 5$ MeV, and even much larger than the Einstain's static mass energy of Bi nucleus $M_{\text{nucleus}}c^2 = 0.3$ erg $= 0.3\times0.624\times10^6$ MeV. $E_{k,up} = 2 \times 10^{10} M_{\text{nucleus}}c^2$. If we rewrite Eq. (6.2.14) as

$$P \equiv \int_0^{E_{k,up}} P'(E_k)dE_k, \tag{6.2.18}$$

then $P'(E_k)$ represents the energy density of transition probability. From the expression of $P'(E_k)$ in Eq. (6.2.14), we know that the larger the energy of ejected electron E_k is, the larger the value of $P'(E_k)$ is, i. e., the ejected electron with higher energy has higher transition probability. Therefore, the internal conversation process is a strong energy non-conversation process.

It is also interesting to note that if the initial excited state of a nucleus transits to final ground state, and a γ photon is emitted, then the transition probability is proportional to (Refer to any text book on nuclear physics for derivation)

$$P_\gamma \propto \int_0^{E_{k,up}} E_k^{2l+1} \frac{sin^2 \frac{(E_k-E_\Delta)t_\circ}{2\hbar}}{\left[\frac{E_k-E_\Delta}{2\hbar}\right]^2} dE_k, \tag{6.2.19}$$

where E_k represents the energy of one photon. Due to that the energy-dependence of factor E_k^{2l+1} in Eq. (6.2.19) is stronger than the factor $E_k^{(2l-1)/2}$ in Eq. (6.2.14), the energy non conservation in the internal conversion of ejecting $\gamma-$rays will be much stronger than that in the internal conversion of electron ejection.

Let us here make some discussions for the above calculations.

(1). Generally speaking, the energy is more or less not conservative in time-dependent and time-independent D-C quantum transition processes.

(2). We can obtain infinite energy from vacuum space in terms of some D-C transitions such as internal conversation of electron ejection. According to our exact calculations for the internal conversation of electron ejection above, the energy can be strongly not conservative.

The internal conversion process can be repeated by three steps:

(2.1). We get the ejected electron in the internal conversion;

(2.2). The ejected electron gives $4 \times 10^{15}/2$ MeV energy to the environment, and then goes back to the original K-shell;

(2.3). The nucleus absorbs a photon with energy 5 Mev, and transits from the state of lower energy Ψ_f to the state of higher energy Ψ_i. After the above three steps the internal conversion process can occur once again, and we can obtain energy $E_{one-ejected-electron} = 4 \times 10^{15}/2$ Mev once again by energy loss of 5 MeV only.

If the volume of element Bi is 1 mm^3, then this volume contains number of Bi atoms

$$N_{Bi} = \frac{9.78}{1000 \times 207 \times 1.6 \times 10^{-24}} = 2.95 \times 10^{19}. \tag{6.2.20}$$

Assume that the number of the Bi nuclei, for which the internal conversion processes of electron ejection occur in the time interval of the lifetime $t_\circ = 10^{-7}$ second, are only $N_{ejected-nuclei} = 10^{-2} \times N_{Bi}$. The energy of oil per gram is $E_{oil} = 2.5 \times 10^{16}$ eV. The energy, obtained in one second from one mm^3 of Bi atoms, is $E = E_{one-ejected-electron}$ $N_{ejected-nuclei}/(t_\circ E_{oil}) = 4.5 \times 10^{23}$ ton of oil. This is a very high rate of energy generation. For example, everybody in our world can share in the energy about 10^7 tons of oil from internal conversation just in one second and in volume mm^3. Because this energy gain process in the internal conversion can be repeated, we can say that one can obtain infinite energy in terms of internal conversion transition process of electron ejection.

If we can obtain infinite energy from vacuum space, then the vacuum space has to have infinite energy.

(3). Although the high rate of energy generation is so inconceivable and unimaginable, but we have to emphasize that our theoretical result on the energy non conservation in internal conversion transition process is absolutely reliable, the reason for which is very simple and obvious. Because our basic physical formula to calculate the internal conversion process is exactly the same as that in AUNRTB, and the sole difference between our method and the methods in AUNRTB is that we make exact numerical calculations without using the two "illegal" approximations, and AUNRTB always make the two "illegal". Here we say two "illegal" approximations because: (a). The $|H_{if}|^2\rho(E_k) \propto E_k^{5/2}$ in Eq. (6.2.14), which is strongly energy-dependent; (b). $t_\circ \to \infty$ is non-physical.

(4). The experimental results in chapter 13 of Ref. [3] for the light absorption in semiconductors are the particular verification for the energy non conservation in D-C quantum transition process.

(5). According to famous Einstein formula, static mass $m_{nucleus}$ can be transformed into energy $m_{nucleus}c^2$, and at the same time $m_{nucleus}$ vanishes into the void. However, according to our theory of internal conversation of electron ejection, one excited Bi (or other) nucleus can produce energy $10^{12} \times m_{Bi-nucleus}c^2 = 1.5 \times 10^{17} m_{free-electron}$, and at the same time $m_{Bi-nucleus}$ still exist after transition process.

(6). The reasons that why the energy non conservation in D-C transition has not yet been reported in a large number are as follows:

(6.1). The until now experiments of the internal conversations are confined to measure the number of the ejected electrons. Therefore, the energy non conservation cannot be discovered;

(6.2). The until now experiment of the internal conversations are confined to measure low angular quantum number, for which the effect of energy non conservation is not large;

(6.3). Although AUNRTB gave also the curve of transition probability versus energy, which has contained small energy non conservation, but AUNRTB think that the second peak in this curve is too low, and therefore can be neglected in any case. Therefore, the present controversial issue is not that the non conservation has not yet been reported in a large number, but is that can one neglects the non conservation just due to its littleness in special case of $\beta = 0$? The obvious answer is: Cannot.

(7). Warning! We have given in this section that the energy, obtained in one second from one mm^3 of Bi atoms contained excited nuclei, is $E = 4.5 \times 10^{21}$ ton of oil. This is a very high rate of energy generation, and absolute value of energy is also too high. It is well known, in energy aspect, one ton of oil $= 10^{-4}$ ton of TNT. Therefore, $E = 4.5 \times 10^{17}$ ton of TNT in energy. From this estimation we know that to use many exited Bi nuclei to verify of our theoretical prediction has to be prohibited at the beginning stage. We give readers an other estimation. The energy of only one ejected electron released in $t_\circ = 10^{-7}$ second by just one exited Bi nucleus is equivalent to 150 Kg of oil. We sincerely hope that the readers consider our estimations. Of course, this might be my unnecessary worrying, because in this process the energy comes from many high energy electrons. The radius of one electron is $< 10^{-18}$ cm. So small electron passes through any object, it might be that there are no even a wee bit of influences for the object.

Chapter 7

Microscopic Superconductivity

7.1 Introduction

At first let me explain what does the (macroscopic) superconductivity mean, and then the microscopic superconductivity mean.

The direct current resistance $R = 0$ at $T \leq T_c$, where T_c is called superconducting transition temperature or critical temperature, was discovered in 1911, and is called (macroscopic) superconductivity. In this textbook, superconductivity means macroscopic superconductivity. The high-T_c superconductivity was discovered in September 1986. Before 1986, the highest value of T_c is 23 K. After 1986, the highest value of T_c reaches 163 K at pressure. From this chapter you will see that in superconducting state there are Cooper pairs. The Cooper pairs are bound state of two free electron or free holes (exactly speaking, quasielectron and quasihole). The value of the so-called gap is nearly equal to half of the bound energy between the two carriers of a Cooper pair. The Cooper pairs are the carriers of supercurrent. If there are long-rang phase coherence, then these Cooper pairs will move together, and form macroscopic supercurrent. However, if there is no the long-range phase coherence between Cooper pairs, then although every Cooper pair move freely without resistance, due to the random motion of every Cooper pairs, the system cannot be in (macroscopic) superconducting state. This textbook calls this state the microscopic state. Considering that (macroscopic) superconductivity does not still appear at room temperature (300 K), and thus its applications are confined much below room temperature. Considering that the microscopic superconductivity already can appear at room temperature, and therefore it might have many applications, this chapter will address only the microscopic superconductivity and the possible applications at room temperature. We just address the conditions forming Cooper pairs, and do not address the long-range phase coherence condition between Cooper pairs. We call the gap, which is nearly equal to the bound energy of a Cooper pair in microscopic superconductivity, the pseudogap. This chapter is written under the bases of Ref. [75]-[80].

According to the distance, which is called coherence length, between the two electrons (or, two holes) forming a Cooper pair, superconductors can be separated into two kinds: conventional and unconventional superconductors. For the conventional superconductor, the coherence length $\xi_\circ \sim 10^3 - 10^4$ $\mathring{A}$, but for unconventional superconductor $\xi_\circ \sim 20 - 30$ $\mathring{A}$. The high temperature superconductivity, discovered in 1986, belongs to the unconventional superconductivity.

AUNRTB (except this textbook and some papers of authors) on the theory and physics of superconductivity have not yet haven the concept on the microscopic superconductivity, and, of course, have not yet pointed out the possible applications of the microscopic superconductivity at room temperature.

7.2 BCS Pairing Theory

The mechanisms forming Cooper pairs in both macroscopic and microscopic superconductivity are exactly the same.

7.2.1 Single Cooper Pair

Cooper showed that if two electrons outside the Fermi surface have a net attractive potential ($-V < 0$), then the two electrons form a bound state so that their total energy is less than $2E_F$. E_F is Fermi energy. We give the derivation of the forming Cooper pair in the following.

For convenience, we make four assumptions: First, $T = 0$; Second, there are only two electrons outside the Fermi surface; Third, the system has translational invariability, so that the mass center momentum of the two electrons, $\mathbf{q}$, is constant; Fourth, the interaction between the two electrons is spin-independent, so that the total spin of the two electrons, S, is a constant. Let the two electron wavefunction be $\psi(\mathbf{r}_1, \mathbf{r}_2) = \phi_{\mathbf{q}}(\mathbf{r})\exp(\mathbf{iq} \cdot \mathbf{R})\exp(\mathbf{i}\phi)$ where $\mathbf{r} = \mathbf{r}_2 - \mathbf{r}_1$, ϕ is the phase of wavefunction of mass-center of the two electrons, and $\mathbf{R} = (\mathbf{r}_1 + \mathbf{r}_2)/2$. From Pauli principle, if $\phi_{\mathbf{q}}(\mathbf{r})$ is symmetric, which means when the two electrons exchange positions, the wavefunction does not change sign, then $S = 0$, i.e., spin singlet state of the two electrons; if $\phi_{\mathbf{q}}(\mathbf{r})$ is antisymmetric, then $S = 1$, i.e., spin triplet state of the two electrons.

The form of $\psi(\mathbf{r}_1, \mathbf{r}_2)$ in case of $\mathbf{q} = 0$ is

$$\psi(\mathbf{r}_1, \mathbf{r}_2) = \phi(\mathbf{r}_2 - \mathbf{r}_1), \tag{7.2.1.1}$$

where $\phi(\mathbf{r}_2 - \mathbf{r}_1)$ includes $\exp(i\phi)$. Considering that the two electrons are outside the Fermi surface, the expansion of $\phi(\mathbf{r})$ in $\mathbf{k}$ space is

$$\phi(\mathbf{r}_2 - \mathbf{r}_1) = \sum_{\mathbf{k}(|\mathbf{k}|>k_F)} a_{\mathbf{k}} e^{i\mathbf{k}\cdot\mathbf{r}_2 - i\mathbf{k}\cdot\mathbf{r}_1}, \tag{7.2.1.2}$$

where $a_{\mathbf{k}}$ includes $\exp(i\phi)$. Since the factors $\exp(i\mathbf{k}\cdot\mathbf{r}_2)$ and $\exp(-i\mathbf{k}\cdot\mathbf{r}_1)$ can be thought of as single-particle states of momentum $\mathbf{k}$ and $-\mathbf{k}$, respectively, we see that the pair wavefunction is a superposition of configurations in each of which a definite pair state ($\mathbf{k}$, $-\mathbf{k}$) is occupied [75]. If $S = 0$, then $\phi(\mathbf{r}_2 - \mathbf{r}_1) = \phi(\mathbf{r}_1 - \mathbf{r}_2)$. From Eq. (7.2.1.2) we obtain $a_{\mathbf{k}} = a_{-\mathbf{k}}$. If $S = 1$, then $\phi(\mathbf{r}_2 - \mathbf{r}_1) = -\phi(\mathbf{r}_1 - \mathbf{r}_2)$. From Eq. (7.2.1.2) we obtain $a_{\mathbf{k}} = -a_{-\mathbf{k}}$.

To find the eigenvalue of $\psi(\mathbf{r}_1, \mathbf{r}_2)$ we write down the Schrödinger equation.

$$(H_\circ + \widehat{V})\psi = W\psi, \tag{7.2.1.3}$$

where $H_\circ$ is the kinetic energy of the two electron system, relative to the Fermi surface, and $\widehat{V}$ the interaction energy between the two electrons. We need to obtain a Schrödinger equation for $a_{\mathbf{k}}$. We do this as follows.

$$\int \frac{d\mathbf{r}_1}{V_1} \int \frac{d\mathbf{r}_2}{V_2} e^{-i\mathbf{k}\cdot(\mathbf{r}_2 - \mathbf{r}_1)} \left\{ \left[-\frac{\hbar^2}{2m^*}\nabla^2_{\mathbf{r}_1} - \frac{\hbar^2}{2m^*}\nabla^2_{\mathbf{r}_2} - 2E_F + V(\mathbf{r}_1, \mathbf{r}_2) \right] \right.$$

$$\left. \sum_{\mathbf{k}'(|\mathbf{k}'|>k_F)} a_{\mathbf{k}'} e^{i\mathbf{k}'\cdot(\mathbf{r}_2 - \mathbf{r}_1)} = W \sum_{\mathbf{k}'(|\mathbf{k}'|>k_F)} a_{\mathbf{k}'} e^{i\mathbf{k}'\cdot(\mathbf{r}_2 - \mathbf{r}_1)} \right\}, \tag{7.2.1.4}$$

where V_1 and V_2 are the system volumes. After integration for Eq. (7.2.1.4) we obtain

$$[W - 2\xi(\mathbf{k})]a_{\mathbf{k}} = \sum_{\mathbf{k}'(|\mathbf{k}'|>k_F)} V_{\mathbf{k}\mathbf{k}'}a_{\mathbf{k}'}, \tag{7.2.1.5}$$

where $\xi(\mathbf{k}) = \hbar^2k^2/(2m^*) - E_F$, $V_{\mathbf{k}\mathbf{k}'} =< \mathbf{k}, -\mathbf{k}|V(\mathbf{r}_1, \mathbf{r}_2)|\mathbf{k}', -\mathbf{k}' >$, and E_F is the Fermi energy.

Assume

$$V_{\mathbf{k}\mathbf{k}'} = \begin{cases} -V(<0), & \text{for} \quad 0 < \xi(\mathbf{k}), \xi(\mathbf{k}') < \hbar\omega_c; \\ 0, & \text{otherwise}, \end{cases} \tag{7.2.1.6}$$

where $\hbar\omega_c$ is the upper boundary of interaction energy. To solve Eq. (7.1.2.5) we use a non-perturbation method, i.e., we look for a non-perturbation solution. From Eq. (7.2.1.5) we have

$$a_{\mathbf{k}} = V \sum_{\mathbf{k}'(|\mathbf{k}'|>k_F)} \frac{a_{\mathbf{k}'}}{-W + 2\xi(\mathbf{k})}. \tag{7.1.2.7}$$

After doing the restrictive summation over $\mathbf{k}$ on both sides of Eq. (7.1.2.7) we obtain

$$\frac{1}{V} = \sum_{\mathbf{k}(|\mathbf{k}|>k_F)} \frac{1}{2\xi(\mathbf{k}) - W}. \tag{7.2.1.8}$$

Transforming the summation in Eq. (7.2.1.8) into integration yields

$$\frac{1}{V} = \int_0^{\hbar\omega_c} d\xi N(\xi) \frac{1}{2\xi - W}. \tag{7.2.1.9}$$

Assuming that the density of states $N(\xi)$ is a slow variation function and performing the integration in Eq. (7.2.1.9) yield

$$W = \frac{-2\hbar\omega_c}{e^{\frac{2}{N(E_F)V}} - 1}, \tag{7.2.1.10}$$

where $N(E_F)$ is the density of states at Fermi surface. Assuming $N(E_F)V < 1/4$, i.e., week coupling, Eq. (7.2.1.10) becomes

$$W = -2\hbar\omega_c e^{\frac{-2}{N(E_F)V}} < 0. \tag{7.2.1.11}$$

$W < 0$ means that the two electron eigenstate forms a bound state so that their total energy is less than $2E_F$. Note that it is a non-perturbation solution because in the above derivations we do not use any perturbation approximation. If one uses perturbation approach, one definitely can not obtain this solution because $V = 0$ is an inherent singular point of the solution.

The two electrons forming a bound state are called a Cooper pair. If we start from a free electron gas, and turn on the interaction $-V$, then the electrons will group themselves in Cooper pairs giving up energy to the external world. The normal state of the electrons is unstable. The largest number of allowed scattering processes, yielding the maximum lowering of energy, is obtained by pairing electrons with equal and opposite momentum [75]. If the total spin of the two electrons $S = 0$, then the Cooper pair consists of a superposition of $\mathbf{k} \uparrow$ and $-\mathbf{k} \downarrow$. In this case we say that the Cooper pair is in time-reversed state. Until now the Cooper pair of a great part of superconductors is in $S = 0$ state. Note that there are no any restriction for the attractive potential $-V < 0$. The $-V < 0$ may come from electron-phonon interaction, electron-magnon interaction, electron-plasmon interaction, electron-local trap interaction, electron-electron gauge force, and so on. The key point to form the bound state of two electrons is that there is net attractive interaction between two electrons. The carriers can be holes instead of electrons. From the above mentioned arbitrary of $-V < 0$, it is not surprised why more than two hundred different mechanisms were proposed until 1990 to interpret the high-T_c superconductivity in cuprates after its discovery in 1986.

7.2.2 The Size of a Cooper Pair

Now we estimate the spatial distance (=coherence length) between the two electrons forming a Cooper pair. The average square distance, ρ^2, is

$$\rho^2 = \frac{\int d(\mathbf{r}_1 - \mathbf{r}_2)|\psi(\mathbf{r}_1 - \mathbf{r}_2)|^2|\mathbf{r}_1 - \mathbf{r}_2|^2}{\int d(\mathbf{r}_1 - \mathbf{r}_2)|\psi(\mathbf{r}_1 - \mathbf{r}_2)|^2}. \quad (7.2.2.1)$$

Inserting Eqs. (7.2.1.1) and (7.2.1.2) into Eq. (7.2.2.1) yields

$$\rho^2 = \frac{\int d(\mathbf{r}_1 - \mathbf{r}_2)|\sum_{\mathbf{k}} a_{\mathbf{k}} \nabla_{\mathbf{k}} e^{i\mathbf{k}\cdot(\mathbf{r}_1-\mathbf{r}_2)}|^2}{\int d(\mathbf{r}_1 - \mathbf{r}_2)|\sum_{\mathbf{k}} a_{\mathbf{k}} e^{i\mathbf{k}\cdot(\mathbf{r}_1-\mathbf{r}_2)}|^2}. \quad (7.2.2.2)$$

Eq. (7.2.2.2) is

$$\rho^2 = \frac{\int d(\mathbf{r}_1 - \mathbf{r}_2)\sum_{\mathbf{k}}\sum_{\mathbf{k}'} a_{\mathbf{k}} a^*_{\mathbf{k}'} \nabla_{\mathbf{k}} e^{i\mathbf{k}\cdot(\mathbf{r}_1-\mathbf{r}_2)} \cdot \nabla_{\mathbf{k}'} e^{-i\mathbf{k}'\cdot(\mathbf{r}_1-\mathbf{r}_2)}}{\int d(\mathbf{r}_1 - \mathbf{r}_2)\sum_{\mathbf{k}}\sum_{\mathbf{k}'} a_{\mathbf{k}} a^*_{\mathbf{k}'} e^{i(\mathbf{k}-\mathbf{k}')\cdot(\mathbf{r}_1-\mathbf{r}_2)}}. \quad (7.2.2.3)$$

Note that the summations over $\mathbf{k}$ and $\mathbf{k}'$ in Eqs. (7.2.2.2-32) are constrained in $|\mathbf{k}| > k_F$ and $|\mathbf{k}'| > k_F$. In Eq. (7.2.2.3) we transform the summations over $\mathbf{k}$ and $\mathbf{k}'$ into $\int d\mathbf{k}V_1/(2\pi)^3$ and $\int d\mathbf{k}'V_2/(2\pi)^3$, respectively, perform integrations by parts for $\mathbf{k}$ and $\mathbf{k}'$, and use

$$\frac{1}{(2\pi)^3}\int d(\mathbf{r}_1 - \mathbf{r}_2)e^{i(\mathbf{k}-\mathbf{k}')\cdot(\mathbf{r}_1-\mathbf{r}_2)} = \delta(\mathbf{k} - \mathbf{k}'), \quad (7.2.2.4)$$

then we have

$$\rho^2 = \frac{\int d\mathbf{k}|\nabla_{\mathbf{k}} a_{\mathbf{k}}|^2}{\int d\mathbf{k}|a_{\mathbf{k}}|^2}. \quad (7.2.2.5)$$

Making variable transformation from $\mathbf{k}$ to ξ $[=\hbar^2k^2/(2m^*) - E_F]$, Eq. (7.2.2.5) becomes

$$\rho^2 \cong \frac{N(E_F)\left(\frac{\partial\xi}{\partial k}\right)_{\xi=0}\int_0^\infty d\xi\left|\frac{\partial a}{\partial\xi}\right|^2}{N(E_F)\int_0^\infty d\xi|a|^2}. \quad (7.2.2.6)$$

We calculate the quantities appearing in Eq. (7.2.2.6) as follows.

$$\left(\frac{\partial\xi}{\partial k}\right)_{\xi=0} = \frac{\partial}{\partial k}\left(\frac{\hbar^2k^2}{2m^*} - E_F\right)_{\xi=0(k=k_F)}$$

$$= \frac{\hbar^2 k_F}{m^*} = \hbar v_F. \quad (7.2.2.7)$$

Using Eq. (7.2.1.7), we have

$$\int_0^\infty d\xi\left|\frac{\partial a}{\partial\xi}\right|^2 = \int_0^\infty d\xi\left|\frac{\partial}{\partial\xi}\frac{C}{(W-2\xi)}\right|^2$$

$$= -\frac{2|C|^2}{3W^3}, \qquad (7.2.2.8)$$

where C is a constant.

$$\int_0^\infty d\xi |a|^2 = \int_0^\infty d\xi \frac{|C|^2}{(W-2\xi)^2} = -\frac{|C|^2}{2W}. \qquad (7.2.2.9)$$

So,

$$\rho = \frac{2}{\sqrt{3}} \frac{\hbar v_F}{W} \cong \hbar \frac{\hbar k_F}{m^*} \frac{1}{W}. \qquad (7.2.2.10)$$

For conventional low-T_c superconductors we take $m^* = m_\circ$ (free electron mass), $k_F = 1/\mathring{A}$, and the bound energy $W = 10$ K, then $\rho \cong 10^{-4}$ cm. The value of ρ is called coherence length $\xi_\circ$ some times.

There is a problem. One knows that Coulomb and phonon interaction between electrons in the normal state lead to a correlation energy of order one eV per electron [75]. However, the bound energy $W \simeq 10$ K. So, even the former's fluctuation can break so weak bound Cooper pair. Especially, the Cooper pairs in the conventional superconductor overlap each other strongly. This long coherence length belongs to strongly overlapping Cooper pair case. Why can the Cooper pair still be formed in this case? The Landau Fermi liquid theory can answer this question [75]. In this theory the electron states of a real normal metal are placed in one-to-one correspondence with those of a free electron gas. The electrons in an interaction system, such as a real normal metal, are quasiparticles which are stable excitations in immediate vicinity of the Fermi surface, and which include all the correlation interaction. The coupling between two quasiparticles of a Cooper pair, mediated by phonon and Coulomb interaction in superconducting state of conventional superconductors, is a residual net attractive interaction between two quasiparticles, neglected within the Fermi liquid approximation. This residual net attractive interaction gives rise to formation of Cooper pair. Exactly speaking, Cooper pair is quasi-electron pairing. On the other hand, for the high-T_c cuprates, such as $YBa_2Cu_3O_{7-\delta}$, $\rho = \xi_\circ \approx 20 - 30$ $\mathring{A}$. This short coherence length belongs to separating pair case. The reason to form Cooper pairs in this case can be interpreted by Landau Fermi liquid theory as well. The formation of Cooper pair in the high-T_c cuprates is caused by the residual attractive interaction between two quasiparticles of Cooper pair.

7.2.3 Pairing Approximation

In this section we expound BCS pairing theory [75]. Assume that the total spin of a Cooper pair is $S = 0$. We consider approximately only the interaction between electrons with $\mathbf{k}\uparrow$ and $-\mathbf{k}\downarrow$, which is called pairing approximation. Then, we can write a so-called reduced Hamiltonian H_{red}.

$$H_{red} = \sum_{\mathbf{k}s} \epsilon_{\mathbf{k}} n_{\mathbf{k}s} + \sum_{\mathbf{k}\mathbf{k}'} V_{\mathbf{k}'\mathbf{k}} b^+_{\mathbf{k}'} b_{\mathbf{k}}, \qquad (7.2.3.1)$$

where

$$V_{\mathbf{k}'\mathbf{k}} = <\mathbf{k}', -\mathbf{k}'|\hat{V}|\mathbf{k}, -\mathbf{k}>, \qquad (7.2.3.2)$$

$$b^+_{\mathbf{k}'} = c^+_{\mathbf{k}'\uparrow} c^+_{-\mathbf{k}'\downarrow}, \qquad (7.2.3.3)$$

$$b_{\mathbf{k}} = c_{-\mathbf{k}\downarrow} c_{\mathbf{k}\uparrow}, \qquad (7.2.3.4)$$

where $b^+_{\mathbf{k}'}$ and $b_{\mathbf{k}}$ are the pair creation and destruction operator, respectively.

At $T = 0$, the system is in ground state, and all electrons are paired. We can rewrite the first term in Eq. (7.2.3.1) as

$$\sum_{\mathbf{k}s} \epsilon_{\mathbf{k}} n_{\mathbf{k}s} = \sum_{\mathbf{k}} 2\epsilon_{\mathbf{k}} b_{\mathbf{k}}^{+} b_{\mathbf{k}}. \tag{7.2.3.5}$$

We prove Eq. (7.2.3.5) as follows.

$$\begin{aligned} 2b_{\mathbf{k}}^{+} b_{\mathbf{k}}|\mathrm{G}> &= 2c_{\mathbf{k}\uparrow}^{+} c_{-\mathbf{k}\downarrow}^{+} c_{-\mathbf{k}\downarrow} c_{\mathbf{k}\uparrow}|\mathrm{G}> \\ &= 2c_{\mathbf{k}\uparrow}^{+} c_{\mathbf{k}\uparrow} c_{-\mathbf{k}\downarrow}^{+} c_{-\mathbf{k}\downarrow}|\mathrm{G}>= 2n_{\mathbf{k}\uparrow} n_{-\mathbf{k}\downarrow}|\mathrm{G}> \\ &= (n_{\mathbf{k}\uparrow} + n_{-\mathbf{k}\downarrow})|\mathrm{G}>, \end{aligned} \tag{7.2.3.6}$$

where $|\mathrm{G}>$ means |ground state>. At $T = 0$, the reduced Hamiltonian , Eq. (7.2.3.1), becomes

$$H_{red}^{(\circ)} = \sum_{\mathbf{k}} 2\epsilon_{\mathbf{k}} b_{\mathbf{k}}^{+} b_{\mathbf{k}} + \sum_{\mathbf{k}'\mathbf{k}} V_{\mathbf{k}'\mathbf{k}} b_{\mathbf{k}'}^{+} b_{\mathbf{k}}. \tag{7.2.3.7}$$

Note that the commutation relations between b^{+} and b are neither Fermi type nor Bose type. For example,

$$[b_{\mathbf{k}}, b_{\mathbf{k}'}^{+}]_{-} = 0, \quad \text{when} \quad \mathbf{k} \neq \mathbf{k}', \tag{7.2.3.8}$$

$$[b_{\mathbf{k}}, b_{\mathbf{k}}^{+}]_{-} = 1 - (n_{\mathbf{k}\uparrow} + n_{-\mathbf{k}\downarrow}). \tag{7.2.3.9}$$

Assume $S = 0$. We look for the BCS ground state wavefunction $\psi_{\circ N}$. Let N denote the number of electrons in a system. Assume that the very large number N is an even number.

$$\psi_{\circ N}(\mathbf{r}_1, \cdots, \mathbf{r}_N) = \widehat{A}\phi(\mathbf{r}_1 - \mathbf{r}_2)\cdots\phi(\mathbf{r}_{N-1} - \mathbf{r}_N)(\uparrow 1)(\downarrow 2)\cdots(\uparrow N-1)(\downarrow N), \tag{7.2.3.10}$$

where $\widehat{A}$ is an antisymmetric operator. The function $\phi(\mathbf{r}_i - \mathbf{r}_{i+1})$ is a relative coordinate wavefunction of a Cooper pair. All Cooper pairs have the same wavefunction, i.e., all Cooper pairs are in the same state in the ground state wavefunction of system. Let $\phi(\mathbf{r}_i - \mathbf{r}_j) = \sum_{\mathbf{k}} g_{\mathbf{k}} \exp[i\mathbf{k} \cdot (\mathbf{r}_i - \mathbf{r}_j)]$ where $g_{\mathbf{k}}$ includes $\exp(i\phi)$. Eq. (7.2.3.102.30) becomes

$$\begin{aligned} \psi_{\circ N}(\mathbf{r}_1, \cdots, \mathbf{r}_N) &= \sum_{\mathbf{k}_1} \cdots \sum_{\mathbf{k}_{N/2}} g_{\mathbf{k}_1} \cdots g_{\mathbf{k}_{N/2}} \widehat{A} e^{i\mathbf{k}_1\cdot(\mathbf{r}_1 - \mathbf{r}_2)} \cdots e^{i\mathbf{k}_{N/2}\cdot(\mathbf{r}_{N-1} - \mathbf{r}_N)} \\ &\cdot (\uparrow 1)(\downarrow 2)\cdots(\uparrow N-1)(\downarrow N) \\ &= \sum_{\mathbf{k}_1} \cdots \sum_{\mathbf{k}_{N/2}} g_{\mathbf{k}_1} \cdots g_{\mathbf{k}_{N/2}} (SD), \end{aligned} \tag{7.2.3.11}$$

where (SD) represents Slater determinant. In second quantization approach Eq. (7.2.3.31) becomes

$$|\psi_{\circ N}>= \sum_{\mathbf{k}_1} \cdots \sum_{\mathbf{k}_{N/2}} g_{\mathbf{k}_1} \cdots g_{\mathbf{k}_{N/2}} c_{\mathbf{k}_1\uparrow}^{+} c_{-\mathbf{k}_1\downarrow}^{+} \cdots c_{\mathbf{k}_{N/2}\uparrow}^{+} c_{-\mathbf{k}_{N/2}\downarrow}^{+}|0> . \tag{7.2.3.12}$$

Note that due to the Pauli principle, there is no same $\mathbf{k}$ in Eq. (7.2.3.12).

For convenience, instead of $|\psi_{\circ N}>$, which has a fixed electron number, BCS takes

$$|\psi_{\circ}>= C\prod_{\mathbf{k}}(1 + g_{\mathbf{k}} c_{\mathbf{k}\uparrow}^{+} c_{-\mathbf{k}\downarrow}^{+})|0>, \tag{7.2.3.13}$$

where $\prod_{\mathbf{k}}$ represents a product of many factors, $|\psi_{\circ}>= |0>_{BCS}$ is called BCS vacuum, $|0>$ is called ordinary vacuum [75]. Set

$$<\psi_{\circ}|\widehat{N}|\psi_{\circ}>= N, \tag{7.2.3.14}$$

where N is the electron number of a real system. The C in Eq. (7.2.3.13) can be determined by the normalization condition, i.e., $<\psi_\circ|\psi_\circ>=1$.

$$C=\frac{1}{\prod_{\mathbf{k}}(1+|g_{\mathbf{k}}|^2)^{1/2}}. \tag{7.2.3.15}$$

Substituting Eq. (7.2.3.15) into Eq. (7.2.3.13) yields

$$|\psi_\circ>=\prod_{\mathbf{k}}\frac{1+g_{\mathbf{k}}b^+_{\mathbf{k}}}{(1+|g_{\mathbf{k}}|^2)^{1/2}}|0>=\prod_{\mathbf{k}}(u_{\mathbf{k}}+v_{\mathbf{k}}b^+_{\mathbf{k}})|0>, \tag{7.2.3.16}$$

where $u_{\mathbf{k}}$ is a real number.

To find the ground state energy, let us look for the variational minimum of the average value of $H^{(\circ)}_{red}-\mu\widehat{N}$ over $|\psi_\circ>$, i.e.,

$$\delta W=\delta<\psi_\circ|H^{(\circ)}_{red}-\mu\widehat{N}|\psi_\circ>=0. \tag{7.2.3.17}$$

Let us at first look for the variational parameter, and then make the variational operation.

$$W=<\psi_\circ|H^{(\circ)}_{red}-\mu\widehat{N}|\psi_\circ>. \tag{7.2.3.18}$$

Substituting Eqs. (7.2.3.13) and (7.2.3.7) into Eq. (7.2.3.18) yields

$$W=\sum_{\mathbf{k}}2(\epsilon_{\mathbf{k}}-\mu)|v_{\mathbf{k}}|^2+\sum_{\mathbf{k}\mathbf{k}'}V_{\mathbf{k}'\mathbf{k}}u_{\mathbf{k}}v_{\mathbf{k}}u_{\mathbf{k}'}v^*_{\mathbf{k}'}. \tag{7.2.3.19}$$

The $u_{\mathbf{k}}$ and $v_{\mathbf{k}}$ in Eq. (7.2.3.16) become

$$u_{\mathbf{k}}=\sin(\vartheta_{\mathbf{k}}), \tag{7.2.3.20}$$

$$v_{\mathbf{k}}=\cos(\vartheta_{\mathbf{k}})\exp(i\phi), \tag{7.2.3.21}$$

where ϕ is the phase of wavefunction of mass-center of a Cooper pair. So, the variational parameter is $\vartheta_{\mathbf{k}}$. Substituting Eqs. (7.2.3.20) and (7.2.3.21) into Eqs. (7.2.3.19) and (7.2.3.17), we have

$$\frac{\partial W}{\partial\vartheta_{\mathbf{k}}}=0, \tag{7.2.3.22}$$

i.e.,

$$-2(\epsilon_{\mathbf{k}}-\mu)\sin(2\vartheta_{\mathbf{k}})+\sum_{\mathbf{k}'}V_{\mathbf{k}'\mathbf{k}}\cos(2\vartheta_{\mathbf{k}})\sin(2\vartheta_{\mathbf{k}'})=0, \tag{7.2.3.23}$$

$$-2(\epsilon_{\mathbf{k}}-\mu)\tan(2\vartheta_{\mathbf{k}})=-\sum_{\mathbf{k}}V_{\mathbf{k}'\mathbf{k}}\sin(2\vartheta_{\mathbf{k}'}). \tag{7.2.3.24}$$

The extra factor of 2 enters in the second sum of Eq. (7.2.3.23) because both $\mathbf{k}$ and $\mathbf{k}'$ in Eq. (7.2.3.19) run over any given value $\mathbf{k}''$. Define

$$\begin{aligned}\Delta_{\mathbf{k}} &= -\sum_{\mathbf{k}'}V_{\mathbf{k}'\mathbf{k}}u_{\mathbf{k}'}v_{\mathbf{k}'}\\ &= -\sum_{\mathbf{k}'}V_{\mathbf{k}'\mathbf{k}}\frac{1}{2}\sin(2\vartheta_{\mathbf{k}'})e^{i\phi}.\end{aligned} \tag{7.2.3.25}$$

From the following derivations we know that $\Delta_{\mathbf{k}}$ represents gap function. Thus from Eq. (7.2.3.25) we see that the gap function contains a phase factor of Cooper pair wavefunction of mass-center. Define

$$E_{\mathbf{k}}=\sqrt{(\epsilon_{\mathbf{k}}-\mu)^2+|\Delta_{\mathbf{k}}|^2}. \tag{7.2.3.26}$$

$E_{\mathbf{k}}$ turns out to be the energy required to create a quasiparticle of momentum $\mathbf{k}$ or $-\mathbf{k}$ in the superconducting state for conventional or unconventional superconductor and in the pseudogap state for unconventional superconductor, and $\Delta_{\mathbf{k}}$ is the energy gap for quasiparticle excitation. From Eqs. (7.2.3.24) and (7.2.3.25) we have

$$\tan(2\vartheta_{\mathbf{k}}) = \frac{|\Delta_{\mathbf{k}}|}{-(\epsilon_{\mathbf{k}} - \mu)}. \tag{7.2.3.27}$$

From Eqs. (7.2.3.20), (7.2.3.21), (7.2.3.26), and (7.2.3.27) we have

$$2u_{\mathbf{k}}v_{\mathbf{k}} = \sin(2\vartheta_{\mathbf{k}})e^{i\phi} = \frac{\Delta_{\mathbf{k}}}{E_{\mathbf{k}}}, \tag{7.2.3.28}$$

$$-u_{\mathbf{k}}^2 + |v_{\mathbf{k}}|^2 = \cos(2\vartheta_{\mathbf{k}}) = \frac{-(\epsilon_{\mathbf{k}} - \mu)}{E_{\mathbf{k}}}. \tag{7.2.3.29}$$

From Eq. (7.2.3.16) we have

$$u_{\mathbf{k}}^2 + |v_{\mathbf{k}}|^2 = 1. \tag{7.2.3.30}$$

From Eqs. (7.2.3.29-30 we have

$$u_{\mathbf{k}}^2 = \frac{1}{2}\left(1 + \frac{\epsilon_{\mathbf{k}} - \mu}{E_{\mathbf{k}}}\right), \tag{7.2.3.31}$$

$$|v_{\mathbf{k}}|^2 = \frac{1}{2}\left(1 - \frac{\epsilon_{\mathbf{k}} - \mu}{E_{\mathbf{k}}}\right). \tag{7.2.3.32}$$

From Eqs. (7.2.3.28) and (7.2.3.25) we have

$$\Delta_{\mathbf{k}} = -\sum_{\mathbf{k}'} V_{\mathbf{k}'\mathbf{k}} \frac{\Delta_{\mathbf{k}'}}{2E_{\mathbf{k}'}}. \tag{7.2.3.33}$$

Eq. (7.2.3.33) is the famous BCS gap equation at 0 K. For convenience, in the following discussions of this section we take $\Delta_{\mathbf{k}}$ and $v_{\mathbf{k}}$ as real numbers. The trivial solution of Eq. (7.2.3.33) is

$$\Delta_{\mathbf{k}} = 0, \tag{7.2.3.34}$$

which corresponds to the normal state.

Let us look for the non-zero energy gap at $T = 0$ K, $\Delta(0)$. Assume

$$V_{\mathbf{k}'\mathbf{k}} = \begin{cases} V < 0, & \text{if } \quad |\epsilon_{\mathbf{k}} - \mu|, |\epsilon_{\mathbf{k}'} - \mu| < \hbar\omega_c, \\ 0, & \text{if } \quad \text{otherwise.} \end{cases} \tag{7.2.3.35}$$

This approximation for $V_{\mathbf{k}'\mathbf{k}}$ is called BCS cut. If $V_{\mathbf{k}'\mathbf{k}}$ comes from phonon-mediated interaction, then ω_c is equal to the maximum phonon frequency, ω_D. Using BCS cut, we have

$$\Delta_{\mathbf{k}} = \begin{cases} \Delta(0), & \text{if } \quad |\epsilon_{\mathbf{k}} - \mu| < \hbar\omega_c; \\ 0, & \text{if } \quad \text{otherwise.} \end{cases} \tag{7.2.3.36}$$

Substituting Eq. (7.2.3.36) into Eq. (7.2.3.33) gives

$$\begin{aligned} \Delta(0) &= V\sum_{\mathbf{k}'} \frac{\Delta(0)}{2\sqrt{(\epsilon_{\mathbf{k}'} - \mu)^2 + \Delta^2(0)}} \\ &= N(E_F)V \int_{-\hbar\omega_c}^{\hbar\omega_c} d\xi \frac{\Delta(0)}{2\sqrt{\xi^2 + \Delta^2(0)}} \\ &= N(E_F)V\Delta(0)\sinh^{-1}\frac{\hbar\omega_c}{\Delta(0)}, \end{aligned} \tag{7.2.3.37}$$

where $\xi = (\epsilon_{\mathbf{k}} - \mu)$. From Eq. (7.2.3.37) we have

$$\Delta(0) = \frac{\hbar\omega_c}{\sinh\frac{1}{N(E_F)V}}. \tag{7.2.3.38}$$

Using weak coupling approximation, $N(E_F)V < 1/4$, Eq. (7.2.3.38) becomes

$$\Delta(0) = 2\hbar\omega_c e^{-\frac{1}{N(E_F)V}}. \tag{7.2.3.39}$$

Let us look for the condensation energy W_{cond}. Its definition is

$$W_{cond} = W_N - W_S, \tag{7.2.3.40}$$

where W_N represents the ground state energy in normal state ($\Delta(0) = 0$), and W_S represents the energy of ground state with $\Delta(0) > 0$. We discuss a unit volume. Substituting Eq. (7.2.3.19) into Eq. (7.2.3.40) gives

$$\begin{aligned} W_{cond} &= \left[\sum_{\mathbf{k}} 2(\epsilon_{\mathbf{k}} - \mu)v_{\mathbf{k}}^2 + \sum_{\mathbf{k}'\mathbf{k}} V_{\mathbf{k}'\mathbf{k}} u_{\mathbf{k}} v_{\mathbf{k}} u_{\mathbf{k}'} v_{\mathbf{k}'}\right]\Bigg|_{\Delta(0)=0} \\ &- \left[\sum_{\mathbf{k}} 2(\epsilon_{\mathbf{k}} - \mu)v_{\mathbf{k}}^2 + \sum_{\mathbf{k}'\mathbf{k}} V_{\mathbf{k}'\mathbf{k}} u_{\mathbf{k}} v_{\mathbf{k}} u_{\mathbf{k}'} v_{\mathbf{k}'}\right]\Bigg|_{\Delta(0)\neq 0}. \end{aligned} \tag{7.2.3.41}$$

From Eqs. (7.2.3.31) and (7.2.3.32) we know that when $\Delta(0) = 0$,

$$v_{\mathbf{k}}^2 = \begin{cases} 1, & \text{if} \quad \epsilon_{\mathbf{k}} - \mu < 0; \\ 0, & \text{if} \quad \epsilon_{\mathbf{k}} - \mu > 0. \end{cases}$$

$$u_{\mathbf{k}}^2 = \begin{cases} 1, & \text{if} \quad \epsilon_{\mathbf{k}} - \mu > 0; \\ 0, & \text{if} \quad \epsilon_{\mathbf{k}} - \mu < 0. \end{cases} \tag{7.2.3.42}$$

Considering Eq. (7.2.3.42) and unit volume, the W_N becomes

$$W_N = \sum_{|\mathbf{k}|<k_F} 2(\epsilon_{\mathbf{k}} - \mu) = N(E_F)\int_{-\hbar\omega_c}^{0} d\xi 2\xi. \tag{7.2.3.43}$$

Substituting Eqs. (7.2.3.31) and (7.2.3.32) into W_S yields

$$W_S = -2N(E_F)\int_0^{\hbar\omega_c} d\xi \frac{\xi^2}{\sqrt{\xi^2 + \Delta^2(0)}} - \Delta^2(0)N(E_F)\int_0^{\hbar\omega_c} d\xi \frac{1}{2\sqrt{\xi^2 + \Delta^2(0)}}. \tag{7.2.3.44}$$

Performing the integrations in Eqs. (7.2.3.43) and (7.2.3.44) yields

$$W_{cond} = -N(E_F)(\hbar\omega_c)^2 + 2N(E_F)\left[\frac{\hbar\omega_c\sqrt{(\hbar\omega_c)^2 + \Delta^2(0)}}{2}\right]. \tag{7.2.3.45}$$

If we assume $\hbar\omega_c \gg \Delta(0)$, then Eq. (7.2.3.45) becomes

$$W_{cond} = \frac{N(E_F)\Delta^2(0)}{2}. \tag{7.2.3.46}$$

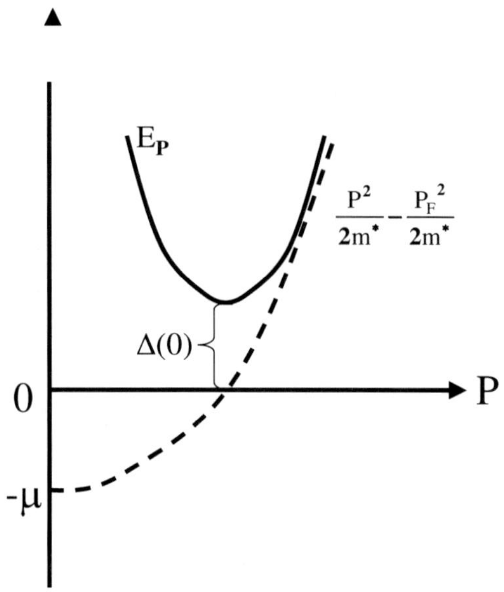

Figure 7.1: The solid line is the schematic plot of $E_\mathbf{p}$ vs. p (=| $\mathbf{p}$ |). The dashed line is the schematic plot of $[p^2/(2m^*) - p_F^2/(2m^*)]$ vs. p. The minimum energy $\Delta(0)$ is required to create each quasiparticle ($\mathbf{p}\uparrow$) or ($-\mathbf{p}\downarrow$) above the chemical potential.

Substituting Eq. (7.2.3.39) into Eq. (7.2.3.46) gives

$$W_{cond} = 2N(E_F)(\hbar\omega_c)^2 e^{-\frac{2}{N(E_F)V}}. \tag{7.2.3.47}$$

From $W_{cond} > 0$, we know that the ground state with $\Delta(0) > 0$ is more stable than the normal state. Here, we have to note that $\Delta(0) > 0$ does not mean superconducting state. The $\Delta(0) > 0$ only means that there are Cooper pairs in the system, and stay in the ground state. If the Cooper pairs have the long-range phase coherence, then the system will be in (macroscopic) superconducting state. For the greater part of low-T_c superconductors, which belong to the conventional superconductors, the amplitude and the long-range phase coherence of Cooper pairs appear nearly at the same temperature. For these conventional superconductors, the $\Delta(0) > 0$ means that the conventional superconductor approaches the superconducting state. However, for unconventional superconductor, even at $T = 0$ K, $\Delta(0) > 0$ does not definitely correspond to superconducting state. For example, for $YBa_2Cu_3O_{7-\delta}$ with $x = 0.05$, $T_c = 0$ K, $T^* = 300$ K, and $\Delta(0) = 34$ meV. Here, T^* represents a characteristic temperature, at which the gap begins to do not equal zero. Section 7.3 will introduce Landau's criterion of microscopic superconductivity. According to this criterion, if pseudogap is not zero, then there is microscopic superconductivity. Therefore, we call T^* the critical temperature of microscopic superconductivity. However, due to the historical misunderstanding, people call the gap at $T_c < T < T^*$ the pseudogap. The x is the carrier's number in one unit cell. The Cooper pair consists of two $O_{p\sigma}$ holes, which are the quasiparticles in the high-T_c cuprates.

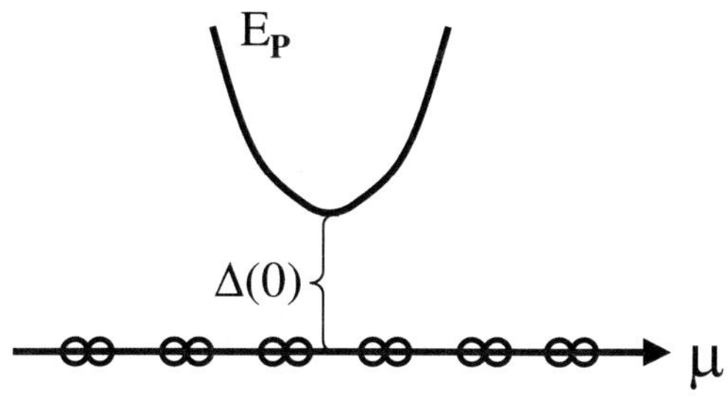

Figure 7.2: BCS model. At BCS ground state all Cooper pairs reside in the chemical potential. The excitation spectrum is shown as a curve above the chemical potential.

7.2.4 Quasiparticle Energy Spectrum

7.2.4.1. Excitation Energy

From Eq. (7.2.3.19), the energy of a pair $(\mathbf{p}\uparrow, -\mathbf{p}\downarrow)$, W_1, is

$$W_1 = 2(\epsilon_{\mathbf{p}} - \mu)v_{\mathbf{p}}^2 + 2\sum_{\mathbf{k}'} V_{\mathbf{pk}'}u_{\mathbf{k}'}v_{\mathbf{k}'}u_{\mathbf{p}}v_{\mathbf{p}}, \tag{7.2.4.1.1}$$

where the second factor "2" comes from double summation. Substituting Eqs. (7.2.3.28) and (7.2.3.33) into Eq. (7.2.4.1.1) yields

$$W_1 = 2(\epsilon_{\mathbf{p}} - \mu)v_{\mathbf{p}}^2 - 2\Delta_{\mathbf{p}}u_{\mathbf{p}}v_{\mathbf{p}}. \tag{7.2.4.1.2}$$

The single quasi-electron $(\mathbf{p}\uparrow)$ excitation energy $W_{\mathbf{p}}$ is defined to be the total excitation energy of the system when the extra electron $(\mathbf{p}\uparrow)$ is added to the system. The only effect of adding the extra electron is to block the pair state $(\mathbf{p}\uparrow, -\mathbf{p}\downarrow)$ from participating in the pairing interaction (due to the Pauli principle). Thus, W_p is equal to the energy change of the system, putting one quasi-electron $(\mathbf{p}\uparrow)$ and taking out a pair $(\mathbf{p}\uparrow, -\mathbf{p}\downarrow)$.

$$W_{\mathbf{p}} = (\epsilon_{\mathbf{p}} - \mu) - \left[2(\epsilon_{\mathbf{p}} - \mu)v_{\mathbf{p}}^2 - 2\Delta_{\mathbf{p}}u_{\mathbf{p}}v_{\mathbf{p}}\right]. \tag{7.2.4.1.3}$$

Substituting Eqs. (7.2.3.32) and (7.2.3.28) into Eq. (7.2.4.1.3) yields

$$W_{\mathbf{p}} = E_{\mathbf{p}}. \tag{7.2.4.1.4}$$

We can simply say that $E_{\mathbf{p}}$ is the energy required to create a quasi-electron in state $(\mathbf{p}\uparrow)$ or $(-\mathbf{p}\downarrow)$. $E_{\mathbf{p}}$ differs from the $(\epsilon_{\mathbf{p}} - \mu)$ clearly in the vicinity of Fermi surface, where $\epsilon_{\mathbf{p}} = p^2/(2m^*)$. A plot of $E_{\mathbf{p}}$ vs. $|\mathbf{p}|$ is given in Fig. 7.1. Note that when a Cooper pair is destroyed, we always get two quasi-electrons $(\mathbf{p}\uparrow)$ and $(-\mathbf{p}\downarrow)$ above the chemical potential. If there is no quasi-electron excitation above the chemical potential, then all Cooper pairs condense into the chemical potential, which is called BCS model and is shown in Fig. 7.2. BCS model means Bose-Einstein condensation of Cooper pairs for which the momentum of mass-center are zero.

7.2.4.2. Momentum Distribution in BCS Ground State

Let us study the average occupation number in the Bloch state $\mathbf{k}\uparrow$, $< n_{\mathbf{k}\uparrow} >$.

$$< n_{\mathbf{k}\uparrow} > = < \psi_\circ | c_{\mathbf{k}\uparrow}^{+} c_{\mathbf{k}\uparrow} | \psi_\circ > . \tag{7.2.4.2.1}$$

Substituting Eq. (7.2.3.16) into Eq. (7.2.4.2.1) yields

$$< n_{\mathbf{k}\uparrow} >= v_{\mathbf{k}}^2. \tag{7.2.4.2.2}$$

$v_{\mathbf{k}}^2$ has been given in Eq. 7.2.3.32). The average occupation number condensed into the Bloch state $\mathbf{k}\uparrow$ is equal to the occupation probability of the pair state $(\mathbf{k}\uparrow, -\mathbf{k}\downarrow)$ because $< n_{\mathbf{k}\uparrow} >=< n_{-\mathbf{k}\downarrow} >$ for BCS ground state. The schematic plot of $< n_{\mathbf{k}\uparrow} >$ vs. k $(=|\mathbf{k}|)$ is shown in Fig. 7.3. Note that, in BCS ground state, even at absolute zero there are vacancies with $k < k_F$ and occupied states with $k > k_F$, i. e., the ground state of normal metal is unstable. According to the Cooper's conclusion in section 7.1, one might at first think that there is no limit to the number of electrons which may be raised from $k < k_F$ to form Cooper pairs with a resultant lowering of the total energy, so that we should end up with all the electrons having $k > k_F$. This, however, is clearly absurd, and the reason is not hard to find. In order that a Cooper pair may be scattered from $(\mathbf{k}\uparrow, -\mathbf{k}\downarrow)$ to $(\mathbf{k}'\uparrow, -\mathbf{k}'\downarrow)$ the state $(\mathbf{k}\uparrow, -\mathbf{k}\downarrow)$ must be first occupied and the state $(\mathbf{k}'\uparrow, -\mathbf{k}'\downarrow)$ must empty. As more and more electrons form Cooper pairs with $k > k_F$, the chance of finding the states $(\mathbf{k}'\uparrow, -\mathbf{k}'\downarrow)$ empty becomes progressively smaller and smaller, so the number of scattering processes which may take place is reduced, with a consequent decrease in the magnitude of the negative potential energy. We know that the kinetic energy increases when two electrons go from $k < k_F$ to $k > k_F$, and the potential energy of the two electrons reduces when the two electrons form a Cooper pair. Eventually a condition is reached in which the lowering of the potential energy is insufficient to outweigh the increase in kinetic energy, and it is no longer possible to lower the total energy of electrons by forming Cooper pairs. In mathematics, the equilibrium occupation probability of the pair state is determined by the variational minimum condition of total energy, i.e., Eq. (7.2.3.22). The transition region, where $|v_{\mathbf{k}}|^2$ goes about from 1 to 0, has an energy width $\Delta(0)$ and a momentum width δk approximately.

$$\frac{\hbar^2}{2m^*}\left[(k_F + \delta k)^2 - k_F^2\right] = \Delta(0). \tag{7.2.4.2.3}$$

$$\delta k = \frac{\Delta(0)}{\hbar v_F}. \tag{7.2.4.2.4}$$

Let us add two electrons to the states $\mathbf{k}\uparrow$ and $-\mathbf{k}\downarrow$ in the condensed state $|\psi_{\circ N} >$. What is the probability amplitude for obtaining the state $|\psi_{\circ N+2} >$? This is the quantity

$$F_{\mathbf{k}} = |< \psi_{\circ N+2}|c_{\mathbf{k}\uparrow}^+ c_{-\mathbf{k}\downarrow}^+|\psi_{\circ N} >| = |< \psi_\circ|c_{\mathbf{k}\uparrow}^+ c_{-\mathbf{k}\downarrow}^+|\psi_\circ >|. \tag{7.2.4.2.5}$$

We call $F_{\mathbf{k}}$ the condensation amplitude in state $\mathbf{k}$. Substituting Eq. (7.2.3.16) into Eq. (7.2.4.2.5) yields

$$F_{\mathbf{k}} = u_{\mathbf{k}}|v_{\mathbf{k}}|. \tag{7.2.4.2.6}$$

$u_{\mathbf{k}}$ and $v_{\mathbf{k}}$ have been given in Eqs. (7.2.3.31) and (7.2.3.32). The schematic plot of $F_{\mathbf{k}}$ vs. k is shown in Fig. 7.4. From Fig. 7.4 we see that $F_{\mathbf{k}}$ does not vanish only in the transition zone $\delta k \cong 1/\xi_\circ$ and has maximum at $k = k_F$. From Figs. 7.3 and 7.4 we see that the Fermi surface of normal state is "smeared" by the pairing correlation.

7.2.5 Quasiparticle Excitation

7.2.5.1. Bogoliubov-Valatin Transformation

This transformation is convenient to describe elementary excitation of superconductors in superconducting or pseudogap state. Define [75]

$$\gamma_{\mathbf{p}\uparrow}^+ = u_{\mathbf{p}} c_{\mathbf{p}\uparrow}^+ - v_{\mathbf{p}}^* c_{-\mathbf{p}\downarrow}, \tag{7.2.5.1.1}$$

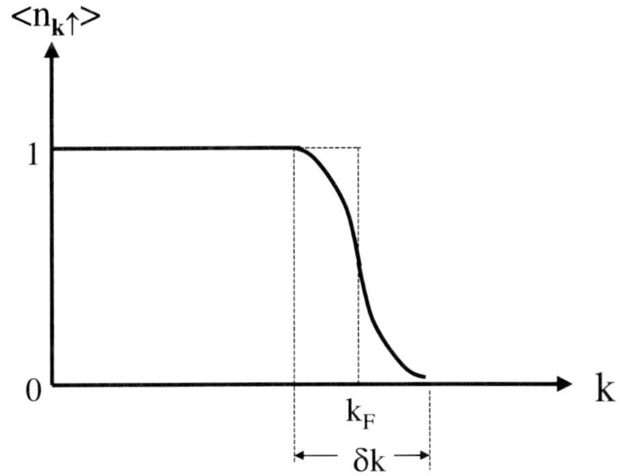

Figure 7.3: The schematic plot of $< n_{\mathbf{k}\uparrow} >$ vs. k. $\delta k = \Delta(0)/(\hbar v_F) \cong 1/\xi_\circ$. The occupation number in the normal state is 1 for $k < k_F$ and 0 for $k > k_F$. The Fermi surface is smeared by the pairing correlation.

$$\gamma_{-\mathbf{p}\downarrow} = u_{\mathbf{p}} c_{-\mathbf{p}\downarrow} + v_{\mathbf{p}} c^+_{\mathbf{p}\uparrow}, \tag{7.2.5.1.2}$$

where $v_{\mathbf{p}}$ includes $\exp(i\phi)$. $\gamma^+_{\mathbf{p}\uparrow}$ and $\gamma_{-\mathbf{p}\downarrow}$ are quasiparticle (elementary excitation), which is not in pairing, creation and annihilation operator, respectively, in both superconducting state or pseudogap state. Eqs. (7.2.5.1.1) and (7.2.5.1.2) are the Bogoliubov-Valatin transformation. The corresponding Hermitian conjugate operators are

$$\gamma_{\mathbf{p}\uparrow} = u_{\mathbf{p}} c_{\mathbf{p}\uparrow} - v_{\mathbf{p}} c^+_{-\mathbf{p}\downarrow}. \tag{7.2.5.1.3}$$

$$\gamma^+_{-\mathbf{p}\downarrow} = u_{\mathbf{p}} c^+_{-\mathbf{p}\downarrow} + v^*_{\mathbf{p}} c_{\mathbf{p}\uparrow}. \tag{7.2.5.1.4}$$

We can show that $\gamma_{-\mathbf{p}\downarrow}$ and $\gamma^+_{\mathbf{p}\uparrow}$ are the annihilation and creation operators of quasiparticles on the BCS ground state. For example,

$$\begin{aligned} \gamma^+_{\mathbf{p}\uparrow}|\psi_\circ > &= (u_{\mathbf{p}} c^+_{\mathbf{p}\uparrow} - v^*_{\mathbf{p}} c_{-\mathbf{p}\downarrow})(u_{\mathbf{p}} + v_{\mathbf{p}} c^+_{\mathbf{p}\uparrow} c^+_{-\mathbf{p}\downarrow}) \prod_{\mathbf{k}(\neq \mathbf{p})} (u_{\mathbf{k}} + v_{\mathbf{k}} c^+_{\mathbf{k}\uparrow} c^+_{-\mathbf{k}\downarrow})|0> \\ &= (u^2_{\mathbf{p}} c^+_{\mathbf{p}\uparrow} + |v_{\mathbf{p}}|^2 c^+_{\mathbf{p}\uparrow}) \prod_{\mathbf{k}(\neq \mathbf{p})} (u_{\mathbf{k}} + v_{\mathbf{k}} c^+_{\mathbf{k}\uparrow} c^+_{-\mathbf{k}\downarrow})|0> \\ &= c^+_{\mathbf{p}\uparrow} \prod_{\mathbf{k}(\neq \mathbf{p})} (u_{\mathbf{k}} + v_{\mathbf{k}} c^+_{\mathbf{k}\uparrow} c^+_{-\mathbf{k}\downarrow})|0> \\ &= |\psi_{\mathbf{p}\uparrow} > . \end{aligned} \tag{7.2.5.1.5}$$

$|\psi_{\mathbf{p}\uparrow} >$ represents a state in which there are the pairs without $(\mathbf{p}\uparrow, -\mathbf{p}\downarrow)$ and a single quasiparticle $(\mathbf{p}\uparrow)$. The single quasiparticle created by $\gamma^+_{\mathbf{p}\uparrow}$ is called a Bogolon. The quasiparticle operators are of Fermion type because it is easy to show that they satisfy the following anticommutation relation.

$$\left[\gamma_{\mathbf{p}s}, \gamma^+_{\mathbf{p}'s'}\right]_+ = \delta_{\mathbf{p}\mathbf{p}'}\delta_{ss'}. \tag{7.2.5.1.6}$$

$$[\gamma_{\mathbf{p}s}, \gamma_{\mathbf{p}'s'}]_+ = \left[\gamma^+_{\mathbf{p}s}, \gamma^+_{\mathbf{p}'s'}\right]_+ = 0. \tag{7.2.5.1.7}$$

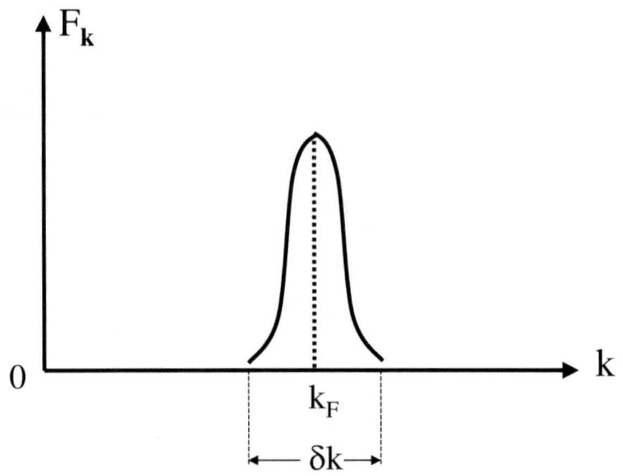

Figure 7.4: The schematic plot of $F_{\mathbf{k}}$ vs. k.

7.2.5.2. Partition Function $\mathbf{Z_{red}}$

The definition of partition function Z_{red} is as follows.

$$Z_{red} \equiv Tr\left[e^{-\beta(H_{red}-\mu\widehat{N})}\right], \tag{7.2.5.2.1}$$

where Tr means to sum the matrix elements in diagonal line of $\exp[-\beta(H_{red}-\mu\widehat{N})]$, and $\beta = 1/(k_B T)$. Using Eq. (7.2.3.1),

$$H_{red} - \mu\widehat{N} = \sum_{\mathbf{k}s} \xi_{\mathbf{k}} n_{\mathbf{k}s} + \sum_{\mathbf{k}\mathbf{k}'} V_{\mathbf{k}'\mathbf{k}} b^{+}_{\mathbf{k}'} b_{\mathbf{k}}, \tag{7.2.5.2.2}$$

where

$$\xi_{\mathbf{k}} = \epsilon_{\mathbf{k}} - \mu. \tag{7.2.5.2.3}$$

Making the Bogoliubov-Valatin transformation, Eq. (7.2.5.2.2) becomes

$$H_{red} - \mu\widehat{N} = H_M + H_I + 2\sum_{\mathbf{k}} \xi_{\mathbf{k}} |v_{\mathbf{k}}|^2 + \sum_{\mathbf{k}\mathbf{k}'} V_{\mathbf{k}'\mathbf{k}} u_{\mathbf{k}} v_{\mathbf{k}} u_{\mathbf{k}'} v^{*}_{\mathbf{k}'}, \tag{7.2.5.2.4}$$

$$H_M = \sum_{\mathbf{k}} E_{\mathbf{k}} (\gamma^{+}_{\mathbf{k}\uparrow}\gamma_{\mathbf{k}\uparrow} + \gamma^{+}_{\mathbf{k}\downarrow}\gamma_{\mathbf{k}\downarrow}). \tag{7.2.5.2.5}$$

H_M is quasiparticle Hamiltonian. $E_{\mathbf{k}}$ is only an introduced parameter at this step.

$$\begin{aligned} H_I &= \sum_{\mathbf{k}} \left[\xi_{\mathbf{k}}(u^2_{\mathbf{k}} - |v_{\mathbf{k}}|^2) - 2u_{\mathbf{k}} v_{\mathbf{k}} \sum_{\mathbf{k}'} V_{\mathbf{k}'\mathbf{k}} u_{\mathbf{k}'} v^{*}_{\mathbf{k}'} - E_{\mathbf{k}}\right] (\gamma^{+}_{\mathbf{k}\uparrow}\gamma_{\mathbf{k}\uparrow} + \gamma^{+}_{\mathbf{k}\downarrow}\gamma_{\mathbf{k}\downarrow}) \\ &+ \sum_{\mathbf{k}\mathbf{k}'} V_{\mathbf{k}'\mathbf{k}} u_{\mathbf{k}} v_{\mathbf{k}} u_{\mathbf{k}'} v^{*}_{\mathbf{k}'} (\gamma^{+}_{\mathbf{k}'\uparrow}\gamma_{\mathbf{k}'\uparrow} + \gamma^{+}_{\mathbf{k}'\downarrow}\gamma_{\mathbf{k}'\downarrow})(\gamma^{+}_{\mathbf{k}\uparrow}\gamma_{\mathbf{k}\uparrow} + \gamma^{+}_{\mathbf{k}\downarrow}\gamma_{\mathbf{k}\downarrow}). \end{aligned} \tag{7.2.5.2.6}$$

To derive Eq. (7.2.5.2.4), we express $c_{\mathbf{p}s}$ and $c^{+}_{\mathbf{p}s}$ by $\gamma_{\mathbf{p}s}$ and $\gamma^{+}_{\mathbf{p}s}$.

$$c^{+}_{\mathbf{p}\uparrow} = u_{\mathbf{p}}\gamma^{+}_{\mathbf{p}\uparrow} + v^{*}_{\mathbf{p}}\gamma_{-\mathbf{p}\downarrow}. \tag{7.2.5.2.7}$$

$$c_{\mathbf{p}\uparrow} = u_{\mathbf{p}}\gamma_{\mathbf{p}\uparrow} + v_{\mathbf{p}}\gamma^{+}_{-\mathbf{p}\downarrow}. \tag{7.2.5.2.8}$$

$$c_{-\mathbf{p}\downarrow} = u_{\mathbf{p}}\gamma_{-\mathbf{p}\downarrow} - v_{\mathbf{p}}\gamma^{+}_{\mathbf{p}\uparrow}. \qquad (7.2.5.2.9)$$

$$c^{+}_{-\mathbf{p}\downarrow} = u_{\mathbf{p}}\gamma^{+}_{-\mathbf{p}\downarrow} - v^{*}_{\mathbf{p}}\gamma_{\mathbf{p}\uparrow}. \qquad (7.2.5.2.10)$$

Substituting Eqs. (7.2.5.2.7) - (7.2.5.2.10) into Eq. (7.2.5.2.2), assuming $\xi_{\mathbf{k}} = \xi_{-\mathbf{k}}$, keeping only the quasiparticle number conservation terms because we will use the $(H_{red} - \mu\widehat{N})$ to get the trace in Eq. (7.2.5.2.1), and using the anticommutation relations, we obtain

$$\sum_{\mathbf{k}s}\xi_{\mathbf{k}}n_{\mathbf{k}s} = \sum_{\mathbf{k}}\xi_{\mathbf{k}}(u^2_{\mathbf{k}} - |v_{\mathbf{k}}|^2)(\gamma^{+}_{\mathbf{k}\uparrow}\gamma_{\mathbf{k}\uparrow} + \gamma^{+}_{-\mathbf{k}\downarrow}\gamma_{-\mathbf{k}\downarrow}) + 2\sum_{\mathbf{k}}\xi_{\mathbf{k}}|v_{\mathbf{k}}|^2. \qquad (7.2.5.2.11)$$

$$\begin{aligned}
&\sum_{\mathbf{k}\mathbf{k}'} V_{\mathbf{k}'\mathbf{k}} c^{+}_{\mathbf{k}'\uparrow} c^{+}_{-\mathbf{k}'\downarrow} c_{-\mathbf{k}\downarrow} c_{\mathbf{k}\uparrow} \\
= & -2\sum_{\mathbf{k}\mathbf{k}'} V_{\mathbf{k}'\mathbf{k}} u_{\mathbf{k}'} v^{*}_{\mathbf{k}'} u_{\mathbf{k}} v_{\mathbf{k}} (\gamma^{+}_{\mathbf{k}\uparrow}\gamma_{\mathbf{k}\uparrow} + \gamma^{+}_{\mathbf{k}\downarrow}\gamma_{\mathbf{k}\downarrow}) \\
+ & \sum_{\mathbf{k}\mathbf{k}'} V_{\mathbf{k}'\mathbf{k}} u_{\mathbf{k}'} v_{\mathbf{k}'} u_{\mathbf{k}} v_{\mathbf{k}} \\
+ & \sum_{\mathbf{k}\mathbf{k}'} V_{\mathbf{k}'\mathbf{k}} u_{\mathbf{k}'} v^{*}_{\mathbf{k}'} u_{\mathbf{k}} v_{\mathbf{k}} (\gamma^{+}_{\mathbf{k}'\uparrow}\gamma_{\mathbf{k}'\uparrow} + \gamma^{+}_{\mathbf{k}'\downarrow}\gamma_{\mathbf{k}'\downarrow})(\gamma^{+}_{\mathbf{k}\uparrow}\gamma_{\mathbf{k}\uparrow} + \gamma^{+}_{\mathbf{k}\downarrow}\gamma_{\mathbf{k}\downarrow}). \qquad (7.2.5.2.12)
\end{aligned}$$

Substituting Eqs. (7.2.5.2.11-12) into Eq. (7.2.5.2.2) yields Eq. (7.2.5.2.84).

H_I is still very complicate, so it is very difficult to get the exact Z_{red}. In the following we look for the average of H_I in terms of the density matrix operator of quasiparticles, $\exp(-\beta H_M)$. Denote the $H_{red} - \mu\widehat{N}$ after averaging of H_I still by $H_{red} - \mu\widehat{N}$, i.e.,

$$H_{red} - \mu\widehat{N} = H_M + C. \qquad (7.2.5.2.13)$$

$$\begin{aligned}
C &= \frac{\mathrm{Tr}\left[e^{-\beta H_M} H_I\right]}{\mathrm{Tr}\left[e^{-\beta H_M}\right]} + 2\sum_{\mathbf{k}}\xi_{\mathbf{k}}|v_{\mathbf{k}}|^2 + \sum_{\mathbf{k}\mathbf{k}'} V_{\mathbf{k}'\mathbf{k}} u_{\mathbf{k}} v_{\mathbf{k}} u_{\mathbf{k}'} v^{*}_{\mathbf{k}'} \\
&= 2\sum_{\mathbf{k}}\xi_{\mathbf{k}}|v_{\mathbf{k}}|^2 + \sum_{\mathbf{k}}\left[\xi_{\mathbf{k}}(u^2_{\mathbf{k}} - |v_{\mathbf{k}}|^2) - E_{\mathbf{k}}\right] 2f(E_{\mathbf{k}}) \\
&+ \sum_{\mathbf{k}\mathbf{k}'} V_{\mathbf{k}'\mathbf{k}} u_{\mathbf{k}'} v^{*}_{\mathbf{k}'} u_{\mathbf{k}} v_{\mathbf{k}} \left[1 - 2f(E_{\mathbf{k}})\right]\left[1 - 2f(E_{\mathbf{k}'})\right], \qquad (7.2.5.2.14)
\end{aligned}$$

where $f(E_{\mathbf{k}})$ is the Fermi distribution of the quasiparticles, $f(E_{\mathbf{k}}) = 1/[\exp(\beta E_{\mathbf{k}}) + 1]$.

The free energy function, F, is

$$F = -k_B T \ln Z_{red}. \qquad (7.2.5.2.15)$$

Substituting Eq. (7.2.5.2.13) into Eq. (7.2.5.2.1) yields

$$\begin{aligned}
F &= -k_B T \ln\mathrm{Tr}\left[e^{-\beta(H_M + C)}\right] = C - k_B T \ln\mathrm{Tr}\left(e^{-\beta\sum_{\mathbf{k}\alpha} E_{\mathbf{k}}\gamma^{+}_{\mathbf{k}\alpha}\gamma_{\mathbf{k}\alpha}}\right) \\
&= C - k_B T \ln\prod_{\mathbf{k}\alpha}\mathrm{Tr}\left(e^{-\beta E_{\mathbf{k}}\gamma^{+}_{\mathbf{k}\alpha}\gamma_{\mathbf{k}\alpha}}\right) = C - k_B T \ln\prod_{\mathbf{k}}\left(1 + e^{-\beta E_{\mathbf{k}}}\right)^2 \\
&= C + 2k_B T\sum_{\mathbf{k}}\ln\left[1 - f(E_{\mathbf{k}})\right]. \qquad (7.2.5.2.16)
\end{aligned}$$

To obtain the real free energy function, we should get the variational minimum value of F. The variational variables are $E_{\mathbf{k}}$ and $|v_{\mathbf{k}}|$. From the two minimum conditions and

$u_{\mathbf{k}}^2 + |v_{\mathbf{k}}|^2 = 1$ we can obtain the equations for $u_{\mathbf{k}}$, $|v_{\mathbf{k}}|$, and energy gap. Substituting Eqs. (7.2.5.2.16) and (7.2.5.2.14) into

$$\frac{\partial F}{\partial E_{\mathbf{k}}} = 0, \tag{7.2.5.2.17}$$

we obtain

$$2k_B T \frac{-\frac{\partial f(E_{\mathbf{k}})}{\partial E_{\mathbf{k}}}}{1 - f(E_{\mathbf{k}})} - 2f(E_{\mathbf{k}}) + 2\left[\xi_{\mathbf{k}}(u_{\mathbf{k}}^2 - |v_{\mathbf{k}}|^2) - E_{\mathbf{k}}\right] \frac{\partial f(E_{\mathbf{k}})}{\partial E_{\mathbf{k}}}$$
$$+2u_{\mathbf{k}}v_{\mathbf{k}} \sum_{\mathbf{k}'} V_{\mathbf{k}'\mathbf{k}} u_{\mathbf{k}'} v_{\mathbf{k}'}^* \left[1 - 2f(E_{\mathbf{k}'})\right] \left[-2\frac{\partial f(E_{\mathbf{k}})}{\partial E_{\mathbf{k}}}\right] = 0. \tag{7.2.5.2.18}$$

Considering $f(E_{\mathbf{k}}) = 1/[\exp(\beta E_{\mathbf{k}}) + 1]$, we have

$$2k_B T \frac{-\frac{\partial f(E_{\mathbf{k}})}{\partial E_{\mathbf{k}}}}{1 - f(E_{\mathbf{k}})} - 2f(E_{\mathbf{k}}) = 0. \tag{7.2.5.2.19}$$

So,

$$E_{\mathbf{k}} = \xi_{\mathbf{k}}(u_{\mathbf{k}}^2 - |v_{\mathbf{k}}|^2) - 2u_{\mathbf{k}}v_{\mathbf{k}} \sum_{\mathbf{k}'} V_{\mathbf{k}'\mathbf{k}} u_{\mathbf{k}'}^* v_{\mathbf{k}'} \left[1 - 2f(E_{\mathbf{k}'})\right]. \tag{7.2.5.2.20}$$

Considering $u_{\mathbf{k}} = \sqrt{1 - |v_{\mathbf{k}}|^2}$ and

$$\frac{\partial F}{\partial |v_{\mathbf{k}}|} = 0, \tag{7.2.5.2.21}$$

we obtain

$$2\xi_{\mathbf{k}} u_{\mathbf{k}} v_{\mathbf{k}} = -(u_{\mathbf{k}}^2 - |v_{\mathbf{k}}|^2) \sum_{\mathbf{k}'} V_{\mathbf{k}'\mathbf{k}} u_{\mathbf{k}'} v_{\mathbf{k}'} \left[1 - 2f(E_{\mathbf{k}'})\right]. \tag{7.2.5.2.22}$$

Let

$$-\sum_{\mathbf{k}'} V_{\mathbf{k}'\mathbf{k}} u_{\mathbf{k}'} v_{\mathbf{k}'} \left[1 - 2f(E_{\mathbf{k}'})\right] = \Delta_{\mathbf{k}}, \tag{7.2.5.2.23}$$

then Eq. (7.2.5.2.20) becomes

$$E_{\mathbf{k}} = \xi_{\mathbf{k}}(u_{\mathbf{k}}^2 - |v_{\mathbf{k}}|^2) + 2u_{\mathbf{k}} v_{\mathbf{k}}^* \Delta_{\mathbf{k}}, \tag{7.2.5.2.24}$$

and Eq. (7.2.5.2.22) becomes

$$2\xi_{\mathbf{k}} u_{\mathbf{k}} v_{\mathbf{k}} = (u_{\mathbf{k}}^2 - |v_{\mathbf{k}}|^2)\Delta_{\mathbf{k}}. \tag{7.2.5.2.25}$$

Substituting Eq. (7.2.5.2.25) into Eq. (7.2.5.2.24) gives

$$E_{\mathbf{k}} = (u_{\mathbf{k}}^2 - |v_{\mathbf{k}}|^2) \frac{\xi_{\mathbf{k}}^2 + |\Delta_{\mathbf{k}}|^2}{\xi_{\mathbf{k}}}. \tag{7.2.5.2.26}$$

Let

$$E_{\mathbf{k}} = \sqrt{\xi_{\mathbf{k}}^2 + |\Delta_{\mathbf{k}}|^2}, \tag{7.2.5.2.27}$$

then Eq. (7.2.5.2.26) becomes

$$u_{\mathbf{k}}^2 - |v_{\mathbf{k}}|^2 = \frac{\xi_{\mathbf{k}}}{E_{\mathbf{k}}}. \tag{7.2.5.2.28}$$

We have already known

$$u_{\mathbf{k}}^2 + |v_{\mathbf{k}}|^2 = 1. \tag{7.2.5.2.29}$$

The solutions of Eqs. (7.2.5.2.28-29) are

$$u_{\mathbf{k}}^2 = \frac{1}{2}\left(1 + \frac{\xi_{\mathbf{k}}}{E_{\mathbf{k}}}\right), \tag{7.2.5.2.30}$$

and

$$|v_{\mathbf{k}}|^2 = \frac{1}{2}\left(1 - \frac{\xi_{\mathbf{k}}}{E_{\mathbf{k}}}\right). \tag{7.2.5.2.31}$$

Substituting Eqs. (7.2.5.2.30-31) into Eq. (7.2.5.2.23) yields

$$\Delta_{\mathbf{k}} = -\sum_{\mathbf{k}'} V_{\mathbf{k}'\mathbf{k}} \frac{\Delta_{\mathbf{k}'}}{2E_{\mathbf{k}'}} \tanh\frac{\beta E_{\mathbf{k}'}}{2}. \tag{7.2.5.2.32}$$

Eq. (7.2.5.2.32) is the famous BCS gap ($\Delta_{\mathbf{k}}$) equation at finite temperature [75]. However, we can use a similar method as to obtain Eq. (7.2.3.33) to show that $\Delta_{\mathbf{k}}$ in Eq. (7.2.5.32) is proportional to $\exp(i\phi)$.

7.2.5.3. Determining T^* from Gap Equation

If $V_{\mathbf{k}'\mathbf{k}}$ is approximated by Eq. (7.2.3.35), then the $\Delta(T)$ in Eq. (7.2.5.2.32) has a simple non-trivial solution as follows. We assume $\Delta(T)$ is a real number. For convenience, we take $\hbar = k_B = 1$.

$$1 = \frac{N(E_F)V}{2} \int_{-\omega_c}^{\omega_c} d\xi \frac{1}{\sqrt{\xi^2 + \Delta^2(T)}} \tanh\frac{\sqrt{\xi^2 + \Delta^2(T)}}{2T}. \tag{7.2.5.3.1}$$

The T^* of unconventional superconductors can be obtained from Eq. (7.2.5.3.1) by using zero-gap condition, i.e., $\Delta(T^*) = 0$. The T_c equation for conventional superconductors can be obtained from Eq. (7.2.5.3.1) by using zero-gap condition, i.e., $\Delta(T_c) = 0$, approximately (The reason for "approximately" is that for conventional superconductor $T_c \approx T^*$.). For example, if $\Delta(T^*) = 0$ in case of unconventional superconductors, then Eq. (7.2.5.3.1) is transformed into

$$\begin{aligned} 1 &= N(E_F)V \int_0^{\omega_c} d\xi \frac{1}{\xi} \tanh\frac{\xi}{2T^*} \\ &= N(E_F)V \int_0^{\omega_c/(2T^*)} \tanh x d\ln x. \end{aligned} \tag{7.2.5.3.2}$$

Assume $\omega_c/(2T^*) \gg 1$, then $\tanh[\omega_c/(2T^*)] \cong 1$. Eq. (7.2.5.3.2) becomes

$$\begin{aligned} 1 &\cong N(E_F)V \left[\ln\frac{\omega_c}{2T^*} - \int_0^{\infty} dx \frac{\ln x}{(\text{ch}x)^2}\right] \\ &= N(E_F)V \left[\ln\frac{\omega_c}{2T^*} + \ln\frac{4\gamma}{\pi}\right], \end{aligned} \tag{7.2.5.3.3}$$

where $\gamma = 1.78$. From Eq. (7.2.5.3.3) we have

$$T^* = 1.13\omega_c e^{-\frac{1}{N(E_F)V}}. \tag{7.2.5.3.4}$$

Reconsidering $\hbar \neq k_B \neq 1$, then

$$T^* = 1.13 \frac{\hbar \omega_c}{k_B} e^{-\frac{1}{N(E_F)V}}. \qquad (7.2.5.3.5)$$

Because we use $\omega_c \gg T^*$, which means $N(E_F)V < 1/4$, we call Eq. (7.2.5.3.5) the formula of T^* in weak coupling approximation.

From Eqs. (7.2.3.39) and (7.2.5.3.5) we have

$$\frac{2\Delta(0)}{T^*} = 3.53, \qquad (7.2.5.3.6)$$

where $\Delta(0)$ is the gap at 0 K. For weak coupling unconventional superconductors the "3.53" is a universal constant. However, for the unconventional superconductors the ratio $2\Delta(0)/T_c$ is meaningless because the value of T_c does not come from the gap equation. (Note that for the unconventional superconductors AUNRTB still seek the meaningless ratio $2\Delta(0)/T_c$.)

For the conventional superconductors the gap, which is proportional to amplitude of wavefunction of a Cooper pair, and the long-range phase coherence between Cooper pairs appear nearly at the same temperature, i. e., $T^* \approx T_c$. The reasons that $T^* \approx T_c$ for conventional superconductor are as follows. The coherence length ξ_o is the average size of a Cooper pair, i.e., the average distance between the two electrons within a Cooper pair. In the conventional superconductors, such as Sn, Hg, and Al, a Cooper pair has an average size, ξ_o, about 10^4 $\AA$. Assume that $\xi_o = 10^4$ $\AA$ and the electron number per cubic centimeter is $n = 10^{22}/\text{cm}^3$. We know that at $T = T^*$ the gap, and thus the number of Cooper pairs, are zero. It is imaginable that, when T is only a little less than $T = T_c$ and T_c is only a little less than T^*, there are 10^{13} electrons in one cm^3 to become Cooper pairs. In this case, within the space of single Cooper pair there exist 10^1 other electrons, each being correlated in Cooper pairs. Because of the strong overlap of Cooper pairs there is a strong coupling between Cooper pairs, and thus the Cooper pairs in static superconducting state have the same unified phase. The meaning of the long-range phase coherence for the conventional superconductors in static superconducting state is that there is the same unified phase for all Cooper pairs. Therefore, the appearance of the Cooper pairs and the long-range phase coherence among Cooper pairs are nearly at the same temperature. Because the gap is proportional to the amplitude of Cooper pair, the $\Delta(T) > 0$ means nearly the appearance of Cooper pairs with the long range phase coherence, i.e., the appearance of superconductivity, for the conventional superconductor. However, strictly speaking or in principle, the BCS pairing theory in this chapter is T_c-independent in mathematics. For example, $\xi_o = 50$ $\AA$ for conventional superconductor Nb_3Sn. For Nb_3Sn the T_c should be clearly less than T^* because it can not be imaginary that at $T = T_c$ which is only a little less than T^*, there are 8×10^{19} electrons in one cm^3 to become Cooper pairs. [$(50 \times 10^{-8}\text{cm})^3 \times 8 \times 10^{19}/\text{cm}^3 = 10^{-2}$ *Cooper pairs*.] For the conventional superconductors with long coherence length ξ_o, repeating the same derivations as to obtain T^* for the unconventional superconductors, we have obviously

$$T_c \cong T^* = 1.13 \omega_c e^{-\frac{1}{N(E_F)V}}, \qquad (7.2.5.3.7)$$

and

$$\frac{2\Delta(0)}{T_c} \cong \frac{2\Delta(0)}{T^*} = 3.53, \qquad (7.2.5.3.8)$$

where $T_c \cong T^*$ for conventional superconductor comes from the above purely physical analysis.

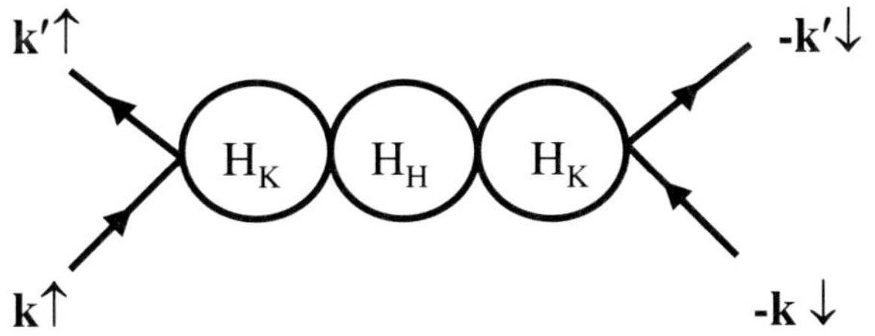

Figure 7.5: The schematic plot of $F_{\mathbf{k}}$ vs. k.

7.2.6 Origins of Attractive Interactions

The formation of Cooper pair of both electrons or holes needs the net attractive force between the two electrons or holes.

According to the work of Bardeen, Cooper, and Schrieffer in 1957, the formation of Cooper pair in conventional superconductivity is caused by exchange of phonon, which is called phonon mechanism [81]. The physics of phonon exchange is as follows. In chapter 11 we introduce phonon and electron-phono interaction. The electron is coupled to the lattice vibrations because the ions are positively charged; this coupling to the lattice means that an electron can emit a phonon, and another electron can absorb this phonon. Inclusion of this intermediate process (i. e., exchange phonon process) means that the energy of the two electrons is altered, so the phonon exchange has the same effect as a direct interaction between the two electrons. Why the phonon exchange causes attractive force between the two electrons? Let us tell you the attractive force origin more directly perceived through senses. When an electron is nearby a positive ion, due to the attractive force between this electron and the ion there is a local excess of positive charge. When the second electron goes through this region of local excess of positive charge, this second electron can feel this excess positive charge, and thus suffers a more strong attractive force. If we do not look at the ion, then there is an effective retardation attractive force between the first and the second electrons.

After discovery of high-temperature superconductivity, which is one kind of unconventional superconductivity, physicists proposed more than 200 mechanisms to explain the origin of the attractive force between the two electrons or the two holes of a Cooper pair. However, until now all the mechanisms cannot explain experimental data well except the mechanism proposed in Ref. [75]. Therefore, let us introduce the mechanism of unconventional superconductivity proposed by Ref. [75].

This mechanism is called two-local-spin-mediated interaction, which is shown in Fig. 7.5. In actual solid, the so called free electron or hole are called quasiparticles. The quasiparticle $\mathbf{k}\uparrow$ has Kondo interaction H_K with the local spin 1. The local spin 1 has Heisenberg interaction with the nearest second local spin. The second local spin has Kondo interaction with another quasiparticle $-\mathbf{k}\downarrow$. If we do not look at intermediate interactions, then there is an effective simultaneous attractive force between the two quasiparticles: $\mathbf{k}\uparrow$ and $-\mathbf{k}\downarrow$.

It can be seen that the two-local-spin-mediated interaction does not belong to the type of exchange of middle phonon (for the case of superconductivity) or middle meson (for the case of particle physics), and belongs to three-fold exchange interaction.

7.3 Microscopic Superconductivity

7.3.1 Introduction

This section will use the same method as that Landau proved the criterion for macroscopic superfluidity of He II [82, 83] to prove two theorems on the microscopic superfluidity and microscopic superconductivity. The present situations of the research on the Landau criterion are as follows. AUNRTB think that:

(1). Landau's criterion is a condition of appearance of macroscopic superfluidity of He II;

(2). The Landau's criterion can be extended to be as a criterion of appearance of macroscopic superconductivity.

On the contrary, in this section we will demonstrate by using the same method as Landau that:

(1)'. Landau's criterion is only a criterion of appearance of microscopic superfluidity of He II. Here, the "microscopic" means that although the liquid does not have macroscopic superfluidity, but some carriers move without resistance and, thus, freely. Due to randomness of the some carrier motions, there is no the macroscopic superfluidity;

(2)'. The Landau's criterion as a criterion of macroscopic superconductivity is not sufficient. The Landau's criterion can be used only to be as a criterion of microscopic superconductivity. Here, the "microscopic" means that although the conductor does not have macroscopic superconductivity, but some Cooper pairs move without resistance and, thus, freely. Due to randomness of the some Cooper pair motions, there is no the macroscopic superconductivity;

(3)'. To produce macroscopic superfluidity and macroscopic superconductivity, the criterion is long-range phase coherence of wavefunctions between Cooper pairs instead of the one condition in Landau's criterion;

(4)'. The Landau's criterion as a criterion of microscopic superconductivity tells us that the microscopic superconductivity can occur in the Fermion system. For example, The electrons in filled bands of all insulators are of microscopic superconductivity.

This section also demonstrates that considering the observed room temperature microscopic superconductivity in the so-called pseudogapbody (, which is a body being of pseudogap,), a room temperature pseudogapbody electronics and electric industry will be established in the nearly future certainly. At that time, many (not all) semiconductors will be replaced by pseudogapbodes being of superior properties, and the strong electricity aspect also will have revolutionary progresses. Sincerely hope readers go ahead in this material revolution, make many inventions, and create miracles.

7.3.2 Criterion for Microscopic Superconductivity and Superfluidity

Theorem XXXXX: Criterion of microscopic superfluidity: The criterion of appearance of microscopic superfluidity is $v_c > 0$. v_c is a critical velocity, at which it is possible that an excitation can be created out of the liquid.

Proof:

Imagine a body of large mass M moving at velocity $\mathbf{V}_i$ through a volume of He II, which is at rest in the laboratory system of coordinates. So long as V_i is small enough, the body will experience no drag from the superfluid (Note that it is enough to have no drag that this superfluid is of microscopic superfluidity!). This will be changed only when the magnitude of V_i reaches a critical value v_c, at which it is possible for an excitation to be created out of the superfluid. This causes a loss of the body's kinetic energy, which is dissipated in the thermal excitation energy, what is as heat. Thus, as soon as its velocity reaches v_c, the body begins to suffer a drag.

In order to calculate v_c, we need to find the minimum value of $\mathbf{V}_i$ at which an excitation can appear. Suppose that the creation of an excitation with energy $\epsilon(\mathbf{p})$ and momentum $\mathbf{p}$ causes the body's velocity to decrease from $\mathbf{V}_i$ to $\mathbf{V}_f$. The process must

$$\frac{1}{2}MV_i^2 = \frac{1}{2}MV_f^2 + \epsilon(\mathbf{p}), \tag{7.3.2.1}$$

and momentum

$$M\mathbf{V}_i = M\mathbf{V}_f + \mathbf{p}. \tag{7.3.2.2}$$

If we eliminate $\mathbf{V}_f$ between Eqs. (7.3.2.1-2), we obtain

$$\epsilon(\mathbf{p}) - \mathbf{p}\cdot\mathbf{V}_i + \frac{p^2}{2M} = 0. \tag{7.3.2.3}$$

We may assume that M is so large that the last term in Eq. (7.3.2.3) can be neglected. If θ is the angle between $\mathbf{p}$ and $\mathbf{V}_i$, we then have

$$pV_i\cos\theta = \epsilon(\mathbf{p}), \tag{7.3.2.4}$$

and since $\cos\theta \leq 1$, the condition

$$V_i \geq \frac{\epsilon(\mathbf{p})}{p} \tag{7.3.2.5}$$

must be satisfied for the excitation to be created. Thus the critical velocity is given by

$$v_c = \left[\frac{\epsilon(\mathbf{p})}{p}\right]_{min}. \tag{7.3.2.6}$$

Microscopic superfluidity can therefore occurs if

$$v_c > 0. \tag{7.3.2.7}$$

This condition is called Landau criterion for macroscopic superfluidity by AUNRTB. However, according to our above derivations, we see that $v_c > 0$ is only a condition of microscopic superfluidity. If the particle-body moves with velocity $v < v_c$, then there is no excitation to be created, which means that the particle-body has its own superfluidity. However, Eq. (7.3.2.7) can not ensure long-range phase coherence between different particle-body. The long-range phase coherence is needed by macroscopic superfluidity. Therefore, Eq. (7.3.2.7) is a criterion of microscopic superfluidity. Because macroscopic superfluidity needs long-range phase coherence [83], and in our above derivations (=Landau's original derivations) there is no condition of long-range phase coherence, thus the Eq. (7.3.2.7) is only a criterion for microscopic superfluidity of He II. QED.

Theorem XXXXXI: Criterion of microscopic superconductivity: The criterion of appearance of microscopic superfluidity is $v_c > 0$. v_c is a critical velocity, at which it is possible that an excitation can be created out of the conductors.

Proof:
The proof is the same as that of theorem XXXXX. QED.

Comments:

Some references think that if the gap is not equal to zero, then the system is in macroscopic superconducting state. From this theorem we see that the non-zero gap just ensures appearance of the microscopic superconductivity. Other some references think that appearance of the macroscopic superconductivity needs both non-zero gap and long-range phase coherence of the wavefunctions of the carriers. This textbook would like to stress that, exactly speaking, the long-range phase coherence of wavefunctions of Cooper pairs is the sole condition to produce macroscopic superconductivity. For example, people have observed gapless superconductivity. Therefore, the long-range phase coherence of wavefunctions of Cooper pairs is more basic. In Ref. [75] there are many studies on the long-range phase coherence condition to produce macroscopic superconductivity. This chapter just addresses microscopic superconductivity.

7.3.3 Systems with Microscopic Superconductivity

Corollary XVIII: Systems with microscopic superconductivity: According to the criterion for microscopic superconductivity Eq. (7.3.2.6-7), the following systems have the microscopic superconductivity:

(1). The electrons within full band.

Assume, for simplicity, that the bands are one dimensional. There is a gap Δ between the empty band bottom and the full band top. The excitation energy of the electron within the full band is

$$\epsilon(\mathbf{p}) = \Delta_1 + \Delta + \frac{p^2}{2m^*} - \frac{p_{e,b}^2}{2m^*}, \tag{7.3.3.1}$$

where Δ_1 is the energy from the electron within the full band to the top of the full band, $p_{e,b}$ is the electron momentum at the bottom of the empty conduction band. According to Eq. (7.3.2.6) the critical velocity of the microscopic superconductivity within the full band is

$$v_c = \frac{\Delta_1 + \Delta}{p_{e,b}}. \tag{7.3.3.2}$$

The velocity v of the electron within full is lower. If $v < v_c$, then the electron within the full band has its own microscopic superconductivity.

This corollary has been verified by many facts. As is well known, the insulators have full bands and one empty conduction band. In the full bands, every electron has a wavenumber. According to the de Bloglie theorem in chapter 1, the electrons in the full bands has momentum $p = \hbar k$. Although the electrons in the full bands are in continuous motions, but the insulators do not become more and more hot, and at last melt.

(2). The electrons in the lower part of Fermi surface of a conduction band.

Assume, for simplicity, that the band is one dimensional, and at 0 K. Let Δ_2 denote the energy from the electron in the lower part of the Fermi surface to the Fermi surface. The excitation energy of the electron in the lower part is

$$\epsilon(\mathbf{p}) = \Delta_2 + \frac{p^2}{2m^*} - \frac{p_F^2}{2m^*}, \tag{7.3.3.3}$$

where p_F is the Fermi momentum. According to Eq. (7.3.2.6) the critical velocity of the microscopic superconductivity of the electron in the lower part of the Fermi surface of the conduction band is

$$v_c = \frac{\Delta_2}{p_F}. \tag{7.3.3.4}$$

(3). The Cooper pairs in pseudogapbody.

The excitation energy of the Cooper pair is

$$\epsilon(\mathbf{p}) = \sqrt{\Delta^2 + (p^2/2m^* - p_F^2/2m^*)^2}, \tag{7.3.3.5}$$

where Δ is the energy gap of the Cooper pair. According to Eq. (7.3.2.6) the critical velocity of the microscopic superconductivity in the Cooper pairs without long-range phase coherence is

$$v_c = \frac{\Delta}{p_F}. \tag{7.3.3.6}$$

The microscopic superconductivity is the general property of the Cooper pairs without long-range phase coherence. We also have to understand that one cannot say that due to non-zero gap or pseudogap the system is of macroscopic superconductivity; we just can say that due to non-zero gap or pseudogap the system is of microscopic superconductivity.

In section 7.2 we have pointed that for unconventional superconductors there is non-zero pseudogap at temperature $T_c < T < T^*$, and we call the unconventional superconductors at $T_c < T < T^*$ the pseudogapbody. The experimentally observed values are as follows: T^* can be 360 K, and the pseudogap at room temperature can be 60 meV, for example. From Eq. (7.3.3.6) we see that these pseudogapbodies have microscopic superconductivity. In subsections 7.3.4-6 we list three aspects of the room temperature applications of the microscopic superconductivity. Actually, the microscopic superconductivity should have many applications. In one word, all thc applications of macroscopic superconductivity, listed in Ref. [84] can also be the applications of microscopic superconductivity.

(4). Carriers in many other bodies, have been observed the pseudogap. However, in these pseudogapbodies the microscopic superconductivity has not yet been observed at any temperature. This corollary also thinks that these pseudogapbodies should be of the microscopic superconductivity. Due to the possibility to have the microscopic superconductivity at room temperature, before entering subsection 7.3.4-6, let me introduce these materials.

(4.1). Colossal Magnetoresistive Oxides.

The phenomenon that an extremely large negative magnetoresistance appears near the Curie temperature of manganese oxides is called colossal magnetoresistance (CMR). The colossal magnetoresistance was observed in many magnetic manganese oxides, such as $La_{1-x}Ca_xMnO_2$ [85], $La_{2/3}Ba_{1/3}MnO_x$ [86], $La_{1.2}Sr_{1.8}Mn_2O_7$, and $La_{0.82}Sr_{0.18}MnO_3$ [87]. Dessau *et al.* performed direct electronic structure measurement of the colossal magnetoresistive oxides using high energy resolution spectroscopy [87, 88]. A key result of these measurements is the discovery of a pseudogap at the Fermi surface below as well as above the magnetic order temperature. There are many possible conjectures on the origin of the pseudogap [87, 88].

We note that the resistance goes down when the temperature goes down from the magnetic order temperature [85, 86]. We guess that the pseudogap belongs to the Cooper pair without the long-range phase coherence. These Cooper pairs have the microscopic superconductivity, and thus lead to the decrease of the whole resistance. It is very interesting to note that Mannella *et al.* reported experimental evidence that a very similar pseudogap state with a nodal-antinodal dichotomous character exists in the ferromagnetic metallic ground-state of the colossal magnetoresistive bilayer manganite $La_{1.2}Sr_{1.8}Mn_2O_7$ [89]. Mannella *et al.*'s findings might imply that the pseudogap in $La_{1.2}Sr_{1.8}Mn_2O_7$ comes from Cooper pairs without the long-range phase coherence as well as the pseudogap in the high-T_c cuprates. For this subsection see also Refs. [90]-[93].

(4.2). Kondo Semiconductors.

Kondo semiconductors, such as CeNiSn, CeRhSb, and CeRhAs, are a class of strongly correlated f-electron materials whose low-energy excitation exhibits a pseudogap at low temperature [94, 95]. The Kondo semiconductors can be distinguished from the conventional semiconductors by the fact that the pseudogap appears below a characteristic temperature T^*, and the pseudogap is very sensitive to any impurity [94]. A number of models have been proposed to account for the unusual pseudogap properties [96].

7.3.4 Room Temperature Electronics of Pseudogapbody

The common characteristic of microscopic superconductivity is the zero-resistance of individual Cooper pair motion. We list two kinds of devises in room temperature electronics of pseudogapbody, and hope the readers to give more room temperature applications of microscopic superconductivity:

(1). Photoconductivity detector [79].

If a photon shines on a pseudogapbody, then the Cooper pair absorbs this photon and becomes two quasiparticles staying up the pseudogap under the condition that $\hbar\omega \geq 2\Delta(T)$. Here, $\hbar\omega$ is the photon's energy, $\Delta(T)$ is the pseudogap at temperature T. Because one Cooper pair absorbs a photon and becomes two quasiparticles, the photon's energy has to larger than $2\Delta(T)$. For some pseudogapbodies, $0 \leq \Delta(300\ \ K) \leq 80\ \ meV$. See Fig. 7.6. The experimental data in Fig. 7.6 come from Ref. [107]

The advantages of pseudogapbody photoconductivity detector are as follows. In order to obtain the best response in the 8 - 14 μm window the detector material must be a semiconductor with its forbidden bandwidth around $E_g = 0.09$ eV. Semiconductor alloy, $Hg_{1-x}Cd_xTe$, is taken to be as an ideal intrinsic infrared radiation detecting materials, and the corresponding detector should be operate at 77 K.

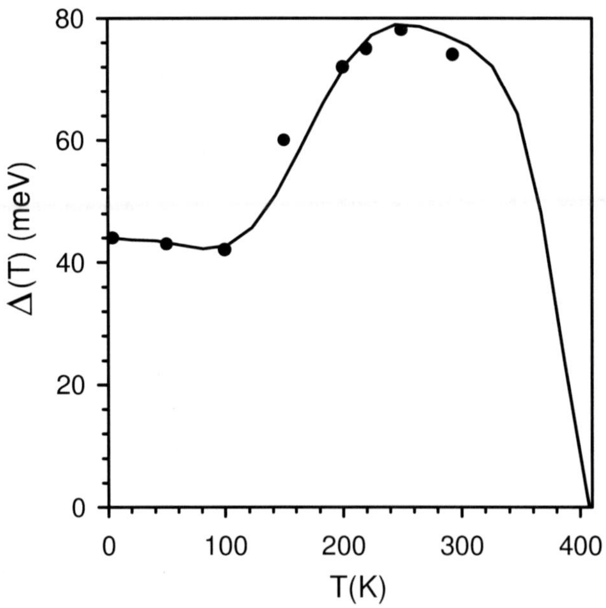

Figure 7.6: Theoretical and experimental pseudogap of Bi2212 sample ($x = 0.124$.). Note that x represents the number of $O_{p\sigma}$ holes in one unit cell of Cu lattice in CuO_2 plane.

However, if we use pseudogapbodies Bi2212 or Y123 (See [79]) with $x = 0.056$ to make the light infrared photoconductivity detector, then it has three advantages over semiconductors: (a). Figs. 7.6, and 7.7 give $E_g=2\times0.025 = 0.05$ eV at 300 K. Therefore, this pseudogapbody device does not need to work ar 77 K; (b). When one Cooper pair absorbs a photon, it can produces two quasiparticles in the conduction band. When one electron in filled band of semiconductor absorbs a photon, it can produces just one quasiparticles in the conduction band; (c). The number density of Cooper pairs of pseudogapbody Y123 with $x = 0.056$ is about $10^{21}cm^3$. The number density of holes in semiconductor is $10^{16}/cm^3$. Therefore, the volume of pseudogapbody device can be much smaller than the volume of semiconductor.

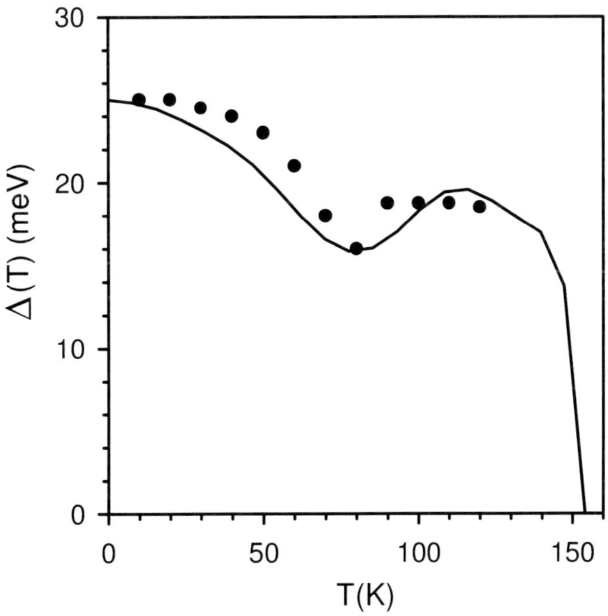

Figure 7.7: Theoretical and experimental pseudogap of Bi2212 sample ($x = 0.188.$). Note that x represents the number of $O_{p\sigma}$ holes in one unit cell of Cu lattice in CuO_2 plane. The experimental data come from Ref. [108].

(2). Light-emitting diode [79].

This is a semiconductor device in which injected carriers recombine by emitting photon. Ref. [79] points out that to obtain light from infrared to ultraviolet more than ten different semiconductors are needed.

The pseudogapbody can be used to make light-emitting diode. The principle is shown in Fig. 7.9.

The Cooper electron pair on the left energy level can tunnel (,or can pass through other weak links,) to the right energy level at $V < 2\Delta/|e|$, where Δ is the pseudogaps on both sides at 300 K. This tunneling does not conserve the energy but takes an energy $2|e|V$ due to the applied static voltage V and microscopic superconductivity. This Cooper pair must tunnel back, since it can not complete a simple motion where energy is not conserved. So long as the voltage exists, the Cooper pair will tunnel to and from between the two pseudogapbody (P) electrodes. This *ac* current can be observed by the external radiation with frequency $\nu = 2|e|V/h < 4\Delta/h$ it produces. If $V < 0.016$ and $V > 5$ eV, then the radiation frequencies are in THz region ($> 80\mu m$) and in deep ultraviolet region ($< 250\mu m$), respec-

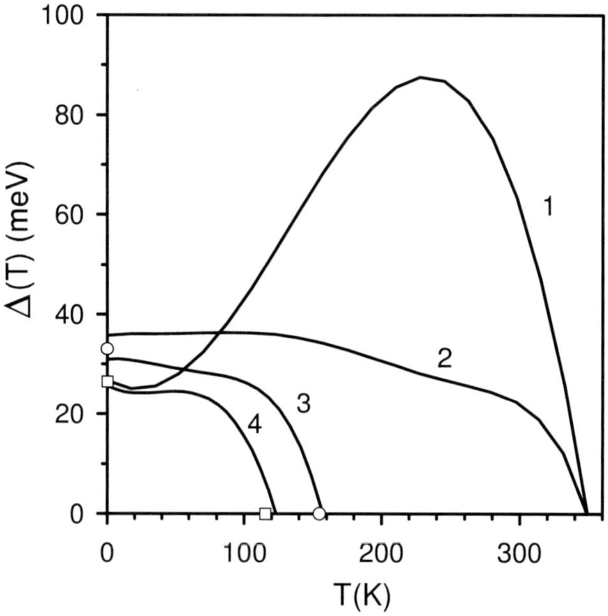

Figure 7.8: Theoretical experimental pseudogap of Bi2212 and Y123. The curve 1, 2, 3, and 4 are for $x = 0.025$, 0.056, 0.16, and 0.208. The circles and squares are for $x = 0.16$ and 0.208, respectively. The experiment data is for $x = 0.056$, and $\Delta \approx 25$ meV, which fits our theoretical curve 2. However, the error of the experimental data in Ref. [79] too large, the experimental data in Ref. [79] does not appear in curve 2.

tively. The pseudogapbody light-emitting diodes can be P(pseudogapbody)-I(insulator)-P(pseudogapbody) tunneling junction, thin film microbridge, point contact, and so on. Therefore, in comparison of pseudogapbody light-emitting diode with semiconductor light-emitting diode the former has two obvious advantages: (a). The device is easy to be made; (b). One can obtain very wide spectrum in terms of only one device and the variations of static voltage. Note: (a). At present the highest breakdown voltage of a tunneling junction is only 2 (Volt). To increase the breakdown voltage should make many basic researches; (b). When the voltage $V(> 2\Delta/|e|)$, there is single particle tunneling besides the Cooper pair tunneling.

(3). Solar cell [79].

7.3.5 Negligible Resistance of Pseudogapbody at Room Temperature

In the previous sections we gave the Landau criterion of microscopic superconductivity, and studied the general theory of microscopic superconductivity. The existence of microscopic superconductivity in pseudogapbody has been confirmed by Ref. [78]. Next we study what is the macroscopic resistance of pseudogapbody?

From subsection 7.2 we know the critical velocity of individual carrier in microscopic superconducting state is

$$v_c = \frac{\Delta(T)}{p_F} = \frac{\Delta(T)}{hk_F}, \tag{7.3.5.1}$$

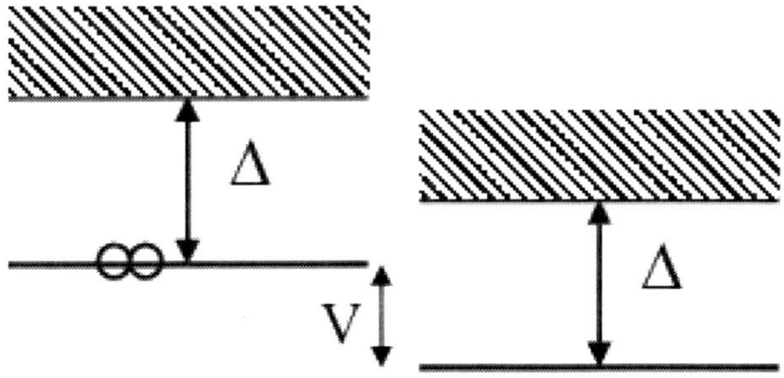

Figure 7.9: The schematic energy level diagram for the two pseudogap-body's (P) electrodes. The shading area denotes the energy levels occupied by quasiparticles. The two circles denote a Cooper pair without the long-range phase coherence. The real line denotes the chemical potential.

where $\Delta(T)$ represents a pseudogap in excitation spectrum, k_F is the wavenumber at Fermi surface. If the velocity v of the any individual Cooper pair surpasses v_c, then the motion of the Cooper pair will suffers resistance. Eq. (7.3.5.1) is called Landau microscopic superconductivity criterion.

If $v < v_c$, then although there is no macroscopic superconductivity, there is microscopic superconductivity in the pseudogapbody so long as $\Delta > 0$. In the pseudogapbody of the high-T_c cuprates, $\Delta > 0$ at $T > T_c$. In microscopic superconducting state the Cooper pairs without the long-range phase coherence move freely, but randomly.

The following numerical calculations demonstrate that the resistance of pseudogapbody as lead or cable at room temperature is only 10^{-10} of that the resistance in copper (Cu) lead.

If a voltage is applied to pseudogapbody, which is taken to be as a lead or cable, then the dc electric field will accelerate the free Cooper pairs. The Cooper pair in electric circuit can sustain a critical voltage V_c, which is determined by $v < v_c$ according to energy conservation.

$$\frac{m^* v_c^2}{2} = 2eV_c, \tag{7.3.5.2}$$

where m^* is the effective mass of a Cooper pair in pseudogapboy.

Assume that the pseudogap at 300 K can be larger than 50 meV. $m* = 10m_e$, where m_e is the mass of a free electron, and $k_F = 3.5 \times 10^7$ cm^{-1} [55]. Let us calculate the values of critical velocity v_c and the critical voltage V_c. From Eq. (7.3.5.1) we obtain $v_c = 4.2 \times 10^6$ cm/sec. Substituting this value into Eq. (7.3.5.2) yields $V_c = 25$ mV. The voltage V_c should be applied on the two sides of the pseudogapbody as lead. Suppose that the length is $L = v_c \times second$ because in this length the maximum velocity of Cooper pairs, accelerated by dc electric field, $v_{max} \lessapprox v_c$. Here, v_{max} comes from the acceleration of dc electric field produced by dc voltage V_c.

Let us calculate the resistance in pseudogapbody as lead (conducting wire) or cable. Assume that the area of cross section is S. The power of the dc electric field, produced by V_c, is

$$W_{pair} = V_c J_{pair} S. \tag{7.3.5.3}$$

If we use the copper (Cu) as lead, with length $L = v_c \times second$ and area of cross section S, and current density $J = J_{pair}$, then the power of the dc electric field in Cu lead has to be

$$W_{Cu} = (JS)^2 R = (J_{pair}S)^2 \rho \frac{L}{S}. \tag{7.3.5.4}$$

Assume that the pseudogapbody is Bi2212 with $x = 0.05$. The volume of an unit cell is $3.9 \times 3.9 \times 37.1$ cm^3. Assume that the pairing probability of a quasiparticle $Q(x = 0.05, 300K) = 0.53$ [80]. Thus, the the number density of Cooper pairs is $n_{pair}(x = 0.05, 300K) = 1.77 \times 10^{20}/\text{cm}^3$. The current density of Cooper pairs is $J_{pair} = ev_c \times n_{pair}(x = 0.05, 300K) = 1.2 \times 10^8$ A/cm^2. The current density J_{pair} is obtained under the applied voltage $V_c = 25$ mV. Assume that the cross section is $S = 1$ cm^2. Thus, from Eq. (7.3.5.3) $W_{pair} = 3.0 \times 10^6$ W. Let us consider copper as lead. The resistivity of copper is $\rho_{Cu} = 10^{-6}\Omega - cm$. Thus, the resistance of copper lead with cross section cm^2 and length L is $R_{Cu} = L\rho/cm^2 = 4.2$ Ω. The power of dc electric field, needed by producing $J = J_{pair}$ in copper lead, is $_{Cu} = I^2 R_{Cu} = (J_{pair} \times cm^2)^2 R = 6.0 \times 10^{16} \times 4.2 = 2.5 \times 10^{17}$ watt. Therefore,

$$\frac{W_{pair}}{W_{Cu}} = \frac{R_{pair}}{R_{Cu}} \approx 5 \times 10^{-10}. \tag{7.3.5.5}$$

From Eqs. (7.3.5.1), and transition theory one obtains

$$Ratio = \frac{W_{pair}}{W_{Cu}} = \frac{R_{pair}}{R_{Cu}} = \frac{m^*}{4e^2 \rho_{Cu} n_{pair}} \times second \propto \frac{1}{n_{pair}}. \tag{7.3.5.6}$$

Eq. (7.3.5.6) shows that the larger the n_{pair}, the less the *Ratio* is.

7.3.6 Macroscopic Superconductivity Induced by Microscopic Superconductivity and Its Room Temperature Applications

7.3.6.1. Induced Macroscopic Superconductivity

Let us consider the ring in Fig. 7.10. The ring is made by pseudogapbody which has microscopic superconductivity. Of course, we have many methods to guarantee that the trapped magnetic flux, Φ_A, is just confined in the region A of Fig. 7.10.

In the ring there is no macroscopic zero resistance superconducting current. The Cooper pairs in the ring make random and free motion. The average velocity of all the Cooper pairs in random motion is equal to zero. However, if we take off the trapped magnetic flux suddenly, then every Cooper pair will suffer the same electromagnetic induced force and get the same additional velocity. In this case the average velocity of all Cooper pairs in the ring will be not zero. Because the Cooper pairs are free, this average velocity will be kept. Therefore, we obtain permanent superconducting current in the ring. The magnitude of this macroscopic permanent supercurrent is same as that of the superconductor ring being of macroscopic superconductivity. We call this zero resistance phenomenon the macroscopic superconductivity induced by microscopic superconductivity.

7.3.6.2. Motion Equation of Cooper Pair in Electromagnetic Field

The pseudogapbody being of microscopic superconductivity does not have (macroscopic) zero-resistance. However, in the special condition such as the case of the ring in subsection 7.3.6.1 we have argued that there is the macroscopic superconducting current. Now, we ask what is the motion equation of the Cooper pairs with induced macroscopic superconductivity in electromagnetic field?

We assume that the wavefunction of $j-th$ Cooper pair is $\psi_j(\mathbf{r})$, neglecting the spin part for brevity. Due to that the Cooper pair motion can be separated into two parts:

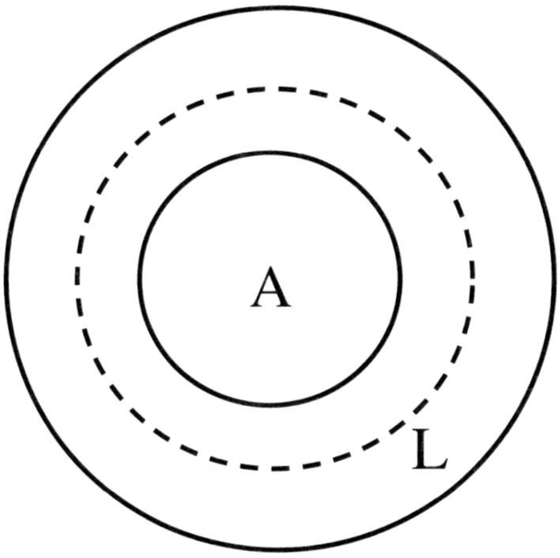

Figure 7.10: The schematic plot of pseudogapbody ring. There is the trapped magnetic flux confined just in region A.

the ordered common motion and the random motion, wavefunction of a Cooper pair can be written as $\psi_j(\mathbf{r}) = |\psi_j(\mathbf{r})| \exp[i(\theta_{common} + \theta_{j,random})]$. Considering that the N Cooper pairs in a system condense into the common ground state, $\sqrt{\psi_j(\mathbf{r})} = \sqrt{\psi(\mathbf{r})}$. Thus, under the condition of neglecting the multiparticle exchange effect, the wavefunction of N Cooper pair's system can be written as

$$\begin{aligned}
\Psi &= \Psi_N = \prod_{j=1}^{N} |\psi_j(\mathbf{r})| e^{[i(\theta_{common}+\theta_{j,random})]} \\
&= \prod_{j=1}^{N} |\psi(\mathbf{r})| e^{[i(\theta_{common}+\theta_{j,random})]} \\
&= \left(|\psi(\mathbf{r})| e^{i\theta_{common}}\right) \prod_{j=1}^{N} e^{\theta_{j,random}} \\
&= \left(|\psi(\mathbf{r})| e^{i\theta_{common}}\right) \exp\left(\sum_{j=1}^{N} \theta_{j,random}\right) \\
&= \left(|\psi(\mathbf{r})|\right) e^{i\theta_{common}} e^{0} \\
&= \sqrt{\rho(\mathbf{r})} e^{i\theta_{common}} \\
&= \sqrt{\rho} e^{i\theta_{common}} \equiv \sqrt{\rho} e^{i\theta}. \qquad (7.3.6.2.1)
\end{aligned}$$

In Eq. (7.3.6.2.1) $|\sqrt{\psi(\mathbf{r})}|^2$ represents the position space probability density distribution of one Cooper Cooper pair. $\left|\left(\sqrt{\psi(\mathbf{r})}\right)^N\right|^2 = \rho(\mathbf{r})$ represents the position space Cooper pair number density distribution. Due to the electrostatic repulsion force between Cooper pairs, $\rho(\mathbf{r}) \approx \rho$ approximately independent of $\mathbf{r}$.

In case that wavefunction represents N particle, Eq. (7.2.16) gives the particle number current density instead of probability current density of one particle. If we multiply the particle number current density by the Cooper pair's charge $2e$, then we obtain the induced macroscopic electric current density is

$$\begin{aligned}\mathbf{J} &= 2e\frac{1}{2\mu}\left[\Psi^*\left(\mathbf{p}-\frac{2e}{c}\mathbf{A}\right)\Psi+\Psi\left(\mathbf{p}-\frac{2e}{c}\mathbf{A}\right)^*\Psi^*\right] \\ &= \frac{2e\rho}{\mu}\left(\hbar\nabla\theta-\frac{2e}{c}\mathbf{A}\right), \qquad (7.2.16)'\end{aligned}$$

where μ is the mass of one Cooper pair.

The θ comes from the common motion of Cooper pairs. Let us discuss the physical meaning of θ further. Substituting Eq. (7.3.6.2.1) into Schrödinger equation

$$i\hbar\frac{\partial}{\partial t}\Psi=\left[\frac{1}{2\mu}\left(-i\hbar\nabla-\frac{2e}{c}\mathbf{A}\right)\cdot\left(-i\hbar\nabla-\frac{2e}{c}\mathbf{A}\right)+2e\phi\right]\Psi. \qquad (7.3.6.2.2)$$

Substituting Eq. (7.3.6.2.1) into Eq. (7.3.6.2.2) yields

$$\frac{\partial}{\partial t}\rho+\nabla\cdot(\rho\mathbf{v})=0, \qquad (7.3.6.2.3)$$

and

$$\frac{\partial}{\partial t}\hbar\theta=-\frac{\mu}{2}v^2-2e\phi+\frac{\hbar^2}{2\mu}\frac{1}{\sqrt{\rho}}\nabla^2\sqrt{\rho}, \qquad (7.3.6.2.4)$$

where

$$\mu\mathbf{v}=\hbar\nabla\theta-\frac{2e}{c}\mathbf{A}. \qquad (7.3.6.2.5)$$

Eq. (7.3.6.2.3) is continuity equation. If we neglect the last term in Eq. (7.3.6.2.4), then $\hbar\theta$, $\mu v^2/2$, and $2e\phi$ are equivalent to velocity potential, kinetic energy, and potential energy of hydrodynamics, respectively. In the interior of the superconductor, $\rho\approx constant$, thus the last term in Eq. (7.3.6.2.4) is nearly equal to zero. From Eq. (7.3.6.2.5) we see that $\hbar\nabla\theta$ is just the canonical momentum of Cooper pair. Next, we demonstrate that if we neglect the last term of Eq. (7.3.6.2.4), then Eq. (7.3.6.2.4) is equivalent to the motion equation of a Cooper pair in electromagnetic field. Taking gradient of Eq. (7.3.6.2.4) yields

$$\frac{\partial}{\partial t}\hbar\nabla\theta=-\frac{\mu}{2}\nabla v^2-2e\nabla\phi. \qquad (7.3.6.2.6)$$

Substituting Eq. (7.3.6.2.5) into Eq. (7.3.6.2.6) yields

$$\mu\frac{\partial}{\partial t}\mathbf{v}=-\frac{\mu}{2}\nabla v^2-2e\nabla\phi-\frac{2e}{c}\frac{\partial}{\partial t}\mathbf{A}. \qquad (7.3.6.2.7)$$

Utilizing

$$\mathbf{v}\times(\nabla\times\mathbf{v})+(\mathbf{v}\cdot\nabla)=\frac{1}{2}\nabla v^2$$

and

$$\mathbf{E}=-\nabla\phi-\frac{1}{c}\frac{\partial}{\partial t}\mathbf{A},$$

Eq. (7.3.6.2.7) is transformed into

$$\mu\left[\frac{\partial}{\partial t}\mathbf{v}+(\mathbf{v}\cdot\nabla)\mathbf{v}\right]=-\mu\mathbf{v}\times(\nabla\times\mathbf{v})+2e\mathbf{E}. \qquad (7.3.6.2.8)$$

Utilizing the curl of Eq. (7.3.6.2.5)

$$\nabla \times \mathbf{v} = -\frac{2e}{\mu c}\nabla \times bfA = -\frac{2e}{\mu c}\mathbf{B}$$

and the relation often used in hydrodynamics

$$\frac{\partial}{\partial t}\mathbf{v} + (\mathbf{v} \cdot \nabla)\mathbf{v} = \frac{d}{dt}\mathbf{v},$$

Eq. (7.3.6.2.8) can be transformed into

$$\mu \frac{d}{dt}\mathbf{v} = 2e\left(\mathbf{E} + \frac{1}{c}\mathbf{v} \times \mathbf{B}\right), \qquad (7.3.6.2.9)$$

which is the motion equation of one Cooper pair with induced macroscopic superconductivity in electromagnetic field.

The profound implication of this demonstration is that Eq. (7.3.6.2.9) is the motion equation in electromagnetic field not only for the Cooper pairs with induced macroscopic superconductivity but also for the Cooper pairs with initial macroscopic superconductivity and for the Cooper pairs just with microscopic superconductivity. In exercises and solutions of section 7.5 we require the readers to answer these questions.

7.3.6.3. Meissner Effect of Induced Macroscopic Superconductivity

The basic characteristics of macroscopic superconductivity are: (a). Zero resistance; (b). Meissner effect or diamagnetism, which says: inside superconductors in macroscopic superconducting state the magnetic induction $B = 0$, and the $B = 0$ is history-independent. This phenomenon is called Meissner effect [75]. Meissner effect is a perfect diamagnetism. This perfect diamagnetism occurs at magnetic field smaller than the thermodynamic critical field H_c for type-I superconductor (See Fig. 7.11.), and at magnetic field smaller than the lower critical magnetic field H_{c1} for type-II superconductor (See Fig. 7.11.). From Meissner effect we realize that superconductors in superconducting state are different from the perfect conductors since the latter tends to trap magnetic flux in, and leads to $B = 0$ history-dependent. The origin of Meissner effect is as follows. The magnetic field parallel to the surface of superconductor decays exponentially, with a penetration depth λ_L, in the interior. $\lambda_L(T = 0 \text{ K})$ ranges from 500 to 10 000 $\mathring{A}$, depending on the material. From Maxwell fourth equation, we know that accompanying this parallel field is a surface current density. This surface macroscopic superconducting current density screens the magnetic field from the interior of superconductor being of macroscopic superconductivity.

The pseudogapbody being of microscopic superconductivity does not have zero-resistance and Meissner effect because Meissner effect needs macroscopic surface superconducting current. However, in the special case such as the ring mentioned in subsection 7.3.6.1, we have argued that there is the induced macroscopic superconducting current. Now we ask if the ring is of Meissner effect? The following demonstration indicates that the induced macroscopic superconductivity also has induced Meissner effect as well.

Substituting Eq. (7.3.6.2.5) into Eq. (7.2.16)' yields

$$\mathbf{J} = 2e\rho\mathbf{v}. \qquad (7.3.6.3.1)$$

From Eqs. (7.3.6.3.1) and (7.3.6.2.9) we obtain

$$\frac{\partial \mathbf{J}}{\partial t} = \frac{\rho(2e)^2}{\mu}\left(\mathbf{E} + \frac{1}{c}\mathbf{v} \times \mathbf{B}\right). \qquad (7.3.6.3.1)'$$

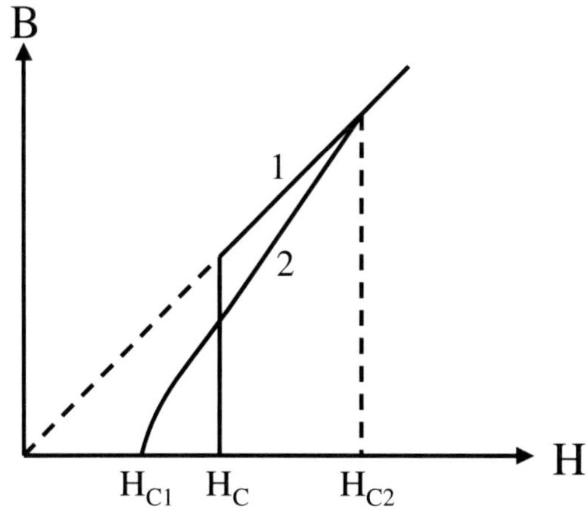

Figure 7.11: The schematic plot of magnetic induction B vs. applied magnetic H for type-I superconductor (curve 1) and type-II superconductor (curve 2).

If $\mathbf{B}$=0, then Eq. (7.3.6.3.1)′ is called the first equation of phenomenological London equation set (1935), which means the acceleration motion of superconducting carrier in electric field.

Taking curl of Eq. (7.3.6.2.6) yields

$$\nabla \times \mathbf{J} = \frac{(2e)^2 \rho}{\mu c} \mathbf{B}, \tag{7.3.6.3.2}$$

which is called the second equation of phenomenological London equation set (1935), which means Meissner effect. The second London equation, when combined with the Maxwell equation $\nabla \times \mathbf{B} = 4\pi \mathbf{J}/c$, leads to

$$\nabla \times (\nabla \times \mathbf{B}) + \lambda_L^{-2} \mathbf{B} = 0, \tag{7.3.6.3.3}$$

$$\lambda_L^2 = \frac{c^2 m^{**}}{4\pi n_s |e^*|^2}. \tag{7.3.6.3.4}$$

Consider a semi-infinite superconducting slab occupying the region $x > 0$ and an applied magnetic field lying parallel to the pseudogapbody with induced macroscopic superconductivity, i.e., $\mathbf{H} \parallel \mathbf{z}$, which we may write as $\mathbf{H} = H(x)\mathbf{z}$. The $\mathbf{H}$ satisfies $\nabla \cdot \mathbf{H} = 0$. Using the vector identity

$$\nabla \times (\nabla \times \mathbf{B}) = \nabla(\nabla \cdot \mathbf{B}) - \nabla^2 \mathbf{B}, \tag{7.3.6.3.5}$$

Eq. (7.3.6.3.3) becomes

$$\frac{\partial^2 B_z(x)}{\partial x^2} = \frac{1}{\lambda_L^2} B_z(x). \tag{7.3.6.3.6}$$

The solution of Eq. (7.3.6.3.6) for pseudogapbody in induced macroscopic superconducting state ($x > 0$) is

$$\mathbf{B}(x) = B_z(0) e^{-x/\lambda_L} \hat{\mathbf{z}}, \tag{7.3.6.3.7}$$

$$B_z(0)\hat{\mathbf{z}} = \mathbf{H}. \tag{7.3.6.3.8}$$

Eq. (7.3.6.3.7) implies that a magnetic field is exponentially screened from the interior of pseudogapbody in macroscopic superconducting state with penetration depth λ_L (See Fig. 7.12.), i.e., the Meissner effect. $\lambda_L(T = 0)$ ranges from 500 to 10 000 $\mathring{A}$, depending on the material.

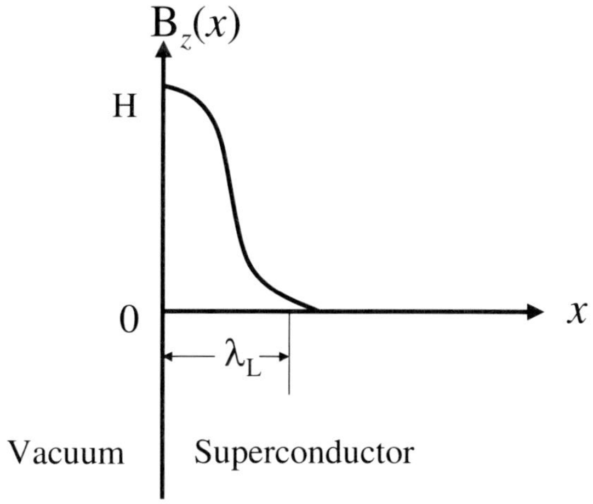

Figure 7.12: The schematic plot of $B_z(x)$ vs. x.

7.3.6.4. Room Temperature Applications of Induced Macroscopic Superconductivity

Because now there are many materials which are of microscopic superconductivity at room temperature, and thus these materials can be of induced macroscopic superconductivity at room temperature as well. Let us list some applications at room temperature.

(1). Quantization of magnetic flux.

Of course, one has method to guaranty that the trapped magnetic flux, Φ_A, is just confined in the region A of Fig. 7.9.

$$\Phi_A = \int \mathbf{B} \cdot d\mathbf{s}. \tag{7.3.6.4.1}$$

Considering $\mathbf{B}(\mathbf{r})$=$\nabla \times \mathbf{A}(\mathbf{r})$ and Stokes theorem yields

$$\Phi_A = \oint_L d\mathbf{l} \cdot \mathbf{A}(\mathbf{r}). \tag{7.3.6.4.2}$$

From Eq. (7.3.6.2.1) the wavefunction of Cooper pairs in induced macroscopic superconductivity is $\Psi(\mathbf{r}) = |\rho| \exp[i\theta]$. Due to the meissner effect the current density is zero on the integration contour L. Thus, Eq. (7.2.16)' gives

$$\mathbf{A}(\mathbf{r}) = \frac{ch}{2|e|} \nabla\theta. \tag{7.3.6.4.3}$$

Substituting Eq. (7.3.6.4.3) into Eq. (7.3.6.4.2) yields

$$\Phi_A = \frac{ch}{2|e|}(\theta_1 - \theta_2), \tag{7.3.6.4.4}$$

where $((\theta_1 - \theta_2)$ is a change of the phase in the round trip of one Cooper pair along the contour L. $((\theta 1 - \theta_2) = 2\pi n$ $(n = \cdots, -2, -1, 0, 1, 2, \cdots)$ due to the single-value property of a wavefunction. Hence, the trapped flux is quantized.

$$\Phi_A = n\Phi_\circ. \tag{7.3.6.4.5}$$

$$\Phi_\circ = \frac{ch}{2|e|} 2.07 \times 10^{-7} Gauss - cm^2. \tag{7.3.6.4.6}$$

The flux quantization has been observed in macroscopic superconducting state in 1961.

(2). Permanent current and strong magnet.

In the first point there is a confined magnetic flux in the region A of Fig. 7.9. If one suddenly takes off the trapped flux, then there is an induced macroscopic superconducting current, which is equivalent to produce a non-zero average velocity of all the Cooper pairs in the pseudogapbody ring at room temperature. Due to the free motion of every Cooper pair in the Cooper pairs without the long-range phase coherence, the average velocity will be kept. Therefore, the permanent current will be formed in the pseudogapbody ring. The estimation value of v_c is $v_c \approx 10^6$ cm/second. Due to the high critical velocity, the ring can have high average current density, and thus the ring or coil can produce strong magnetic field, and can be used as permanent strong magnet at room temperature.

7.4 Spontaneous Symmetry Breaking

7.4.1 Introduction

In section 7.4.2 we will demonstrate that the second order phase transformation occurs at T^*. Generally speaking, a second-order phase transition is characterized by spontaneous symmetry breaking. Spontaneous symmetry breaking is a spontaneous process by which a system in a high symmetrical state ends up in a low asymmetrical state.

Spontaneous symmetry breaking in condensed matter physics:

The picture of the second order phase transitions is as follows. There are also two phases, differing by a certain symmetry. One phase possesses this symmetry, and in the other phase, this symmetry is broken. For instance, for the ferromagnetic phase transition at $T < T^*$, that is the spontaneous rotational symmetry breaking: It exists in the non-magnetic phase, in which phase there is of a rotational symmetry, but is broken in the magnetic phase. For the pyroelectric transition, this is inversion symmetry that is broken in the pyroelectric phase. For structural phase transitions, the symmetry is encoded in the symmetry group of the crystal. For microscopically superconducting phase transition, the gauge invariance ($U(1)$ symmetry) is broken in the macroscopically superconducting phase. $U(1)$ transformation means: $\psi \rightarrow \exp(-i\theta)\psi$ and $\psi^+ \rightarrow \exp(i\theta)\psi^+$. If θ does not depend on $(\mathbf{r}, t)$, then one calls it a global transformation. If θ depends on $(\mathbf{r}, t)$, then we would be dealing with the so-called local transformation. For a body being of microscopic superconductivity, its Lagrangian density and the equation of motion of Ψ do not change under the $U(1)$ (gauge) transformation.

Second order phase transition is a spontaneously symmetry breaking. Mathematically, this is described by the symmetry groups of the phases involved. If the low symmetry phase is described by the symmetry group $\mathbf{L}$, and the high symmetry phase by the symmetry group $\mathbf{H}$, $\mathbf{L}$ has to be a subgroup of $\mathbf{H}$. If this is not the case, second order phase transition is impossible, and the transition almost always is of the first order. Because of this, the two phases do not coexist at any temperature except for T_{charac}. What happens we describe first with an example of the ferromagnetic phase transition, where the low-symmetry-ferromagnetic-phase is characterized by the magnetization. The magnetization decreases with the temperature and at the temperature T_{charac} turns to zero. Thus, at the transition temperature the ferromagnetic phase becomes paramagnetic above the transition temperature, the magnetic moment equals zero. The low temperature phase transforms into the high temperature phase continuously. This continuous transformation can be generalized to any species of the second order phase transition. One defines second order phase transitions: two phases with different symmetry.

Spontaneous symmetry breaking examples: Structure phase transitions: Space symmetry group; Pyroelectricity: Inversion symmetry; Magnetism: Rotational symmetry; Superconductivity: Gauge invariance (U(1) symmetry).

To describe the second order phase transition, the order parameter is introduced. This is a quantity that characterizes the symmetry of the phases. The order parameter is zero in the high symmetry phase, finite in the low symmetry phase, and as the function of temperature turns to zero at the transition point, to account for the continuous nature of the transition. Since the order parameter is continuous at the transition and is zero above the transition, its derivative must be discontinuous. This discontinuity is responsible for the singularities in the thermodynamic properties. Our definition of the order parameter is very general. Indeed, it can be anything. For the structural phase transition, the order parameter is an atomic displacement, sometimes a scalar, sometimes a vector. For the ferromagnetic (Curie) transition the magnetization as an order parameter is a vector. For the microscopically superconducting phase transition, the order parameter is the pseudogap function which is equal to the square of absolute value of wavefunction of Cooper pair. There are transitions, for instance, in liquid crystals, where the order parameter is a tensor, or in spin glasses even a functional.

For the pseudogapbody (i. e., a body which is of microscopic superconductivity at $T_c < T < T^*$, and in it there are Cooper pairs making free and random motion. The more general definition of pseudogapbody should be a body which is of pseudogap.), next we make some phenomenological discussions on the order parameter.

Ginzburg and Landau (1950) introduced phenomenologically for the macroscopic superconductivity a complex pseudo-wavefunction (or effective wavefunction) ψ, and took $|\psi|^2$ as order parameter to describe the second order phase transition at T_c for superconductor [75]. We find that the Ginzburg and Landau's work can also be used to pseudogapbody being of microscopic superconductivity. We do not make normalization for ψ. Thus, the local number density of Cooper pairs in pseudogapbody can be

$$n_s(\mathbf{r}) = |\psi(\mathbf{r})|^2. \tag{7.4.1.1}$$

Since the pseudo-wavefunction is complex, only the form $\psi\psi^*$ may enter the expansion of free energy F. We therefore write in cases of $T \lesssim T^*$, applied field H = 0, and infinite superconductor, i.e., no boundary effect,

$$F = F_n(T) + \int d\mathbf{r} \left\{ \alpha|\psi|^2 + \frac{1}{2}\beta|\psi|^4 \right\}, \tag{7.4.1.2}$$

where $F_n(T)$ is the free energy of the normal state in the absence of a magnetic field. In Eq. (7.4.1.2) there are no terms linear and cubic in ψ. These terms contain an arbitrary phase of ψ, and should be excluded because F is physically observable. Thermodynamic stability of microscopic superconducting state requires $\beta > 0$ (otherwise the system would seek a divergently large $|\psi|$). However, $\alpha(T)$ may have either sign. If $\alpha(T) > 0$, then the minimum of Eq. (1.22) occurs at $|\psi| = 0$. If $\alpha(T) < 0$, then the minimum is at $\psi \neq 0$. Hence we can model a second order phase transition simply by arranging for the sign of $\alpha(T)$ to change at T^*, which is easily done by writing $\alpha(T) = d(T - T^*)$. Thermodynamic stability of microscopic superconducting state requires $d > 0$. It is sufficient to regard β as a constant independent of temperature. Setting $\partial F/\partial|\psi| = 0$ yields

$$|\psi| = 0, \qquad T > T^*, \tag{7.4.1.3}$$

$$|\psi| = \left(-\frac{\alpha}{\beta}\right)^{\frac{1}{2}}, \qquad T \lesssim T^*, \tag{7.4.1.4}$$

i.e., in this case $n_s = -\alpha/\beta$. $|\psi|$ is independent of space, which describes a homogeneous microscopic superconducting state. In cases of that the applied field is not equal to zero and/or there is a boundary effect, the F should contain kinetic term, and ψ depends on space.

$$\begin{aligned} F &\equiv \int d\mathbf{r} f \equiv \int d\mathbf{r}(f_n + f_p) \\ &= F_n(T) + \int d\mathbf{r} \left\{ \alpha|\psi(\mathbf{r})|^2 + \frac{\beta}{2}|\psi(\mathbf{r})|^4 \right. \\ &+ \left. \frac{1}{2m^{**}} \left| \left[\hbar\nabla - \frac{ie^*}{c}\mathbf{A}(\mathbf{r}) \right] \psi(\mathbf{r}) \right|^2 + \frac{B^2(\mathbf{r})}{8\pi} \right\}, \end{aligned} \tag{7.4.1.5}$$

where m^{**} and e^* are the effective mass and charge of one Cooper pair. $e^* = 2e$. $m^{**} = 2m^*$. Varying $\psi^*(\mathbf{r})$ by $\delta\psi^*(\mathbf{r})$ and integrating by parts, we obtain

$$\begin{aligned} \delta F &= \int d\mathbf{r}\delta\psi^*(\mathbf{r}) \left\{ \alpha\psi(\mathbf{r}) + \beta|\psi(\mathbf{r})|^2\psi(\mathbf{r}) - \frac{1}{2m^{**}} \left[\hbar\nabla - i\frac{e^*}{c}\mathbf{A}(\mathbf{r}) \right]^2 \psi(\mathbf{r}) \right\} \\ &+ \frac{1}{2m^{**}} \int\int \delta\psi^*(\mathbf{r})d\mathbf{s} \cdot \left[\hbar\nabla - \frac{ie^*}{c}\mathbf{A}(\mathbf{r}) \right] \psi(\mathbf{r}), \end{aligned} \tag{7.4.1.6}$$

where the second integral is over the surface of the superconductor. In equilibrium, $\delta F = 0$ and both integrals should vanish for any $\delta\psi^*(\mathbf{r})$. As a result, we obtain an equation

$$-\frac{1}{2m^{**}} \left[-i\hbar\nabla - \frac{e^*}{\hbar c}\mathbf{A}(\mathbf{r}) \right]^2 \psi(\mathbf{r}) + \alpha\psi(\mathbf{r}) + \beta|\psi(\mathbf{r})|^2\psi(\mathbf{r}) = 0, \tag{7.4.1.7}$$

which is called the first equation of GL equation set. (Note that, in AUNRTB GL equation set is an equation set for macroscopic superconductivity.) The equation obtained from the surface integral in Eq. (7.4.1.6) corresponds to the boundary condition at the surface, $\mathbf{n} \cdot [\hbar\nabla - ie^*/c\mathbf{A}(\mathbf{r})]\psi(\mathbf{r}) = 0$, where $\mathbf{n}$ is unit vector normal to the surface.

Varying $\mathbf{A}(\mathbf{r})$ by $\delta\mathbf{A}(\mathbf{r})$ in the free energy, Eq. (7.4.1.5), and setting $\delta F = 0$, we obtain the Maxwell equation $\nabla \times \nabla \times \mathbf{A}(\mathbf{r}) = (4\pi/c)\mathbf{J}_s(\mathbf{r})$, and the current density $\mathbf{J}_s(\mathbf{r})$ is

$$\mathbf{J}_s(\mathbf{r}) = \frac{e^*\hbar}{i2m^{**}}[\psi^*(\mathbf{r})\nabla\psi(\mathbf{r}) - \psi(\mathbf{r})\nabla\psi^*(\mathbf{r})] - \frac{e^{*2}}{m^{**}c}|\psi(\mathbf{r})|^2\mathbf{A}(\mathbf{r}), \tag{7.4.1.8}$$

which is called the second equation of GL equation set. Using above boundary condition at the surface and the current density equation, Eq. (7.4.1.8), we can obtain $\mathbf{n} \cdot \mathbf{J}_s(\mathbf{r}) = 0$, i.e., the component of the current density should be zero on the surface.

It is easy to verify that the free energy density in Eq. (7.4.1.5) and equation of motion Eq. (7.4.1.7) do not change under the $U(1)$ gauge transformation:

$$\psi \to \psi' = e^{i\varphi}\psi, \quad \mathbf{A} \to \mathbf{A}' = \mathbf{A} + \frac{\hbar}{2e}\nabla\varphi, \tag{7.4.1.9}$$

where φ is a scalar function of $\mathbf{r}$. It is easy to prove that

$$(-i\hbar\nabla - 2e\mathbf{A}')\psi' = e^{i\varphi}(-i\hbar\nabla - 2e\mathbf{A})\psi. \tag{7.4.1.10}$$

From Eq. (7.4.1.10) we know that the Eqs. (7.4.1.5) and (7.4.1.7) are of $U(1)$ gauge invariance.

The existence of $U(1)$ gauge invariance in pseudogapbody indicates that the gauge field (here it is electromagnetic field $\mathbf{A}$) and the phase field ϕ degree of freedom of the order parameter in the pseudogapbody $\psi = \exp(i\phi)|\psi|$ can transform each other. The phase field ϕ is called Goldstone boson field. Of course, we can choose $\varphi \equiv \phi$. In this choice, the free energy density in Eq. (7.4.1.5) becomes

$$f = f_n + \frac{\hbar^2}{2m^{**}}(\nabla|\psi|)^2 + \frac{m_A^2}{2}\mathbf{A}^2 + \alpha|\psi|^2 + \frac{\beta}{2}|\psi|^4 + \frac{\mathbf{H}^2}{2\mu_\circ}, \tag{7.4.1.11}$$

$$m_A^2 = \frac{4e^4}{m^{**}}|\psi|^2. \tag{7.4.1.12}$$

In Eq. (7.4.1.11) we use the the SI system. SI system is also known as MKSA (for meter, kilogram, second, ampere) [14, 110]. The gauge $\varphi \to \phi$ is called unitary gauge. In unitary gauge, the phase field ϕ of ψ is absorbed completely by gauge field $\mathbf{A}$, and thus does not appear in the expression of free energy density or equation of motion. However, we find that the corresponding cost is that there is a term $m_A^2/2\mathbf{A}^2$ in Eq. (7.4.1.11). What does this term mean? Due to spontaneous gauge symmetry breaking in microscopic superconducting state $|\psi|^2 > 0$, $|m_A^2| > 0$, and thus the gauge field $\mathbf{A}$ obtains an effective mass m_A formally and apparently. This is the so called Anderson-Higgs mechanism in macroscopic superconducting state [109, 110]. Here we proved that Anderson-Higgs mechanism holds in case of microscopic superconducting state as well (For more discussions, please see the eight point in the section 9.7 of chapter 9).

Utilizing the formula $\mathbf{H} = \nabla \times \mathbf{A}$ and the variation of Eq. (7.4.1.11) over A, one obtains that the gauge field satisfies

$$\nabla \times (\nabla \times \mathbf{A}) + 2\mu_\circ m_A^2\mathbf{A} = 0. \tag{7.4.1.13}$$

Further, using the equality

$$\nabla \times (\nabla \times \mathbf{H}) = -\nabla^2\mathbf{H}, \tag{7.4.1.14}$$

one obtains

$$\nabla^2\mathbf{H} = 2\mu_\circ m_A^2\mathbf{H}. \tag{7.4.1.15}$$

From Eq. (7.4.1.15), in the simple case,

$$\mathbf{H}(x) = H_z(0)e^{-x/\lambda_L}\hat{\mathbf{z}}. \tag{7.3.6.3.7$'$}$$

The $H \sim x$ is shown in Fig. 7.11. This antimagnetic phenomenon is called Meissner effect. In Eq. (7.3.6.7)$'$ the penetration is

$$\frac{1}{\lambda_L^2} = 2\mu_\circ m_A^2. \tag{7.4.1.16}$$

If we make comparison of Eq. (7.4.1.16) with Eq. (7.3.6.3.4), then we obtain

$$n_S = 4|\psi|^2. \tag{7.4.1.17}$$

The Eq. (7.4.1.11) shows that under the unitary gauge, the phase field φ (or, generally speaking, Goldstone boson field) of the order parameter ψ is absorbed by the electromagnetic field $\mathbf{A}$ (or, generally speaking, gauge field) completely, and thus phase field ϕ does not have influence on the free energy. It is looks like formally that the pseudogapbody reduces a degree of freedom ϕ. Actually, it is not. The massless $U(1)$ gauge field (here, it is the electromagnetic field $\mathbf{A}$) is a transverse field, has no longitudinal component, and thus has only two components. But, after absorbing the Goldstone bosons from the phase field (Goldstone boson field) the gauge field gets mass formally, and thus its longitudinal component is not zero, and has three degrees of freedom. Therefore, the total degrees of freedom do not reduce, and just appears in different form. Just from this aspect, we can say that Meissner effect is a consequence of spontaneous $U(1)$ symmetry breaking, and is a manifestation of Anderson-Higgs mechanism. However, the physical origin of producing the Meissner effect is as follows. Under condition of the applied magnetic field, the phase field ϕ of the order parameter ψ will vary in position space, induces a macroscopic superconducting current, which gives an opposite magnetic field to screen the applied magnetic field. Equivalently, the gauge field (the magnetic field) gets mass, and thus the propagation of the gauge field (magnetic field) in space becomes shorter range. However, we have to remember that this acquirement of mass is only apparent.

7.4.2 Second Order Phase Transition at T^*

The characteristic of second order phase transition is that the entropy is continuous at T^*, but the derivative of entropy at T^* is discontinuous. We prove that the specific heat, which is the first order derivative of entropy, jump occurrs at T^* for both the unconventional and the conventional superconductors. Let us study, as an example, the unconventional superconductors.

From section 7.2 we know that in the unconventional superconductors there is a quasiparticle excitation with energy $E_{\mathbf{k}} = \sqrt{\xi_{\mathbf{k}}^2 + \Delta_{\mathbf{k}}^2(T)}$. For simplicity, in the following we assume that $\Delta_{\mathbf{k}}(T)$ is independent of $\mathbf{k}$. Of course, in ordinary case the quasiparticles satisfy the free Fermion distribution $f(E_{\mathbf{k}}) = 1/[\exp(\beta E_{\mathbf{k}})+1]$. However, now we just consider the Cooper pairs without the long-range phase coherence at $T > T_c$. Thus we consider just these Cooper pair's excitation through the gap at $T > T_c$. The entropy of this ideal Fermion gas in pseudogap state is

$$S_{es} = -2k_B \sum_{\mathbf{k}} \{[1 - f(E_{\mathbf{k}})] \ln [1 - f(E_{\mathbf{k}})] + f(E_{\mathbf{k}}) \ln f(E_{\mathbf{k}})\} . \tag{7.4.2.1}$$

The specific heat can be written as

$$C_{es} = T\frac{dS_{es}}{dT} = -\beta\frac{dS_{es}}{d\beta}. \tag{7.4.2.2}$$

Substituting Eq. (7.4.2.1) into Eq. (7.4.2.2) yields

$$C_{es} = 2\beta k_B \sum_{\mathbf{k}} -\frac{\partial f(E_{\mathbf{k}})}{\partial E_{\mathbf{k}}} \left(E_{\mathbf{k}}^2 + \frac{1}{2}\beta\frac{d\Delta^2}{d\beta} \right). \tag{7.4.2.3}$$

At $T = T^*$, the $\Delta(T) \to 0$, and thus the first term of Eq. (7.4.2.3) reduces to the electronic specific heat of usual normal state

$$C_{en} = \frac{2\pi^2}{3} N(E_F) k_B^2 T. \tag{7.4.2.4}$$

The second term is finite just below T^*, where $d\Delta^2/dT$ is large, but it is zero above T^*, giving rise to a discontinuity of the quasiparticle specific heat at T^*. The size of the discontinuity is

$$(C_{es} - C_{en})|_{T=T^*} = N(E_F) \left(-\frac{d\Delta^2(T)}{dT} \right)\bigg|_{T=T^*} . \tag{7.4.2.5}$$

From the solution of BCS gap equation, we obtain:

$$\Delta(T) = 3.06 T^* \sqrt{1 - \frac{T}{T^*}}. \tag{7.4.2.6}$$

Substituting Eq. (7.4.2.6) into Eq. (7.4.2.5) yields

$$(C_{es} - C_{en})|_{T=T^*} = (3.06k_B)^2 N(E_F)T^*. \tag{7.4.2.7}$$

Combining Eqs. (7.4.2.4) and (7.4.2.7), we have

$$\frac{(C_{es} - C_{en})|_{T=T^*}}{C_{en}|_{T=T^*}} = \frac{(3.06)^2}{2\pi^2/3} = 1.43. \tag{7.4.2.8}$$

AUNRTB think that the specific heat jump given by Eq. (7.4.2.8) takes place just at T_c, related to the formation of Cooper pairs in macroscopically superconducting state, and that there is no specific heat at T^*, related to the formation of Cooper pairs in microscopically superconducting state. Even when people discovered the specific heat jump at $T > T_c$ for the high-T_c cuprates, which belong to one kind of the unconventional superconductors, they always attributed it to some causes other than that related to the formulation of Cooper pair without the long-range phase coherence, or just simply neglected it.

For the unconventional superconductors, such as the high-T_c cuprates, at $T = T_c < T^*$ the long-range phase difference locking among Cooper pairs begins to appear, i.e., the macroscopically supercurrent Cooper pairs begin to appear. Therefore, there is another phase transition at $T = T_c$, and the specific heat jump at T_c for the high-T_c cuprates is observed in experiments.

7.4.3 Spontaneous Symmetry Breaking at T^*

From Euler-lagrange equation, and the following Lagrangian density,

$$\mathcal{L}(\psi, \psi^+) = i\hbar\psi^+ \frac{\partial}{\partial t}\psi + \psi^+ \left(\frac{\hbar^2\nabla^2}{2m} + \mu\right)\psi + \frac{g}{2}\psi^+\psi_\beta^+\psi_\beta\psi, \tag{7.4.3.1}$$

we can obtain the non-relativistic Schrödinger field equation Eq. (7.4.3.2). In Eq. (7.4.3.1), $g > 0$ represents the attractive potential between two carriers, and $\psi = \psi_\beta$, $\beta = 1, 2$ represent the spins of carriers $\uparrow, \downarrow$, respectively. g leads to the formulation of Cooper pair, i. e., leads to what is called a "condensate" in Cooper pairs. All Cooper pairs are in ground state with the same energy level. One calls all Cooper pairs vividly "condensate".

The non-relativistic Schrödinger field equation is:

$$i\hbar\frac{\partial}{\partial t}\psi_\alpha(x) + \left(\frac{\hbar^2\nabla^2}{2m} + \mu\right)\psi_\alpha(x) + g\psi_\beta^+(x)\psi_\beta(x)\psi_\alpha(x) = 0. \tag{7.4.3.2}$$

The Lagrangian density Eq. (7.4.3.1) and motion equation Eq. (7.4.3.2) are of $U(1)$ gauge invariance.

BCS pairing approximation of the previous section will correspond here to replacing the bilinear field operators $\psi_\beta\psi_\alpha$ by its BCS vacuum expectation value:

$$g\psi_\beta(x)\psi_\alpha(x) \quad \rightarrow \quad g\{_{BCS}\!< 0|\psi_\beta(x)\psi_\alpha(x)|0 >_{BCS}\} \equiv M\epsilon_{\alpha\beta}, \tag{7.4.3.3}$$

where $\epsilon_{\alpha\beta}$ is an antisymmetric tensor to reflect the fact that $\psi_\alpha(x)$ and $\psi_\beta(x)$ anticommute. Since one can write $\psi(x) = \exp(-iPx)\psi(0)\exp(iPx)$, using the translational operator P, and since vacuum is invariant under translation, M is a constant. M is called the order parameter, because its definition tells us that it is proportional to the amplitude of Cooper pair. M is a complex number generally. [Here, we would like to provide a little more knowledges on M. The Eq. (2.166) in Ref. [75] shows clearly that $M \equiv$ *energy function* , *i.e.*, $\Delta(\mathbf{r})$. The Eq. (2.229) in Ref. [75] shows clearly that $M \propto \psi(\mathbf{r})$ in Eq. (7.4.1.1). In superconductivity physics, people call both the M and the $\psi(\mathbf{r})$ the order parameter.]

Under BCS pairing approximation, Eq. (7.4.3.2) becomes

$$i\hbar\frac{\partial}{\partial t}\psi_\alpha(x) + \left(\frac{\hbar^2\nabla^2}{2m} + \mu\right)\psi_\alpha(x) + M\epsilon_{\alpha\beta}\psi_\beta^+(x) = 0. \tag{7.4.3.4}$$

Since M is just a number and not an operator, it cannot be affected by the $U(1)$ transformation, and therefore, Eq. (7.4.3.4) is no longer invariant under the $U(1)$ transformation. The culprit is clearly Eq. (7.4.3.3) where we replaced a composite operator by its vacuum expectation value. Here, we would like to say a little more about the physics. Because M in Eq. (7.4.3.3) is related to the Cooper pair, $M \neq 0$ means that energy gap and the free Cooper pair appear in the system. Or, equivalently speaking, the condensate appears in the ground state (=vacuum state). In this case if the $U(1)$ gauge symmetry is broken, then we do not feel strange.

If we examine Eq. (7.4.3.3), we find that if the vacuum is assumed to be invariant under the $U(1)$ transformation, then we must have

$$_{BCS} < 0|\psi_\beta(x)\psi_\alpha(x)|0 >_{BCS} = e^{-i2\theta} \; _{BCS} < 0|\psi_\beta(x)\psi_\alpha(x)|0 >_{BCS} . \tag{7.4.3.6}$$

Since θ is an arbitrary variable, $_{BCS} < 0|\psi_\beta(x)\psi_\alpha(x)|0 >_{BCS} = 0$. The only alternative available to us is to assume that vacuum does not respect the $U(1)$ symmetry. Thus the vacuum must break the $U(1)$ symmetry. This type of symmetry breaking of the condensate in the pseudogapbody which is of he microscopic superconductivity is called spontaneous symmetry breaking where the Lagrangian density satisfies the symmetry but not the ground state, i. e., not the BCS vacuum (=ground) state $|0 >_{BCS}$.

Let us consider the relation between the BCS and ordinary vacuum state in Eq. (7.2.3.16)'

$$|0 >_{BCS} = \prod_{\mathbf{k}}(u_{\mathbf{k}} + v_{\mathbf{k}}b_{\mathbf{k}}^+)|0 > \tag*{(7.2.3.16)$'$}$$

By implementing the $U(1)$ transformation for the creation operators we find that

$$U(\theta)|0 >_{BCS} = \prod_{\mathbf{k}}(u_{\mathbf{k}} + v_{\mathbf{k}}e^{-2i\theta}b_{\mathbf{k}}^+)|0 >, \tag{7.4.3.7}$$

where $U(\theta)$ designates the $U(1)$ transformation. Clearly, the BCS vacuum state is not invariant under the $U(\theta)$ transformation. If we write

$$|\theta > = U(\theta)|0 >_{BCS}, \tag{7.4.3.8}$$

then $|\theta >$ is a different BCS vacuum state. Furthermore,

$$U(\theta')|\theta >_{BCS} = |\theta + \theta' > . \tag{7.4.3.9}$$

Thus, since θ' is continuous, we have an infinite number of states that are all BCS vacuum states, and they are degenerate.

This example illustrates the fact that the presence of condensate leads to the spontaneous symmetry breaking, and vice versa. Since spontaneous symmetry breaking occurs due to the formation of condensates, which are dynamical entities, the spontaneous symmetry breaking such as that found in microscopic superconductivity is called dynamical spontaneous symmetry breaking.

AUNRTB think that the order parameter in the second order phase transition is a long range order parameter. Or, equivalently speaking, AUNRTB think that there is long-range phase coherence between any Cooper pairs. On the contrary, we have proven that the Cooper pairs in microscopic superconductivity is not of the long-range phase coherence. Therefore, the order parameter, which is proportional to the number of Cooper pairs can be short-range order parameter as well. Actually, Eq. (7.2.3.21) has indicated that the every Cooper pair can have own phase.

7.5 Exercises and Solutions

(1). There are many systems in which there are the observed pseudogaps in the quasiparticle excitation spectrums: heavy-Fermion system, some organic materials, many antiferromagnetic materials, Kondo semiconductors, colossal magnetoresistive oxides [75]. Please argue that, at least, these systems are of microscopic superconductivity certainly.

Solution:

The Landau criterion of microscopic superconductivity comes from exact derivation. If there are pseudogap in the quasiparticle excitation spectrum, then the corresponding materials have to be of microscopic superconductivity.

(2). Prove that Eq. (7.3.6.2.9) is the motion equation in electromagnetic field not only for the Cooper pairs with induced macroscopic superconductivity but also for the Cooper pairs with macroscopic superconductivity and for the Cooper pairs just with microscopic superconductivity.

Solution:

(3). Design an experiment to detect the microscopic superconductivity.

Solution:

Suppose that there is no the trapped flux in Fig. 7.9, which is made from the pseudogap-body with room temperature microscopic superconductivity. Let the ring turn. Considering that the Cooper pairs in the ring are free and random, the Cooper pairs do not follow the turning movement of the ring, and thus an average current, in opposite direction of the ring movement, coming from the contributions of all the Cooper pairs will be produced. This phenomenon can occur at room temperature.

This experiment will become an effective method to measure the number density of Cooper pairs.

(4). Give the maximum value of T^* and corresponding value of x (the definitions are given in subsection 7.2.3) from Fig. 7.13 (The data are from [75].), and explain the physical meaning of T^*.

Solution:

The solid line is a solution of BCS gap equation at zero gap, and clearly shows the relation between T^* and x for $Bi2212$. Of course, using the same method given in Ref. [75] one can obtain the relations between T^* and x for Hg2213, Y123, and La214. Fig. 7.13 clearly shows that the best $Bi2212$ used as room-temperature-pseudogapbody is $x = 0.04$ because in this case T^* takes maximum value 372 K.

(5). Give the values of the pseudogaps at 300 K for $x = 0.056$ and 0.025 from Fig. 7.8.

Solution:

$\Delta(300\ \ K) = 25$ meV for $x = 0.056$. $\Delta(300\ \ K) = 70$ meV for $x = 0.025$.

(6). In subsection 7.3.4 we demonstrate that in comparison with the resistance of Cu lead, the effective resistance at room temperature of pseudogapbody is negligible. Please argue that besides the negligible resistance, the pseudogapbody as lead or cable has another advantage in comparison with the conventional conductors, i. e., the power provided by

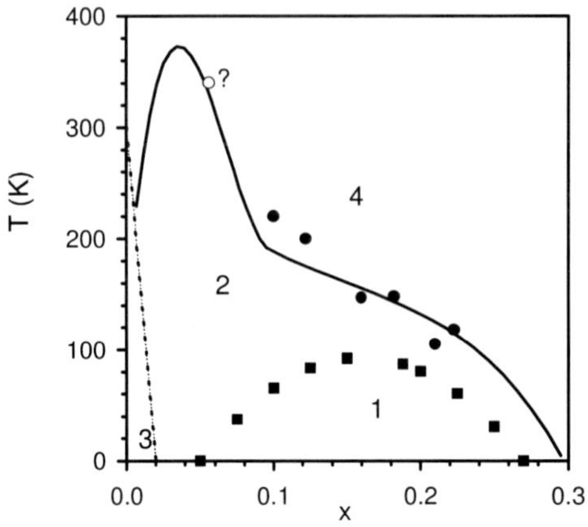

Figure 7.13: Phase diagram. The states in ranges 1, 2, 3, and 4 are macroscopic superconducting state, pseudogap state, antiferromagnetic insulating state, and normal state, respectively. The data of T^* vs. x are for $Bi2212$. The solid squares indicate the experimental values of T_c vs. x. The solid line is the theoretical curve of T^* vs. x.

applied dc voltage on the pseudogapbody does not transform to Jour heat energy.

Solution:
When a pseudogapbody is in the microscopic superconductive state, the power is used only for acceleration of freely random Cooper pair.

(7). Refs. [110, 111] give a formula to determine the transition temperature T_c of macroscopic superconductivity. They think that: the more larger the phase fluctuation energy $V_\circ$ is, the higher the value of T_c is, and from this idea obtain the value of T_c is determined by the condition $k_B T_c = V_\circ$. From this condition, Refs. [110, 111] obtain that the value of T_c is proportional to the supercurrent Cooper pair density $n_s(T)$ at 0 K $n_s(0\ K)$. Their numerical results for the high-T_c superconductor are: $T_{c,theory} \approx (30 \to 100)T_{c,experiment}$. Please give comments for theirs theory and analyze the reasons of significant difference between theoretical calculations and experimental date. Even you have not yet known many about macroscopic superconductivity.

Solution:
The theories in Refs. [110, 111] look just like OK formally. Actually, these theories are not correct. Our reasons are:

(A). The supercurrent Cooper pair number at 0 K is only one factor to benefit the appearance of macroscopic superconductivity at $T = T_c > 0$ K. Refs. [110, 111] just consider this one factor. Actually, we still need to consider other factors. We should consider the situation in the $T_c > 0$ K. At $T \lesssim T_c$ many factors can have actions on the value of T_c. Such as, the resistance can make the coupling between Cooper pairs weaker, different materials have different dependence of superconducting Cooper pair on the temperature. Therefore, the reliable method to determine the value of T_c should be $n_s(T_c) = 0$. The idea to determine the value on T^* in this chapter is $\Delta(T = T^*) = 0$. If you have interest for the correct method to determine the value of T_c, then please look at Eq. (3.124) in Ref. [75].

(B). It is not strange that theirs theoretical numerical results are much larger than the experimental data because their theories just consider a non important factor, and completely neglect the most important factors.

(8).

(A). Use the Eqs. (7.4.1.16) and (7.4.1.12) to estimate the formally equivalent dynamical mass of the photon in the Meissner effect at 0 K. ($n_S(T = 0\ K) = 10^{27}/M^3$.)

(B). Use the Eq. (7.4.1.16) and experimental value at 0 K $\lambda_L = 4000 \times 10^{-8}$ cm to estimate the value of M_A.

(C). Make comparison of the two values.

(D). Do you belief that the photon in Meissner effect really obtain mass?

Solution:

(A). The dimension of M_A is 1/second. $M_A = 0.37 \times 10^{10}/\text{second}, \quad 1.5 \times 10^{-5}$ eV, 2.7×10^{-39} g. $\lambda_L(0\ K) = 2000 \times 10^{-8}$ cm.

(B). $M_A = 0.37/2 \times 10^{10}/\text{second}, \ 1.5/2 \times 10^{-5}$ eV, $2.7/2 \times 10^{-39}$ g.

(C). Nearly, the same.

(D). Although the acquirement of mass of the photon is very small, but we still absolutely do not belief that in Meissner effect the photon obtains mass. Our reason is simple. Experimentalists already many times observed the Meissner antimagnetic supercurrent in a thin layer of the surface. From this example, we infer that the so-called Higgs mechanism, which says that the gauge field particle masses come from the Higgs bosons, is a false appearance, or, at last, is a phenomenological mechanism. One should go further into the mechanism of the acquisition of the mass of gauge field particles.

(9). Please find more possible applications of the microscopic superconductivity at room temperature from Refs. [112, 113].

Chapter 8

Relativistic Quantum Mechanics

8.1 Introduction

Before this chapter we only proved the nonrelativistic quantum mechanics by many demonstrations. For example, we proved

$$E = \frac{p^2}{2m} + V. \tag{8.1.1}$$

$$E \to i\hbar\frac{\partial}{\partial t}, \quad \mathbf{p} \to \ - i\hbar\nabla. \tag{8.1.2}$$

$$i\hbar\frac{\partial}{\partial t}\Psi = \left[-\frac{\hbar^2}{2m}\triangle + V\right]\Psi. \tag{8.1.3}$$

Eq. (8.1.3) is the Schrödinger equation, which is a nonrelativistic single-particle equation.

Now we would like to extend our wavepacket-only theory to the relativistic domain. The natural thing is to search a relativistic single-particle wave equation to replace the Schrödinger equation. It turns out that the form of the relativistic equation can be proved to depend on the spin of the particle,

spin-0: Klein-Gordon equation

spin-1/2: Dirac equation

massive spin-1: Proca equation etc

You will find that for the zero spin particle the canonical quantization method and basic concepts are same as that in nonrelativistic quantum mechanics except that when we write the Hamiltonian, we use the special theory of relativity other than Newton's mechanics. The relativistic quantummechanical equation of spin zero particle is called Klein-Gorden equation.

As to the spin 1/2 particle, let us at first tell the interesting historical story.

The first direct experimental evidence of the electron spin was the Stern-Gerlach experiment of 1922. The electron spin was discovered in the context of the emission spectrum of alkali metals. In 1924 Pauli introduced what he called a "two-valued quantum degree of freedom" associated with the electron in the outermost shell. This allowed him to formulate the Pauli exclusion principle, stating that no two electrons can share the same quantum state at the same time. The physical interpretation of Pauli's "degree of freedom" was initially unknown. Konig, one of Land's assistants, suggested in early 1925 that it was produced by the self-rotation of the electron. When Pauli heard about this idea, he criticized it severely, noting that the electron's hypothetical surface would have to be moving faster than the speed of light in order for it to rotate quickly enough to produce the necessary angular momentum. This would violate the special theory of relativity. Largely due to Pauli's criticism,

Kronig decided not to publish his idea. In the autumn of 1925, the same thought came to two Dutch physicists, Uhlenbeck and Goudsmit. Under the advice of Ehrenfest, they published their results. It met a favorable response, especially after Thomas managed to resolve a factor-of-two discrepancy between experimental results and Uhlenbeck and Goudsmit's calculations (and Kronig's unpublished ones). This discrepancy was due to the orientation of the electron's tangent frame, in addition to its position. Mathematically speaking, a fiber bundle description is needed. The tangent bundle effect is additive and relativistic; that is, it vanishes if c goes to infinity. It is one half of the value obtained without regard for the tangent space orientation, but with opposite sign. Thus the combined effect differs from the latter by a factor two (Thomas precession). The mathematical theory was worked out in depth by Pauli in 1927. Despite his initial objections, Pauli formalized the mathematical theory of spin in 1927, using the theory of quantum mechanics discovered by Schrödinger and Heisenberg. He pioneered the use of Pauli matrices as a representation of the spin operators, and introduced a two-component spinor wave-function. Pauli's theory of spin was nonrelativistic. However, in 1928, Dirac published the Dirac equation, which described the relativistic electron. In the Dirac equation, a four-component spinor (known as a "Dirac spinor") was used for the electron wave-function. In 1940, Pauli proved the spin-statistics theorem, which states that fermions have half-integer spin and bosons integer spin.

This chapter introduces the relativistic quantum theory for the spin 1/2 electron along the line given by our wavepacket-only theory other than the historical line. At first we derive the Dirac equation, based on the canonical quantization method and the concept of our wavepacket-only theory, essentially the same method as that of establishment of Klein-Gorden equation.

It is useful to study these single-particle equations (Klein-Gorden and Dirac) and their solutions for certain problems. However, at certain point these single particle relativistic quantum theory encounter fatal inconsistencies and break down. Essentially, this is because while energy is conserved in special relativity but mass is not. Particles with mass can be created and destroyed in real physical processes. For example, pair annihilation $e^+e^- \rightarrow 2\gamma$, muon decay $\mu^- \rightarrow e^-\nu_e^-\nu_\mu$. At that stage we are forced to abandon single-particle relativistic wave equations and go to quantum field theory, in which particles can be created and destroyed, which is the subject of chapter 9.

8.2 Related Recalls from Chapter 1

In Eq. (1.5.1) of chapter 1, the wavepacket $\Psi(x,t)$ representing a massive particle is of the form

$$\begin{aligned}\Psi(x,t) &= \mathrm{e}^{ik_{xo}x-i\omega(k_{xo})t} \times \mathrm{e}^{-ik_{xo}x'} \times \Psi(x',0) \\ &\equiv I \times II \times III,\end{aligned} \tag{8.2.1}$$

where I, II, III are called guide-factor (or, often say, wavefunction of a particle), front-factor, and body-factor, respectively. The motion of body-factor III is determined by the guide-factor I according to the spontaneous equiprobability symmetry breaking rule. Therefore, the guide-factor is an important factor. It is easy to verify that the guide-factor I satisfies the following differential equation.

$$\begin{aligned}i\hbar\frac{\partial}{\partial t}e^{ik_{xo}x-i\omega(k_{xo})t} &= \hbar\omega(k_{xo})e^{ik_{xo}x-i\omega(k_{xo})t} \\ &= \frac{\hbar^2k_{xo}^2}{2m}e^{ik_{xo}x-i\omega(k_{xo})t} \\ &= -\frac{\hbar^2}{2m}\frac{\partial^2}{\partial x^2}e^{ik_{xo}x-i\omega(k_{xo})t}.\end{aligned} \tag{8.2.2}$$

Eq. (8.2.2) can be written as

$$i\hbar\frac{\partial}{\partial t}\Psi(x,t) = -\frac{\hbar^2}{2m}\frac{\partial^2}{\partial x^2}\Psi(x,t). \tag{8.2.3}$$

Note that the meanings in $\Psi(x,t)$ in Eqs. (8.2.3) and (8.2.1) are different. The Ψ in Eq. (8.2.3) represents the quide-factor (wavefunction) other than the particle itself. Do not confuse the meanings of Ψ in different cases, because we use the same notation Ψ to represent both wavepacket and wavefunction.

Eq. (8.2.3) is an one dimensional time-dependent guide-factor equation of free wavepacket. This equation is called one dimensional time-dependent Schrödinger (wave) equation of free particle. Schrödinger equation is a requirement of wavepacket-only theory, and can be proved by wavepacket-only theory. The another method to "prove" Eq. (8.2.3) can be found in chapter 5 of Ref. [3].

In relativistic case

$$\begin{aligned} E^2 &= p_{x\circ}^2 c^2 + m_\circ^2 c^4 = (\hbar k_{x\circ})^2 c^2 + m_\circ^2 c^4 \\ &= \hbar^2\omega^2(k_{x\circ}) \neq \frac{\hbar^2 k_{x\circ}^2}{2m}, \end{aligned} \tag{8.2.4}$$

$$\omega^2(k_{x\circ}) = c^2(k_{x\circ}^2 + k_c^2), \tag{8.2.5}$$

where $k_c = m_\circ c/\hbar = 1/\lambda_c$, and λ_c is the Compton wavelength.

It is easy to see that although in relativistic case the guide-factor of a free wavepacket still is a plane wave, but this plane wave does not satisfy the Schrödinger equation. It turns that there are other two differential equations which are satisfied by the guide-factor. The two equations are called Klein-Gorden equation and Dirac equation.

8.3 Klein-Gorden Equation

In three dimensional case, the Schrödinger equation can be written as

$$i\hbar\frac{\partial}{\partial t}\Psi(\mathbf{x},t) = -\frac{\hbar^2}{2m}\nabla^2\Psi(\mathbf{x},t). \tag{8.3.1}$$

Due to the different orders of its temporal and spatial derivatives, this equation is not Lorentz-covariant (See the appendix). This means that, passing from one inertia system to another, the equation changes its structure, thus contradicting the principle of relativity.

In relativistic case, the wavefunction ($\equiv$ guide-factor) is

$$\Psi(\mathbf{x},t) = e^{i\mathbf{k}_\circ\cdot\mathbf{x} - i\omega(\mathbf{k}_\circ)t}, \tag{8.3.2}$$

where $\mathbf{x}$ represents the space vector. It is easy to verify that the wavefunction in Eq. (8.3.2) satisfies the following equation

$$i\hbar\frac{\partial}{\partial t}\Psi(\underline{x}) = \left(-c^2\hbar^2\nabla^2 + m_\circ^2 c^4\right)^{1/2}\Psi(\underline{x}), \quad \underline{x} = (x^\mu) = (x^\circ, x^1, x^2, x^3), \quad (x^\circ = ct). \tag{8.3.3}$$

However, this equation has two grave flaws. On the one hand, due to the unsymmetrical appearance of space and time derivatives, the relativistic invariance of this equation is not apparent. On the other hand, the operator on the right hand side is square root whose expansion leads to highly nonlocal theory.

It is easy to verify that due to $\hbar^2k^2c^2 + m_\circ^2c^4 = \hbar^2\omega^2(k) = E^2$ the wavefunction in Eq. (8.3.2) satisfies the following equation

$$-\hbar^2\frac{\partial^2}{\partial t^2}\Psi(\underline{x}) = \left(-c^2\hbar^2\nabla^2 + m_\circ^2c^4\right)\Psi(\underline{x}), \quad \underline{x} = (x^\mu) = (x^\circ, x^1, x^2, x^3), \quad (x^\circ = ct). \quad (8.3.4)$$

Eq. (8.3.4) is called Klein-Gorden equation. We will see that both the problems of Eq. (8.3.3) are circumvented by Eq. (8.3.4). Using the corresponding rule:

$$p^\circ \equiv \frac{E}{c} \rightarrow i\hbar\frac{\partial}{\partial ct}, \ \mathbf{p} \rightarrow -i\hbar\nabla \Leftrightarrow p^\mu \rightarrow i\hbar\partial^\mu, \ (four\ momentum) \quad (8.3.5)$$

Eq. (8.3.4) can be immediately brought into Lorentz-covariant form,

$$(p_\mu p^\mu - m_\circ c^2)\Psi(\underline{x}) = 0. \quad (8.3.6)$$

Eq. (8.3.6) is called Klein-Gorden equation, which determines the guide-factor of a wavepacket.

Next let us derive the Klein-Gorden equation from the canonical commutation relation theorem in chapter 1. Taking $[x, p] = i\hbar$ as a starting point, one can derive

$$|\Psi(t) >= e^{-\frac{i}{\hbar}H(t-t_\circ)}|\Psi(t_\circ) > . \quad (8.3.7)$$

Taking its first order time derivative yields

$$i\hbar\frac{\partial}{\partial t}|\Psi(t) >= H|\Psi(t) > . \quad (8.3.8)$$

Now we take its second order time derivative, we will obtain

$$-\hbar^2\frac{\partial^2}{\partial t^2}|\Psi(t) >= H^2|\Psi(t) > . \quad (8.3.9)$$

From Eq. (8.3.9) we can also obtain Klein-Gorden equation Eq. (8.3.4).

If we do not use the relativistic notations in appendix 1, then Eq. (8.3.4) is

$$\left(\frac{1}{c^2}\frac{\partial^2}{\partial t^2} - \nabla^2 + \frac{m_\circ^2c^4}{\hbar^2}\right)\Psi = 0. \quad (8.3.10)$$

The complex conjugate of Eq. (8.3.10) is given by

$$\left(\frac{1}{c^2}\frac{\partial^2}{\partial t^2} - \nabla^2 + \frac{m_\circ^2c^4}{\hbar^2}\right)\Psi^* = 0. \quad (8.3.11)$$

Multiplying Eq. (8.3.10) by Ψ^* on the left and Eq. (8.3.11) by Ψ on the left and subtracting, we obtain

$$\frac{1}{c^2}\left(\Psi^*\frac{\partial^2\Psi}{\partial t^2} - \Psi\frac{\partial^2\Psi^*}{\partial t^2}\right) - \Psi^*\nabla^2\Psi + \Psi\nabla^2\Psi^* = 0. \quad (8.3.12)$$

Since

$$\frac{\partial}{\partial t}\left(\Psi^*\frac{\partial\Psi}{\partial t} - \Psi\frac{\partial\Psi^*}{\partial t}\right) = \Psi^*\frac{\partial^2\Psi}{\partial t^2} - \Psi\frac{\partial^2\Psi^*}{\partial t^2}, \quad (8.3.13)$$

and

$$\nabla\cdot(\Psi^*\nabla\Psi - \Psi\nabla\Psi^*) = \Psi^*\nabla^2\Psi - \Psi\nabla^2\Psi^*, \quad (8.3.14)$$

Eq. (8.3.12) can be written in the form of continuity ($or, say, conservation$) equation of probability current

$$\frac{\partial\rho}{\partial t} + \nabla\cdot\mathbf{j} = 0, \quad (8.3.15)$$

choosing for the current density $\mathbf{j}$ the same expression as that given in chapter 1 for the nonrelativistic case

$$\begin{aligned} \mathbf{j} &= \frac{\hbar}{2mi}[\Psi^* \cdot \nabla\Psi - (\nabla\Psi^*)\Psi] \\ &= \frac{\mathbf{p}}{m}\Psi^*\Psi. \end{aligned} \tag{8.3.16}$$

At the same time, the probability density ρ in Eq. (8.3.15) is

$$\begin{aligned} \rho &= \frac{i\hbar}{2mc^2}\left(\Psi^*\frac{\partial\Psi}{\partial t} - \Psi\frac{\partial\Psi^*}{\partial t}\right) \\ &= \frac{E_k}{mc^2}\Psi^*\Psi, \end{aligned} \tag{8.3.17}$$

where the second equality comes from that the Ψ is a plane wave. Here we recall that for the nonrelativistic Schrödinger equation

$$\rho_s = \Psi^*\Psi \geq 0. \tag{8.3.18}$$

The second equality of Eq. (8.3.17) is much different from the nonrelativistic Eq. (8.3.18). The probability density ρ given by the first equality of Eq. (8.3.17) involves both Ψ and $\partial\Psi/\partial t$ which can be fixed arbitrary and hence admits both positive and negative values. The probability density ρ given by the second equality of Eq. (8.3.17) involves E, which means that the sign of ρ dependent on the sign of $E \lesseqgtr 0$. Since the probability density should be a positive-definite quantity, the Klein-Gorden equation was not accepted as a correct relativistic equation until Pauli and Weiskopf reinterpreted it as a quantum field equation in the same sense as Maxwell's quantum field equation for electromagnetic field [114]. Interpretation of the Klein-Gorden equation as a quantum field equation has shown the seed for the development of the quantum field theory.

We would like to mention here that the negative probability difficulty does not come from relativistic quantum mechanics. On the contrary, this difficulty comes from classical relativistic mechanics. The presence of E_k in ρ of Eq. (8.3.17) is not surprising since ρ is the first component of the current four-vector $(j^\mu = [c\rho, \mathbf{j}])$ [Eq. (A2.25)], just like as E is the first component of momentum four-vector $(p^\mu) = [E/c, \mathbf{p}]$. Indeed, one can write

$$j^\mu = \frac{p^\mu}{m}\Psi^*\Psi. \tag{8.3.19}$$

In appendix 1 we have pointed out that the classical relativistic mechanics already requires the existence of the negative energy particle.

$$E_k = \pm\sqrt{p^2c^2 + m_\circ^2c^4}. \tag{8.3.20}$$

Therefore, if we consider to introduce the special theory of relativity into quantum theory, then the guide-factor (wavefunction) itself has to allow to correspond to the negative energy, i. e., to antiparticle. The probability density is no longer positive definite.

The quantum field theory in chapter 9 allows the existence of antiparticles. The idea and procedures of field quantization are same as the Heisenberg canonical quantization, which will be given in detail in chapter 9. At this stage, we make the following understanding: We treat the Eq. (8.3.15) as a charge conservation equation. The charge density is $Q\rho$. The charge of antiparticle is $-|Q|$, which leads to that the charge density of antiparticle is $Q\rho < 0$. The antiparticle has negative energy.

8.4 Dirac Equation

The physical idea of Dirac equation is to avoid the negative probability density difficulty. It is easy to see that if we use the first order time derivative to replace the second order time derivative in Klein-Gorden equation, then we can avoid this difficulty.

Starting from the relativistic energy $E^2 = p^2c^2 + m_\circ^2 c^4$ of a free particle, Dirac took the following linear relation

$$E = c\boldsymbol{\alpha} \cdot \mathbf{p} + \beta m_\circ c^2, \tag{8.4.1}$$

where $\boldsymbol{\alpha}$, β are independent of $\mathbf{p}$ and $\mathbf{x}$=(x, y, z) = (x^1, x^2, x^3). In relativistic case, for convenience, people like to use $\mathbf{x}$ instead of $\mathbf{r}$.

It is obvious that Eq. (8.4.1) really makes the energy E is simple instead of $E = \pm\sqrt{c^2p^2 + m_\circ c^4}$. However, the cost is that the coefficients $\alpha_x, \alpha_y, \alpha_z, \beta$ become the matrices of dimension 4.

The guide-factor of a free particle in our wavepacket-only theory still can be written as

$$\begin{aligned} \Psi(\mathbf{x}, t) &= ue^{i\mathbf{k}_\circ \cdot \mathbf{x} - i\omega(\mathbf{k}_\circ)t} \\ &\equiv ue^{i\mathbf{k} \cdot \mathbf{x} - i\omega(\mathbf{k})t} = ue^{i(\mathbf{p} \cdot \mathbf{x} - Et)/\hbar}, \end{aligned} \tag{8.4.2}$$

where u is independent of $\mathbf{r}$ and $\mathbf{p}$, is an allowed arbitrary column vector by wavepacket-only theory, and $E = E_\pm = \pm\sqrt{p^2c^2 + m_\circ^2 c^4}$. Note that in subsection 8.5.2 we will argue one can also has other form of solution of plane wave. In Eq. (8.4.2) we write $\mathbf{k} = \mathbf{k}_\circ$ for convenience. It is obvious that $\Psi(\mathbf{x}, t)$ in Eq. (8.4.2) satisfies the following time-dependent Dirac equation:

$$i\hbar \frac{\partial}{\partial t} \Psi(\mathbf{x}, t) = (-ic\hbar \boldsymbol{\alpha} \cdot \nabla + \beta m_\circ c^2)\Psi(\mathbf{x}, t) = H\Psi(\mathbf{x}, t), \tag{8.4.3}$$

where H is called Dirac Hamiltonian. Substituting Eq. (8.4.2) into Eq. (8.4.3) yields a time-independent Dirac equation:

$$(c\boldsymbol{\alpha} \cdot \mathbf{p} + \beta m_\circ c^2)u = Eu. \tag{8.4.4}$$

In Eq. (8.4.4) $\mathbf{p}$ is only a momentum vector and not an operator because we have not yet use the commutation relations.

Let us determine the $\boldsymbol{\alpha}$ and β. Squaring Eq. (8.4.1), we obtain

$$E^2 = c^2(\boldsymbol{\alpha} \cdot \mathbf{p})^2 + \beta^2 m_\circ^2 c^4 + (\boldsymbol{\alpha} \cdot \mathbf{p})\beta m_\circ c^3 + \beta(\boldsymbol{\alpha} \cdot \mathbf{p}) m_\circ c^3. \tag{8.4.5}$$

If we require $E^2 = c^2p^2 + m_\circ^2 c^4$, then we need

$$(\boldsymbol{\alpha} \cdot \mathbf{p})^2 = p^2, \quad \beta^2 = 1, \quad (\boldsymbol{\alpha}\beta = -\beta(\boldsymbol{\alpha}. \tag{8.4.6}$$

The relation

$$(\boldsymbol{\alpha} \cdot \mathbf{p})^2 = (\alpha_x p_x + \alpha_y p_y + \alpha_z p_z)^2 = p^2, \tag{8.4.7}$$

implies

$$\alpha_x^2 = \alpha_y^2 = \alpha_z^2 = 1$$

and

$$\alpha_x\alpha_y = -\alpha_y\alpha_x, \quad \alpha_y\alpha_z = -\alpha_z\alpha_y, \quad \alpha_z\alpha_x = -\alpha_x\alpha_z. \tag{8.4.8}$$

The Dirac Hamiltonian should be Hermitian, i. e., $H = H^+$. Since $\mathbf{p}$ is Hermitian,

$$H = c(\hbar\boldsymbol{\alpha} \cdot \nabla + \beta m_\circ c), \quad H^+ = c(\hbar\boldsymbol{\alpha}^+ \cdot \nabla + \beta^+ m_\circ c). \tag{8.4.9}$$

Eq. (8.4.9) means that α_i $(i = x,\ y,\ z)$ and β matrices are Hermitian, i. e.,

$$\alpha_i^+ = \alpha_i, \quad (i = x,\ y,\ z), \quad \beta^+ = \beta. \tag{8.4.10}$$

Since we have

$$\alpha_i^2 = 1, \quad (i = x,\ y,\ z) \quad and \quad \beta^2 = 1,$$

from Eq. (8.4.10) we derive

$$\alpha_i = \alpha_i^{-1}, \quad (i = x,\ y,\ z) \quad and \quad \beta = \beta^{-1}, \tag{8.4.11}$$

which means that matrices α_i and β are non-singular and consequently their determinant is non-zero.

$$\det(\alpha_i) \neq 0, \qquad \det\beta \neq 0. \tag{8.4.12}$$

Next we neglect the tedious derivations, and give directly the last results:

$$\alpha_x = \begin{bmatrix} 0 & 0 & 0 & 1 \\ 0 & 0 & 1 & 0 \\ 0 & 1 & 0 & 0 \\ 1 & 0 & 0 & 0 \end{bmatrix}, \quad \alpha_y = \begin{bmatrix} 0 & 0 & 0 & -i \\ 0 & 0 & i & 0 \\ 0 & -i & 0 & 0 \\ i & 0 & 0 & 0 \end{bmatrix}, \tag{8.4.13}$$

$$\alpha_z = \begin{bmatrix} 0 & 0 & 1 & 0 \\ 0 & 0 & 0 & -1 \\ 1 & 0 & 0 & 0 \\ 0 & -1 & 0 & 0 \end{bmatrix}, \quad \beta = \begin{bmatrix} 1 & 0 & 0 & 0 \\ 0 & 1 & 0 & 0 \\ 0 & 0 & -1 & 0 \\ 0 & 0 & 0 & -1 \end{bmatrix}. \tag{8.4.14}$$

Because $\alpha_i,\ \beta$ are matrices of dimension 4×4, we have to set

$$\psi = \begin{bmatrix} \bar{a} \\ \bar{b} \\ \bar{c} \\ \bar{d} \end{bmatrix}, \quad \psi^+ = [\bar{a}^* \ \ \bar{b}^* \ \ \bar{c}^* \ \ \bar{d}^*]. \tag{8.4.15}$$

Eqs. (8.4.13-14) can also be written as

$$\alpha_x = \begin{bmatrix} 0 & \sigma_x \\ \sigma_x & 0 \end{bmatrix}, \quad \alpha_y = \begin{bmatrix} 0 & \sigma_y \\ \sigma_y & 0 \end{bmatrix}, \tag{8.4.13$'$}$$

$$\alpha_z = \begin{bmatrix} 0 & \sigma_z \\ \sigma_z & 0 \end{bmatrix}, \quad \beta = \begin{bmatrix} I & 0 \\ 0 & -I \end{bmatrix}, \tag{8.4.14$'$}$$

where $\boldsymbol{\sigma}$ is the Pauli matrix vector, I is a unit 2×2 matrix. Using Eqs. (8.4.13-14)$'$. Setting

$$\psi = \begin{bmatrix} \phi \\ \chi \end{bmatrix},$$

where ϕ and χ are two simple spinors, then the free particle Dirac equation becomes

$$\begin{bmatrix} (-i\partial_\circ + k_c) & -i\boldsymbol{\sigma} \cdot \nabla \\ i\boldsymbol{\sigma} \cdot \nabla & i\partial_\circ + k_c \end{bmatrix} \begin{bmatrix} \phi \\ \chi \end{bmatrix} = 0, \tag{8.4.10$'$}$$

where $k_c = m_\circ c/\hbar$ is the Compton wavenumber. (Some references like to use natural units: $\hbar = c = k_B = 1$. Refer to Ref. [14] for detail.)

Dirac equation can also be written as relativistic covariant form. Multiplying the two sides of Eq. (8.4.3) by $\beta/\hbar c$ and defining new Hermitian matrixes, Gamma matrices (See appendix A for detail.):

$$\gamma^j = \beta\alpha_j, \quad j = 1, 2, 3, \ or \ j = x, y, z, \quad \alpha_j = \beta\gamma^j,$$

$$\beta = \gamma^\circ,$$

we obtain instead of Eq. (8.4.3)

$$(i\gamma^\mu \partial_\mu - k_c)\Psi(\mathbf{x}, t) = 0. \tag{8.4.3$'$}$$

For convenience, we write the form of Eq. (8.4.3)′ in natural units

$$(i\gamma^\mu \partial_\mu - m_\circ)\Psi(\mathbf{x}, t) = 0. \tag{8.4.3$''$}$$

Note that in the following discussions, for convenience, if we use gamma matrixes, then we always write the form in natural units. Eqs. $(8.4.3' - 3'')$ are called the form of Dirac equation in Pauli-Dirac representation.

The continuity equation derived from Dirac equation is as follows. Multiplying the Dirac equation Eq. (8.4.3) on the left by Ψ^+ yields

$$i\hbar\Psi^+ \frac{\partial}{\partial t}\Psi = -ic\hbar\Psi^+ \boldsymbol{\alpha} \cdot \nabla\Psi + \beta m_\circ c^2 \Psi^+\Psi. \tag{8.4.16}$$

Taking the Hermitian conjugate of the Dirac equation Eq. (8.4.3) and multiplying on the right by Ψ, we get

$$-i\hbar\Psi^+ \frac{\partial}{\partial t}\Psi(\mathbf{r}, t) = ic\hbar(\nabla\Psi^+) \cdot \boldsymbol{\alpha}\Psi + \beta m_\circ c^2 \Psi^+\Psi. \tag{8.4.17}$$

Due to that H is Hermitian, α_i, β are Hermitian matrices. Subtracting Eq. (8.4.17) from Eq. (8.4.16) yields

$$\begin{aligned} i\hbar\frac{\partial}{\partial t}(\Psi^+\Psi) &= -ic\hbar\left(\Psi^+\boldsymbol{\alpha} \cdot \nabla\Psi + (\nabla\Psi^+) \cdot \boldsymbol{\alpha}\Psi\right) \\ &= -ic\hbar\nabla \cdot (\Psi^+\boldsymbol{\alpha}\Psi), \end{aligned} \tag{8.4.18}$$

which can be rewritten as a continuous equation

$$\frac{\partial\rho}{\partial t} + \nabla \cdot \mathbf{j} = 0, \tag{8.4.19}$$

with

$$\rho = \Psi^+\Psi = \psi^+\psi, \quad \mathbf{j} = c\Psi^+\boldsymbol{\alpha}\Psi = c\psi^+\boldsymbol{\alpha}\psi, \tag{8.4.20}$$

where the ψ is given by Eq. (8.4.2), and is spacetime independent, and we consider a unit volume of a system.

Using Eq. (8.4.15), the probability density becomes

$$\rho = \Psi^+\Psi = \psi^+\psi = \bar{a}^*\bar{a} + \bar{b}^*\bar{b} + \bar{c}^*\bar{c} + \bar{d}^*\bar{d}, \tag{8.4.21}$$

which is positive-definite quantity, overcoming the negative probability difficulty encountered in the Klein-Gorden equation. The probability current density and the probability density obeys the continuity equation. So, Dirac equation has become a physical acceptable relativistic wave function.

Considering the Eq. (8.4.18) and the boundary conditions (Ψ is confined in a box or approaches zero at infinity), and making integrations by parts, one obtains

$$\frac{\partial}{\partial t}\int \Psi^{+}\Psi d\mathbf{r} = 0, \tag{8.4.22}$$

which means that one can normalize the $\Psi(\mathbf{r}, t)$ (or, say, $\Psi \in \mathcal{L}^2$.) Thus we can give a probability density amplitude interpretation for Ψ, and use the Dirac wavefunction to find the average values of any operator A, i. e.,

$$\langle A\rangle = \int \Psi^{+}A\Psi d\mathbf{r}. \tag{8.4.23}$$

However, we should remember that in appendix A we have already pointed out that the classical relativistic mechanics predicts the existence of antiparticles. Although the Dirac equation is a relativistic quantummechanical equation, but it is still a single body equation, and meets with a negative energy difficulty, which will be seen in the next section. This negative energy difficulty is met by Klein-Gorden equation as well. This difficulty is naturally explained by the existence of antiparticles in quantum field theory.

Instead of Ψ^{+}, one usually uses $\overline{\Psi} \equiv \Psi^{+}\gamma^{\circ}$. In this case, $\rho = \Psi^{+}\Psi = \overline{\Psi}\gamma^{\circ}\Psi$ and $\mathbf{j} = \overline{\Psi}\boldsymbol{\gamma}\Psi$.

$$\frac{\partial}{\partial t}\left(\overline{\Psi}\gamma^{\circ}\Psi\right) + \nabla\cdot\left(\overline{\Psi}\boldsymbol{\gamma}\Psi\right) = \partial_\mu\overline{\Psi}\gamma^\mu\Psi \equiv \partial_\mu j^\mu = 0, \tag{8.4.24}$$

where j^μ is called four-vector current.

It is easy to prove, by using the following Eq. (8.4.28) or referring to Ref. [14],

$$j^\mu = \frac{p^\mu}{m_\circ}\overline{\Psi}\Psi. \tag{8.4.25}$$

We use

$$j \ \ and \ \ \mathbf{j}$$

to express four- or three-vector current, respectively. (See appendix A.)

Let us use Dirac equation to deduce that an electron has to have an additional degree of freedom, i. e., spin. In other worlds, we demonstrate that Dirac equation predicts the existence of new freedom of electron.

The Dirac Hamiltonian H in Eq. (8.4.4) indicates

$$[\mathbf{p}, H] = 0, \tag{8.4.26}$$

which means the conservation of momentum of electron. In physics, due to the homogeneity of position space, the momentum conservation is naturally.

Considering

$$\frac{d}{dt}x = \frac{1}{i\hbar}[x, H] = \frac{1}{i\hbar}[x, c\boldsymbol{\alpha}\cdot\mathbf{p} + \beta m_\circ c^2] = \frac{1}{i\hbar}[x, c\alpha_x p_x] = c\alpha_x, \tag{8.4.27}$$

we have the relativistic velocity operator of electron is

$$\mathbf{v} \equiv \dot{\mathbf{r}} = c\boldsymbol{\alpha}. \tag{8.4.28}$$

Let us consider the time dependence of the orbital angular momentum $\mathbf{L} = (L_x, L_y, L_z)$ of an electron.

$$\begin{aligned}\frac{d}{dt}L_x &= \frac{1}{i\hbar}[L_x, H] = \frac{c}{i\hbar}[L_x, \alpha_x p_x + \alpha_y p_y + \alpha_z p_z] \\ &= \frac{c}{i\hbar}\{\alpha_x[L_x, p_x] + \alpha_y[L_x, p_y] + \alpha_z[L_x, p_z]\} \\ &= c(\alpha_y p_z - \alpha_z p_y) = c(\vec{\alpha}\times\mathbf{p})_x.\end{aligned} \tag{8.4.29}$$

Therefore,

$$\frac{d}{dt}\mathbf{L} = c(\boldsymbol{\alpha} \times \mathbf{p}), \tag{8.4.30}$$

or

$$[\mathbf{L}, H] = i\hbar c(\boldsymbol{\alpha} \times \mathbf{p}), \tag{8.4.31}$$

Eq. (8.4.31) indicates that the orbital angular momentum of an electron is not conservative. However, for a free electron , the position space should be isotropic. This is a very sharp contradiction on quantum mechanics consistency. In this case people naturally remember that the experiment observed the spin of electron, and try if the total angular momentum $\mathbf{J} = \mathbf{L} + \mathbf{S}$ is conservative.

Let us calculate $[\mathbf{J}, H] =?$, and look for the condition that it is equal to zero. Suppose that an operator $\mathbf{W}$ satisfies the following algebraic relations:

$$[\mathbf{W}, \beta] = 0, \quad [W_i, \alpha_i] = 0, \quad i = x,\ y,\ z$$

$$[W_i, \alpha_j] = 2i\varepsilon_{ijk}\alpha_k. \tag{8.4.32}$$

Using Eq. (8.4.32) one can demonstrate

$$\left[\frac{1}{2}\mathbf{W}, H\right] = -ic(\boldsymbol{\alpha} \times \mathbf{p}).(8.4.33)$$

Therefore, if set

$$\mathbf{S} = \frac{\hbar}{2}\mathbf{W}, \tag{8.4.34}$$

then

$$[\mathbf{S}, H] = -i\hbar c(\boldsymbol{\alpha} \cdot \mathbf{p}). \tag{8.4.35}$$

Substituting Eq. (8.4.35) into Eq. (8.4.31) one obtains $[\mathbf{J}, H] = 0$, and thus proves that $\mathbf{J}$ is a conservative quantity. To fit the experiment data for the electron magnetic moment, we should set

$$W_x = \pm 1, \quad W_y = \pm 1, \quad W_z = \pm 1. \tag{8.4.36}$$

Due to that the $\mathbf{S}$ is of property of angular momentum, according to the algebra of angular momentum, we require

$$[S_i, S_j] = i\hbar\varepsilon_{ijk}s_k. \tag{8.4.37}$$

From Eqs. (8.4.34) and (8.4.37) one obtains

$$[W_i, W_j] = 2i\varepsilon_{ijk}W_k. \tag{8.4.38}$$

From Eqs. (8.4.38) and (8.4.36) we see that algebra properties of $(W_x,\ W_y,\ W_z)$ are same as the properties of Pauli operators $(\sigma_x,\ \sigma_y,\ \sigma_z)$. Therefore, we conclude that Dirac equation describes the particle with intrinsic angular momentum $\hbar/2$.

8.5 Free Particle Solution of Dirac Equation

8.5.1 Free Particle at Rest

We are now in a position to study the properties of wavefunction (≡ guide-factor) determined by the Dirac equation. Already we notice, however, that the guide-factor has four components, and is called a spinor.

Look for the free particle guide-factors of the form:

$$\Psi(\mathbf{x}, t) = u(E, \mathbf{p})e^{i(\mathbf{p}\cdot\mathbf{x} - Et)/\hbar}, \tag{8.5.1.1}$$

where $u(E, \mathbf{p})$ is a constant four-component spinor. Substituting Eq. (8.5.1.1) into Dirac equation is

$$(i\gamma^\mu \partial_\mu - m_\circ)\Psi = 0 \tag{8.5.1.2}$$

yields

$$(\gamma^\circ E - \sum_{i=1}^{3} \gamma^i p_i - m_\circ)u = (\gamma^\mu_\mu - m_\circ)u = 0. \tag{8.5.1.3}$$

For particle at rest

$$(\gamma^\circ E - m_\circ)u = 0. \tag{8.5.1.4}$$

Eq. (8.5.1.4) has solution

$$E \begin{pmatrix} 1 & 0 & 0 & 0 \\ 0 & 1 & 0 & 0 \\ 0 & 0 & -1 & \\ 0 & 0 & 0 & -1 \end{pmatrix} \begin{pmatrix} \phi_1 \\ \phi_2 \\ \phi_3 \\ \phi_4 \end{pmatrix} = m_\circ \begin{pmatrix} \phi_1 \\ \phi_2 \\ \phi_3 \\ \phi_4 \end{pmatrix}. \tag{8.5.1.5}$$

Eq. (8.5.1.5) has four orthogonal solutions:

$$u_1(m_\circ, 0) = \begin{pmatrix} 1 \\ 0 \\ 0 \\ 0 \end{pmatrix}, \; u_2(m_\circ, 0) = \begin{pmatrix} 0 \\ 1 \\ 0 \\ 0 \end{pmatrix}, \; u_3(m_\circ, 0) = \begin{pmatrix} 0 \\ 0 \\ 1 \\ 0 \end{pmatrix}, \; u_4(m_\circ, 0) = \begin{pmatrix} 0 \\ 0 \\ 0 \\ 1 \end{pmatrix}, \tag{8.5.1.6}$$

where $E = m_\circ$ for u_1 and u_2, and $E = -m_\circ$ for u_3 and u_4, i. e., there is negative energy solution. Dirac equation cannot just discard the $E < 0$ solutions as unphysical as we require a complete set of states, i. e., four solutions. Substituting Eq. (8.5.1.6) into Eq. (8.5.1.1) yields

$$\Psi_1 = \begin{pmatrix} 1 \\ 0 \\ 0 \\ 0 \end{pmatrix} e^{-im_\circ t}, \; \Psi_2 = \begin{pmatrix} 0 \\ 1 \\ 0 \\ 0 \end{pmatrix} e^{-im_\circ t}, \; \Psi_3 = \begin{pmatrix} 0 \\ 0 \\ 1 \\ 0 \end{pmatrix} e^{im_\circ t}, \; \Psi_4 = \begin{pmatrix} 0 \\ 0 \\ 0 \\ 1 \end{pmatrix} e^{im_\circ t}. \tag{8.5.1.7}$$

Note that if we we do not use the natural units, then $m_\circ \to m_\circ c^2/\hbar$.

We will need to deal with the interpretation of these solutions later. Both the upper (Ψ_1 and Ψ_2) and the lower (Ψ_3 and Ψ_4) halves of the Dirac spinor wavefunction, however, have only one nonzero component that it is tempting to call "spin up" and "spin down", respectively. This interpretation is in fact correct, but we need to be somewhat more ambitions before we can state this with confidence. The interpretation of the negative solutions also must be postulated until later.

8.5.2 General Plane Wave Solution

Eq. (8.4.2) is a plane wave solution of free particle, the coefficient u is a column vector given by Eq. (8.4.15). Substituting Eq. (8.4.15) into the Dirac equation Eq. (8.4.4) we obtain

$$(E - m_\circ c^2)\bar{a} - cp_z\bar{c} - c(p_x - ip_y)\bar{d} = 0$$

$$(E - m_\circ c^2)\bar{b} - c(p_x + ip_y)\bar{c} + cp_z\bar{d} = 0$$

$$(E + m_\circ c^2)\bar{c} - cp_z\bar{a} - c(p_x - ip_y)\bar{b} = 0$$

$$(E + m_\circ c^2)\bar{d} - c(p_x + ip_y)\bar{a} + cp_z\bar{b} = 0. \tag{8.5.2.1}$$

These equations have a nontrivial solution only if the determinant of the coefficient matrix vanishes. These results

$$(E^2 - m_\circ^2 c^4 - c^2 p^2)^2 = 0, \tag{8.5.2.2}$$

which is noted to be in agreement with the fundamental relativistic energy relation.

Explicit forms of the $\bar{a}_j$ coefficients correspond to choices of the sign of the energy. For each such choice there are two linearly independent solutions.

$$E_+ = +\sqrt{c^2p^2 + m_\circ^2 c^4} \equiv +E$$

$$u_1 = N_1 \begin{bmatrix} 1 \\ 0 \\ cp_z/\Delta_+ \\ c(p_x + ip_y)/\Delta_+ \end{bmatrix}, \quad u_2 = N_2 \begin{bmatrix} 0 \\ 1 \\ c(p_x - ip_y)/\Delta_+ \\ -cp_z/\Delta_+ \end{bmatrix},$$

$$E_- = -\sqrt{c^2p^2 + m_\circ^2 c^4} \equiv -E$$

$$u_3 = N_3 \begin{bmatrix} cp_z/\Delta_- \\ c(p_x + ip_y)/\Delta_- \\ 1 \\ 0 \end{bmatrix}, \quad u_4 = N_4 \begin{bmatrix} c(p_x - ip_y)/\Delta_- \\ -cp_z/\Delta_- \\ 0 \\ 1 \end{bmatrix}, \tag{8.5.2.3}$$

where

$$\Delta_\pm \equiv (E_\pm \pm m_\circ c^2). \tag{8.5.2.4}$$

Eqs. (8.5.2.3-4) can also be expressed as

$$u_i = N_i \begin{bmatrix} \phi_i \\ c/\Delta_+ \boldsymbol{\sigma}\cdot\mathbf{p}\phi_i \end{bmatrix}, \; i = 1, 2 \;\; \phi_1 = \begin{bmatrix} 1 \\ 0 \end{bmatrix}, \;\; \phi_2 = \begin{bmatrix} 0 \\ 1 \end{bmatrix}.$$

$$u_j = N_j \begin{bmatrix} c/\Delta_- \boldsymbol{\sigma}\cdot\mathbf{p}\phi_j \\ \phi_j \end{bmatrix}, \; j = 3, 4 \;\; \phi_3 = \begin{bmatrix} 1 \\ 0 \end{bmatrix}, \;\; \phi_4 = \begin{bmatrix} 0 \\ 1 \end{bmatrix}. \tag{8.5.2.3$'$}$$

Note that in natural unit system, $c = \hbar = k_B = 1$.

The free particle spinor solutions of Dirac equation are given by Eq. (8.4.2):

$$\begin{aligned} \Psi(\mathbf{x}, t) &= u(E, \mathbf{p})e^{i\mathbf{k}_\circ\cdot\mathbf{x} - i\omega(\mathbf{k}_\circ)t} \\ &\equiv u(E, \mathbf{p})e^{i\mathbf{k}\cdot\mathbf{x} - i\omega(\mathbf{k})t} = u(E, \mathbf{p})e^{i(\mathbf{p}\cdot\mathbf{x} - Et)/\hbar}, \end{aligned} \tag{8.5.2.5)(8.4.2}$$

which satisfy

$$(\gamma^\mu p_\mu - m_\circ)u = 0. \tag{8.5.2.5$'$}$$

Actually, the free particle spinor solution of Dirac equation can also take the following two forms other than the form of Eq. (8.5.2.5) [115]. The antiparticle solutions in terms of the physical energy and momentum are:

$$\begin{aligned} \Psi(\mathbf{x}, t) &= \nu(E, \mathbf{p})e^{-i\mathbf{k}_\circ\cdot\mathbf{x} + i\omega(\mathbf{k}_\circ)t} \\ &\equiv \nu(E, \mathbf{p})e^{-i\mathbf{k}\cdot\mathbf{x} + i\omega(\mathbf{k})t} = \nu(E, \mathbf{p})e^{-i(\mathbf{p}\cdot\mathbf{x} - Et)/\hbar}, \end{aligned} \tag{8.5.2.6)(8.4.2$'$}$$

which satisfy

$$(\gamma^\mu p_\mu + m_\circ)\nu = 0. \tag{8.5.2.6$'$}$$

For simplicity, we set $\mathbf{k}_\circ = \mathbf{k}$. The spinor wavefunctions are

$$\nu_1 = N_1' \begin{bmatrix} c(p_x - ip_y)/\Delta_+ \\ -cp_z/\Delta_+ \\ 0 \\ 1 \end{bmatrix}, \quad \nu_2 = N_2' \begin{bmatrix} cp_z/\Delta_+ \\ c(p_x + ip_y)/\Delta_+ \\ 1 \\ 0 \end{bmatrix},$$

$$\nu_3 = N_3' \begin{bmatrix} 1 \\ 0 \\ cp_z/\Delta_- \\ c(p_x + ip_y)/\Delta_- \end{bmatrix}, \quad \nu_4 = N_4' \begin{bmatrix} 0 \\ 1 \\ c(p_x - ip_y)/\Delta_- \\ -cp_z/\Delta_- \end{bmatrix}, \tag{8.5.2.7}$$

For these spinor wavefunctions the spins are given by $S_\nu = -S$, for both particle and antiparticle solutions: $E = \sqrt{|\mathbf{p}|^2 + m_\circ^2}$ [115]. It is easy to see that one can simply define anti-particle spinor wavefunction by flipping the sign of E and $\mathbf{p}$ following the Feynman-Stückelurg interpretation ($\hbar = 1$): [115]

$$\nu_1(E, \mathbf{p})e^{-i(\mathbf{p}\cdot\mathbf{x}-Et)} = u_4(-E, -\mathbf{p})e^{i(\mathbf{p}\cdot\mathbf{x}-Et)}. \tag{8.5.2.7$'$}$$

$$\nu_2(E, \mathbf{p})e^{-i(\mathbf{p}\cdot\mathbf{x}-Et)} = u_3(-E, -\mathbf{p})e^{i(\mathbf{p}\cdot\mathbf{x}-Et)}. \tag{8.5.2.7$''$}$$

Next we study the normalization of the spinor wavefunction. We consider

$$\psi_1(\mathbf{x}, t) = u_1 e^{i(\mathbf{p}\cdot\mathbf{x}-Et)}. \tag{8.5.2.8}$$

$$\begin{aligned}
1 &= \langle\psi_1|\psi_1\rangle \\
&= |N_1|^2 \frac{1}{V} \left[1, 0, \frac{cp_z}{\Delta_+}, \frac{c(p_x - ip_y)}{\Delta_+}\right] \begin{bmatrix} 1 \\ 0 \\ cp_z/\Delta_+ \\ c(p_x + ip_y)/\Delta_+ \end{bmatrix} \\
&= |N_1|^2 \frac{1}{V} \left(1 + \frac{c^2p^2}{(E + m_\circ c^2)^2}\right) \\
&= |N_1|^2 \frac{1}{V} \left(\frac{(E + m_\circ c^2)^2 + c^2p^2}{(E + m_\circ c^2)^2}\right) \\
&= |N_1|^2 \frac{1}{V} \left(\frac{(E + m_\circ c^2)^2 + E^2 - m_\circ^2 c^4}{(E + m_\circ c^2)^2}\right) \\
&= |N_1|^2 \frac{1}{V} \left(\frac{2E^2 + 2Em_\circ^2}{(E + m_\circ c^2)^2}\right) \\
&= |N_1|^2 \frac{1}{V} \left(\frac{2E}{E + m_\circ c^2}\right),
\end{aligned} \tag{8.5.2.9}$$

which for the desired $2E$ particles per unit volume $V = 1$, requires that [115]

$$N_1 = \sqrt{E + m_\circ c^2} \rightarrow \sqrt{E + m_\circ}. \tag{8.5.2.10}$$

Using the same method yields $N_1 = N_2 = N_3 = N_4 = N_1' = N_2' = N_3' = N_4' =$.

Note that in our derivation of the normalization factors we assume that $\mathbf{p}$ take discrete values. If we assume that $\mathbf{p}$ take continuous values, then we will obtain different normalization factors, i. e., $\sqrt{V} \rightarrow 1/\sqrt{(2\pi)^3}$. If we define Dirac adjoint operation (or, "bar" operation) $\overline{\Psi} \equiv \Psi^+\gamma^\circ$, and the normalization is for $\overline{\Psi}\Psi$, then the normalization factors are also different from Eq. (8.5.3.10) [116, 7].

8.5.3 Helicity Operator

In this subsection we argue the the free particle solutions indeed behave appropriately according to the spin up/down assignment that we have conjectured in subsection 8.5.1. To do this, we introduce helicity operator

The definition of helicity operator $\mathfrak{H}$ is [117, 118, 119, 120]

$$\mathfrak{H} = \mathbf{\Sigma} \cdot \frac{\mathbf{p}}{|\mathbf{p}|}, \tag{8.5.3.1}$$

$$\mathbf{\Sigma} = \begin{bmatrix} \boldsymbol{\sigma} & 0 \\ 0 & \boldsymbol{\sigma} \end{bmatrix}. \tag{8.5.3.2}$$

The helicity operator $\mathfrak{H}$ is a projective operator to project the spin operator $\mathbf{S} = \mathbf{\Sigma}/2$ on the direction of $\mathbf{p}$. Using Eq. (8.4.35), we can prove

$$\frac{d\mathbf{S}\cdot\mathbf{p}}{dt} = i[H, \mathbf{S}\cdot\mathbf{p}] = -\boldsymbol{\alpha} \times \mathbf{p} \cdot \mathbf{p} = 0. \tag{8.5.3.3}$$

Eq. (8.5.3.3) indicates that $\mathfrak{H}$ and H have common eigenstates and the eigenvlues of $\mathfrak{H}$ are conserved.

From

$$\mathfrak{H}^2 = \left\{ \frac{1}{|\mathbf{p}|} \begin{bmatrix} \boldsymbol{\sigma}\cdot\mathbf{p} & 0 \\ 0 & \boldsymbol{\sigma}\cdot\mathbf{p} \end{bmatrix} \right\}^2 = \begin{bmatrix} 1 & 0 \\ 0 & 1 \end{bmatrix}, \tag{8.5.3.4}$$

we obtain the eigenvalues of helicity operator are +1 and -1. This implies that the spin of the particle is 1/2, because only that value has exactly two values along an axis of quantization (here the direction of $\mathbf{p}$.) We call $\mathfrak{H} = +1$ and -1 the spin up and spin down, respectively.

To determine which of the four spinors u_i, $i = 1, 2, 3, 4$ has spin up and down, we do as follows. Assume that the particle's momentum $\mathbf{p} = (0, 0, p_z)$. Then

$$\mathfrak{H} = \frac{1}{p_z} \begin{bmatrix} \sigma_z p_z & 0 \\ 0 & \sigma_z p_z \end{bmatrix} = \begin{bmatrix} \sigma_z & 0 \\ 0 & \sigma_z \end{bmatrix} = \begin{bmatrix} 1 & 0 & 0 & 0 \\ 0 & -1 & 0 & 0 \\ 0 & 0 & 1 & 0 \\ 0 & 0 & 0 & -1 \end{bmatrix}. \tag{8.5.3.5}$$

We have for positive energy

$$\mathfrak{H} u_1 = N_1 \begin{bmatrix} \sigma_z & 0 \\ 0 & \sigma_z \end{bmatrix} \begin{bmatrix} \phi_1 \\ c/\Delta_+ \sigma_z p_z \phi_1 \end{bmatrix} = +u_1, \tag{8.5.3.6}$$

$$\mathfrak{H} u_2 = N_2 \begin{bmatrix} \sigma_z & 0 \\ 0 & \sigma_z \end{bmatrix} \begin{bmatrix} \sigma_z & 0 \\ 0 & \sigma_z \end{bmatrix} \begin{bmatrix} \phi_1 \\ c/\Delta_+ \sigma_z p_z \phi_2 \end{bmatrix} = -u_2, \tag{8.5.3.7}$$

and for negative energy

$$\mathfrak{H} u_3 = N_3 \begin{bmatrix} \sigma_z & 0 \\ 0 & \sigma_z \end{bmatrix} \begin{bmatrix} \phi_3 \\ c/\Delta_- \sigma_z p_z \phi_3 \end{bmatrix} = +u_3, \tag{8.5.3.8}$$

$$\mathfrak{H} u_4 = N_4 \begin{bmatrix} \sigma_z & 0 \\ 0 & \sigma_z \end{bmatrix} \begin{bmatrix} \sigma_z & 0 \\ 0 & \sigma_z \end{bmatrix} \begin{bmatrix} \phi_4 \\ c/\Delta_+ \sigma_z p_z \phi_4 \end{bmatrix} = -u_2, \tag{8.5.3.9}$$

Thus we conclude that u_1 and u_3 have positive helicity and represent the object being of right handness (= chirality), and that u_2 and u_4 have positive helicity and represent left handness. We call the two components of spinor wavefunction with $E < 0$ the antiparticle. Then, the four components of the spinor wavefunction in Eq. (8.5.1.1) correspond to spin up and spin down of particle and antiparticle.

In the limit $E >> m_\circ$, the eigenstates of helicity operator are also the eigenstates of γ^5

$$\gamma^5 u_\uparrow = +u_\uparrow,\ \gamma^5 u_\downarrow = -u_\downarrow,\ \gamma^5 \nu_\uparrow = -\nu_\uparrow,\ \gamma^5 \nu_\downarrow = +\nu_\downarrow. \tag{8.5.3.10}$$

In general, define the eigenstates of γ^5 as left and right handed chiral states: u_R, u_L, ν_R, ν_L, i. e.,

$$\gamma^5 u_R = +u_R,\ \gamma^5 u_L = -u_L,\ ,\gamma^5 \nu_R = -\nu_R,\ ,\gamma^5 \nu_L = +u_L. \qquad (8.5.3.11)$$

Only in the limit $E >> m_\circ$

$$u_R \equiv u_\uparrow,\ u_L \equiv u_\downarrow,\ \nu_R \equiv \nu_\uparrow,\ \nu_L \equiv \nu_\downarrow. \qquad (8.5.3.12)$$

It is now known that all particles either have an antiparticle or they are their own antiparticle. For the latter to be the case, the particle has to be neutral [120].

To demonstrate the consistency of the present spinor wavefunctions of spin up and spin down with the nonrelativistic forms in quantum mechanics we do follows. This nonrelativistic limit is described by the condition $v/c \to 0$. The two lower entries of u_1 behave as $cp/\Delta_+ \approx v/c \to 0$. It follows that in this nonrelativistic limit one may write for spin up particle

$$\psi_1(\mathbf{x}, t) \sim \begin{pmatrix} 1 \\ 0 \end{pmatrix} e^{i(\mathbf{p}\cdot\mathbf{x}-Et)/\hbar}$$

Similarly, for spin down particle

$$\psi_2(\mathbf{x}, t) \sim \begin{pmatrix} 0 \\ 1 \end{pmatrix} e^{i(\mathbf{p}\cdot\mathbf{x}-Et)/\hbar}. \qquad (8.5.3.10)$$

These plane waves were encountered previously, relevant to the propagating of free spin 1/2 electron with "spin up" and "spin down", respectively.

8.6 Symmetries of Dirac Objects

Generally speaking, the last purpose to study the so-called symmetries is to give the properties of the objects, including particle, antiparticle, phonon, photon, and so on. For convenience, we call the object obeying Dirac equation, Dirac object. Here, the symmetries do not confined in the conservative properties of some quantities of that object. For example, the charge conjugation in subsection 8.6.3 belongs to a symmetry of Dirac object, but this symmetry does not have a wee bit of connection with so called conservation; on the contrary, charge conjugation represents some property of Dirac particle-antiparticle conjugation under some operation. Therefore, one cannot always connect conservation with symmetry.

8.6.1 Spatial Isotropy

Chapter 1 tells us that the spatial isotropy leads to angular momentum operator does not vary with time, which means that the quantummechanical average of angular momentum operator is conservative. Let us look at that if this conservation holds for Dirac equation.

Theorem XXXXXII: Total angular momentum conservation: The spatial isotropy leads to conservation of quantummechanical average value of total angular momentum of Dirac spinor objects.

Proof:

This theorem does not need to prove. After Eq. (8.4.31), we say that if position space is of spatial isotropy, then this isotropy property has to lead to conservation of some kind angular momentum. Since Eq. (8.4.31) clearly tells us that the orbital angular momentum by oneself is not conservative, we have to add another term to the angular momentum, and set the total angular momentum to be conservative. This means that in terms of the requirement of conservation of some kind of angular momentum, people predicts spin angular momentum upon Dirac Hamiltonian. In fact, section 8.4 has proved

$$[\mathbf{J}, H_{Dirac}] = [\mathbf{J}+\mathbf{S}, H_{Dirac}] = 0, \quad \frac{d\mathbf{J}}{dt} = \frac{\partial \mathbf{J}}{\partial t} + [H_{Dirac}, H_{Dirac}] = 0. \tag{8.6.1.1}$$

Refer to Ref. [116] for other method to prove this theorem

8.6.2 Spatial Inversion

8.6.2.1. Parity Conservation

At first we study the spatial inversion transformation on the ket state $|\alpha>$, which is not a spinor. We call $\pi|\alpha>$ the the spatial reversion (= reflection) state of $|\alpha>$, where π is a unitary operator, and is called parity operator, which make spatial inversion operation on $|\alpha>$. Spatial inversion transformation belongs to discrete transformation other than continuous transformation like spatial rotation. (1). Definition of parity operator

$$|\alpha> \rightarrow \pi|\alpha> \equiv |\alpha'>, \; in\ coordinate\ representation\ \pi\Psi(\mathbf{x}) = e^{i\delta}\Psi(-\mathbf{x}). \tag{8.6.2.1.1}$$

For simplicity, take $\delta = 0$. So,

$$\pi\Psi(\mathbf{x}) = \Psi(-\mathbf{x}). \tag{8.6.2.1.2}$$

(2). Properties of parity operator

(2.1). Linearity.

$$\pi[c_1|\alpha>_1 + c_2|\alpha>_2] = c_1\pi|\alpha>_1 + c_2\pi|\alpha>_2 . \tag{8.6.2.1.3}$$

(2.2). Eigenvalues of π.

Applying π twice

$$\pi\pi\Psi(\mathbf{x},t) = \pi e^{i\delta}\Psi(-\mathbf{x},t) = e^{i2\delta}\Psi(\mathbf{x},t) = e^{i2\delta}\Psi(\mathbf{x},t). \tag{8.6.2.1.4}$$

$$\pi\pi = 1 \Rightarrow \pi = \pm 1, \quad and \quad \pi^{-1} = \pi. \tag{8.6.2.1.5}$$

The eigenstates of parity operator with $\pi = 1$, *and* -1 are called even (or positive) and odd (or negative) parity.

(2.3). Unitarity

To preserve the normalization of the wavefunction

$$<\Psi|\Psi> = <\Psi'|\Psi'> = <\Psi|\pi^{+}\pi\Psi> \Rightarrow \pi^{+}\pi = 1, \; \pi^{-1} = \pi \; i.\ e.,\ unitary. \tag{8.6.2.1.6}$$

(2.4). Hermitian

$$From\ \pi\pi = 1 \ \ and\ \ \pi^{+}\pi = 1 \Rightarrow \pi = \pi^{+}, \tag{8.6.2.1.7}$$

which implies Parity is an observable quantity. If the interaction Hamiltonian commutes with π, parity is an observable conserved quantity.

(2.5). Parity of operator.

(2.5.1).

$$\pi\mathbf{x}\pi^{+} = -\mathbf{x}, \quad \mathbf{x}\ \ negative\ parity. \tag{8.6.2.1.8}$$

(2.5.2).

$$\pi\mathbf{p}\pi^{+} = -\mathbf{p}, \quad \mathbf{p}\ \ negative\ parity. \tag*{(8.6.2.1.8)$'$}$$

We call a vector being of negative parity (ordinary) vector or polar operator.

(2.5.3).

$$\pi\mathbf{L}\pi^{+} = \pi\mathbf{x}\times\mathbf{p}\pi^{+} = \pi\mathbf{x}\pi^{+}\times\pi\mathbf{p}\pi^{+} = \mathbf{L}. \tag*{(8.6.2.1.8)$''$}$$

We call a vector being of positive parity axialvector or pseudovector operator.

(2.5.4).

$$\pi\mathbf{p}\cdot\mathbf{x}\pi^{+}=\pi\mathbf{p}\pi^{+}\cdot\pi\mathbf{x}\pi^{+}=\mathbf{p}\cdot\mathbf{x}. \tag{8.6.2.1.9}$$

We call a scalar being of positive parity scalar operator a scalar.

(2.5.5).

$$\pi\mathbf{p}\cdot\mathbf{S}\pi^{+}=\pi\mathbf{p}\pi^{+}\cdot\pi\mathbf{S}\pi^{+}=-\mathbf{p}\cdot\mathbf{S}. \tag{8.6.2.1.10}$$

We call a scalar being of negative parity a pseudoscalar operator.

For the Dirac equation being of form with 4×4 matrix, we do not know the form of parity operator. However, we hope it commute with the Dirac Hamiltonian. Thus, at first we express the full parity operator by $\mathcal{P}=\pi U_P$, which renders the Dirac Hamiltonian invariant under $\mathcal{P}$, i. e., we hope to still keep the Dirac Hamiltonian invariant under operation of $\mathcal{P}$. To keep the correctness of Dirac equation, we look for the necessary corresponding variance of the spinor wavefunction. In $\mathcal{P}=\pi U_P$, the π just causes the spatial reflection $\mathbf{x}\to-\mathbf{x}$. The U_p should be a 4×4 matrix in spinor space.

Theorem XXXXXIII: Parity conservation: In case of interaction Hamiltonian $V(\mathbf{x})=V(|\mathbf{x}|)$, Dirac Hamiltonian has invariance of spatial inversion. Define $\mathcal{P}=\pi\gamma^{\circ}$ as full spatial inversion operator, where π is spatial inversion operator. We have

$$[\mathcal{P},H_{Dirsac}]=0\Rightarrow\frac{d\mathcal{P}}{dt}=\frac{\partial\mathcal{P}}{\partial t}+[\mathcal{P},H_{Dirsac}]=0. \tag{8.6.2.1.11}$$

Parity is a conserved quantity.

Proof:

$$\begin{aligned}
\mathcal{P}i\hbar\frac{\partial}{\partial t}\Psi(\mathbf{x},t) &= i\hbar\frac{\partial}{\partial t}U_P\Psi(-\mathbf{x},t)\\
&= (-ic\hbar\gamma^{\circ}\boldsymbol{\alpha}\cdot(-)\nabla+\beta m_{\circ}c^2)\Psi(-\mathbf{x},t)\\
&= (-ic\hbar\gamma^{\circ}\boldsymbol{\alpha}\cdot(-)\nabla+\beta m_{\circ}c^2)\gamma^{\circ+}\gamma^{\circ}\Psi(-\mathbf{x},t)\\
&= (-ic\circ\boldsymbol{\alpha}\cdot\nabla+\beta m_{\circ}c^2)\gamma^{\circ}\Psi(-\mathbf{x},t).
\end{aligned} \tag{8.6.2.1.12}$$

Note that the spinor wavefunction has to have corresponding variation. For a multiparticle system, the general expression of parity conservation should consider the particle's intrinsic parity besides its orbital angular momentum. Set the intrinsic parities of electron and positron are ±1. The particle's number of initial and final states of system can be different. The general parity conservation expression is

$$\Pi_{i=1}^{N_{initial}}(-1)^{l_i}\pi_i=\Pi_{j=1}^{N_{final}}(-1)^{l_j}\pi_j. \tag{8.6.2.1.13}$$

8.6.2.2. Parity Nonconservation of Weak Interaction

We emphasize that one has to distinguish the parity conservation from left-right asymmetry of Dirac spinor wavefunction itself. In the following we will take the electron as an example of Dirac object to derive the asymmetry of left-right by solving the Dirac equation. You will see really that if the velocity of electrons is not equal to zero, then the left-handness and right-handness (=chirality) of electron spinor wavefunction are asymmetry. Then we explain that what does the parity nonconservation in weak interaction mean.

Eqs. (8.4.2) and (8.5.2.3) gives the Dirac spinor wavefunction of electron at rest is

$$\Psi_1=e^{-im_{\circ}t}\begin{bmatrix}1\\0\\0\\0\end{bmatrix},\quad \Psi_2=e^{-im_{\circ}t}\begin{bmatrix}0\\1\\0\\0\end{bmatrix}. \tag{8.6.2.2.1}$$

Eqs. (8.4.2) and (8.5.2.4) gives the Dirac spinor wavefunction of antielectron at rest is [118]

$$\Psi_3 = e^{im_\circ t} \begin{bmatrix} 0 \\ 0 \\ 1 \\ 0 \end{bmatrix}, \quad \Psi_4 = e^{im_\circ t} \begin{bmatrix} 0 \\ 0 \\ 0 \\ 1 \end{bmatrix}. \qquad (8.6.2.2.2)$$

Therefore, parity transformations of the Dirac spinor wavefunctions of electron and anti-electron at rest are

$$\mathcal{P}\Psi = \pi\gamma^\circ \Psi_i = u_i, \quad i = 1, 2, \quad \mathcal{P}\Psi_j = -\Psi_j, \quad j = 3, 4. \qquad (8.6.2.2.3)$$

The spinor wavefunctions of particle and antiparticle at rest have an definite opposite parity, particle and antiparticle have positive and negative parity, respectively. More general speaking, from the matrix form of γ°, one sees that in the rest frame of the particle, the spinor wavefunctions of positive and negative that are eigenstatesn of γ°-with opposite eigenvalues, i. e., opposite parity. This means that the intrinsic parities of particles and antiparticles are opposite.

For example, in the quark model:

$$quarks \quad u, d, s, c, b \quad have \quad a \ parity \quad +1$$

$$antiquarks \quad \overline{u}, \overline{d}, \overline{s}, \overline{c}, \overline{b} \quad have \quad a \ parity \quad -1$$

The full parity transformation of Dirac spinor wavefunction of free electron is

$$\begin{aligned} \Psi_i &= \mathcal{P}\Psi_i(\mathbf{x}, t) \\ &= \pi\gamma^\circ \Psi_i(\mathbf{x}, t) \\ &= \mathcal{N}^{(+)}_{\uparrow,\downarrow} \gamma^\circ \begin{bmatrix} \phi \\ c/\Delta_+ \boldsymbol{\sigma} \cdot \mathbf{p}\phi \end{bmatrix} \pi e^{-iEt + i\mathbf{p}\cdot\mathbf{x}} \\ &= N \begin{bmatrix} \phi \\ -c/\Delta_+ \boldsymbol{\sigma} \cdot \mathbf{p}\phi \end{bmatrix} e^{-iEt + i(-)\mathbf{p}\cdot(-)\mathbf{x}}, \end{aligned} \qquad (8.6.2.2.4)$$

where

$$\phi = \begin{bmatrix} 1 \\ 0 \end{bmatrix} \quad for \ spinup, \quad \phi = \begin{bmatrix} 0 \\ 1 \end{bmatrix} \quad for \ spindown, \qquad (8.6.2.2.5)$$

The parity transformation of Dirac spinor wavefunction of antielecton is

$$\begin{aligned} \Psi_j &= \mathcal{P}\Psi_j(\mathbf{x}, t) \\ &= \pi\gamma^\circ \Psi_j(\mathbf{x}, t) \\ &= N\gamma^\circ \begin{bmatrix} c/\Delta_- \boldsymbol{\sigma} \cdot \mathbf{p}\phi \\ \phi \end{bmatrix} \pi e^{iEt + i\mathbf{p}\cdot\mathbf{x}} \\ &= N \begin{bmatrix} -c/\Delta_- \boldsymbol{\sigma} \cdot \mathbf{P}\phi \\ \phi \end{bmatrix} e^{-iE_- t + i(-)\mathbf{p}\cdot(-)\mathbf{x}}. \end{aligned} \qquad (8.6.2.2.6)$$

Let us prove the asymmetry of left-handed and right-handed electron spinor wavefunction in Eqs. (8.6.2.2.5-6).

To do this, define helicity λ as the projection of spin on the direction of motion,

$$\lambda = \mathbf{S} \cdot \hat{\mathbf{p}}, \quad \hat{\mathbf{p}} = \frac{\mathbf{p}}{|\mathbf{p}|}. \qquad (8.6.2.2.7)$$

$\lambda > 0$ and $\lambda < 0$ represent right- and left-handed electron spinor wavefunction in special case of $\mathbf{p} = (0, 0, p_z)$ and $E >> m_\circ$, respectively. Dirac spinor wavefunctions of particle and antiparticle have helicity $\lambda = \pm 1/2$, respectively.

From Eq. (8.5.2.3)′, we obtain the Dirac spinor wavefunction of positive and negative helicity are, respectively, (Use natural units.)

$$u_1 = \sqrt{E + m_\circ}\begin{bmatrix} 1 \\ 0 \\ p/(E + m_\circ) \\ 0 \end{bmatrix} \; has\ \lambda = +\frac{1}{2}\ positive\ helicity\ \mathbf{S} \uparrow\uparrow \mathbf{p}. \tag{8.6.2.2.8}$$

$$u_2 = \sqrt{E + m_\circ}\begin{bmatrix} 0 \\ 1 \\ 0 \\ -p/(E + m_\circ) \end{bmatrix} \; has\ \lambda = -\frac{1}{2}\ negative\ helicity\ \mathbf{S} \downarrow\uparrow \mathbf{p}. \tag{8.6.2.2.9}$$

For $E >> m_\circ$:

$$u_1 = \sqrt{E + m_\circ}\begin{bmatrix} 1 \\ 0 \\ 1 \\ 0 \end{bmatrix}, \quad u_2 = \sqrt{E + m_\circ}\begin{bmatrix} 0 \\ 1 \\ 0 \\ -1 \end{bmatrix}. \tag{8.6.2.2.10}$$

$$\frac{1}{2}(1 - \gamma^5)u_1 = 0, \quad \frac{1}{2}(1 - \gamma^5)u_2 = u_2. \tag{8.6.2.2.11}$$

$$\frac{1}{2}(1 + \gamma^5)u_1 = u_1, \quad \frac{1}{2}(1 + \gamma^5)u_2 = 0. \tag{8.6.2.2.12}$$

$$P_L = \frac{1}{2}(1 - \gamma^5)\ is\ projection\ operator\ for\ left\ handness. \tag{8.6.2.2.13}$$

$$P_R = \frac{1}{2}(1 + \gamma^5)\ is\ projection\ operator\ for\ right\ handness. \tag{8.6.2.2.14}$$

General properties of projection operators:

$$P_L^2 = P_L, \quad P_R^2 = P_R, \quad P_R P_L = P_L P_R = 0. \tag{8.6.2.2.15}$$

With the help of the projection operators P_R and P_L the spinor wavefunction can be decomposed into right-handed part R and left-handed part L, because

$$u = \frac{1}{2}(1 + \gamma^5)u + \frac{1}{2}(1 - \gamma^5)u = u_R + u_L. \tag{8.6.2.2.16}$$

$$P_R u_R = u_R, \ P_R u_L = 0, \ P_L u_R = 0, \ P_L u_L = u_L. \tag{8.6.2.2.17}$$

Eq. (8.6.2.2.8) gives the spinor wavefunction with right handness with $\lambda = +1/2$. We have

$$\frac{1}{2}(1 - \gamma^5)u_1(p) = \sqrt{E + m_\circ}\frac{1}{2}\left(1 - \frac{p}{(E + m_\circ)}\right)\begin{bmatrix} 1 \\ 0 \\ -1 \\ 0 \end{bmatrix}. \tag{8.6.2.2.18}$$

Eq. (8.6.2.2.19 gives the spinor wavefunction with negative helicity with $\lambda = -1/2$. We have

$$\frac{1}{2}(1 - \gamma^5)u_2(p) = \sqrt{E + m_\circ}\frac{1}{2}\left(1 + \frac{p}{(E + m_\circ)}\right)\begin{bmatrix} 1 \\ 0 \\ 0 \\ -1 \end{bmatrix}. \tag{8.6.2.2.19}$$

Set $P(\lambda = -1/2)$ and $P(\lambda = +1/2)$ represent the probability of appearances of left handness spinor wavefunction in $\lambda = -1/2$ and $\lambda = +1/2$, respectively. The asymmetry in left-right-handed electron production probability is

$$\begin{aligned}
\frac{P(\lambda=+1/2)-P(\lambda=-1/2)}{P(\lambda=+1/2)+P(\lambda=-1/2)} &= \frac{\left(1-\frac{p}{E+m_\circ}\right)^2-\left(1+\frac{p}{E+m_\circ}\right)^2}{\left(1-\frac{p}{E+m_\circ}\right)^2+\left(1-\frac{p}{E+m_\circ}\right)^2} \\
&= -\frac{2p}{E+m_\circ}\frac{(E+m_\circ)^2}{(E+m_\circ)^2+p^2} \\
&\overset{E>\geq m_\circ}{=} -\frac{p}{E}=-\frac{v}{c},
\end{aligned} \tag{8.6.2.2.20}$$

where v is the velocity of emitted electron. Eq. (8.6.2.28) tells us that electron with negative helicity has preferentially a left-handed parity.

In actual motion of electron, the body-factor of electron will obey the guide of the spinor wavefunction (or spinor guide-factor) according to the spontaneous equiprobability symmetry breaking rule. Generally speaking, an elecron spinor wavefunction is a linear superposition of many spinor wavefunctions with different parity.

To this day, one of the clearest demonstration of parity asummetry in weak interaction is the experiment that first revealed it (1957 Wu). The experiment studied weak interaction β-decay of polarized cobalt nuclei:

$$^{60}Co \rightarrow^{*} + e^- + \overline{\nu}_e. \tag{8.6.2.2.21}$$

The decay electrons are detected in the direction parallel or antiparallel to the nuclei spin **S**, depending on the sign of the external polarized magnetic field. If parity were asymmetry, then expect equal rate for producing electron in directions along and opposite to the nuclear spin. However, an asymmetry in the direction of emitted electrons was observed. The asymmetry was found to change sign upon reversal of the magnetic field such that electrons prefer to be emitted in a direction opposite to that of the nuclei spin **S**, which shows that electron emitted in weak interaction has preferentially a left-handed parity. This phenomenon is only found in weak interaction.

In 1934, before the discovery of parity violation, Fermi proposed, in analogy with QED, that the invariant element for β-decay, which belongs to weak interaction, was of the form:

$$M_{ji} = G_F g_{\mu\nu}[\overline{\Psi}\gamma^\mu\Psi][\overline{\Psi}\gamma^\nu\Psi], \tag{8.6.2.2.22}$$

where $G_F = 1.166 \times 10{-}5GeV^{-2}$. After the discovery of parity violation in 1956 by Yang and Lee and in 1957 by Wu, M_{ji} was modified to

$$M_{ji} = \frac{G_F}{\sqrt{2}} g_{\mu\nu}[\overline{\Psi}\gamma^\mu(1-\gamma^5)\Psi][\overline{\Psi}\gamma^\nu(1-\gamma^5)\Psi]. \tag{8.6.2.2.23}$$

The weak interaction is of the form Vector-Axialvector (V-A)

$$H' = \frac{-ig_w}{\sqrt{2}}\frac{1}{2}\gamma^\mu(1-\gamma^5) = \frac{-ig_w}{\sqrt{2}}\frac{1}{2}\gamma^\mu - \frac{-ig_w}{\sqrt{2}}\frac{1}{2}\gamma^\mu\gamma^5 \equiv Vector - Axialvector. \tag{8.6.2.2.24}$$

Due to the factor $(1-\gamma^\circ)$ in H', only left-handed parity particle states and right-handed parity anti-particle states participate in weak interaction. Therefore, the weak interaction has maximal parity violation.

8.6.3 Charge Conjugation

Charge Conjugation ($\mathcal{C}$) is a symmetry transformation of Dirac equation. The charge conjugation operator is defined as$\mathcal{C} = i\gamma^2 K$, where K is a complex conjugation operator.

Theorem XXXXXIV: Charge conjugation: Under $\mathcal{C}$ operation *particle* $\Leftrightarrow$ *antiparticle* for Dirac equation

Proof:

Let us demonstrate the above property of Dirac equation. Quantum mechanics shows that the motion of a charged particle in an electromagnetic potential $A^\mu = (\phi, \mathbf{A})$ can be obtained by the following substitution

$$\mathbf{p} \to \mathbf{p} - e\mathbf{A}; \quad E \to E - e\phi, \; where \; \mathbf{p} = -i\nabla; \;\; E = i\frac{\partial}{\partial t} \;\; (\hbar = c = 1). \tag{8.6.3.1}$$

Eq. (8.6.3.1) can be written as

$$\partial_\mu \to \partial_\mu + ieA_\mu, \tag{8.6.3.2}$$

and thus Dirac equation becomes

$$\gamma^\mu(\partial_\mu + ieA_\mu)\Psi(\mathbf{x}, t) + im_\circ \Psi(\mathbf{x}, t) = 0. \tag{8.6.3.3}$$

Taking the complex conjugate and pre-multiplying by $-i\gamma^2$ yields

$$-i\gamma^2\gamma^{\mu *}(\partial_\mu - ieA_\mu)\Psi^*(\mathbf{x}, t) - m_\circ\gamma^2\Psi^*(\mathbf{x}, t) = 0. \tag{8.6.3.4}$$

Considering $\gamma^{\circ *} = \gamma^\circ$, $\gamma^{1*} = \gamma^1$, $\gamma^{2*} = -\gamma^2$, $\gamma^{3*} = \gamma^3$ $\gamma^2\gamma^{\mu *} = -\gamma^\mu\gamma^2$ yields

$$\gamma^\mu(\partial_\mu - ieA_\mu)\Psi'(\mathbf{x}, t) + im_\circ\Psi'(\mathbf{x}, t) = 0, \tag{8.6.3.5}$$

where

$$\Psi'(\mathbf{x}, t) = \mathcal{C}\Psi = i\gamma^2 K\Psi(\mathbf{x}, t), \tag{8.6.3.6}$$

Making comparison of Eq. (8.6.3.5) with Eq. (8.6.3.3), we see that the spinor wavefunction $\Psi'(\mathbf{x}, t)$ of charge conjugation transformation of $\Psi(\mathbf{x}, t)$ describes a particle with the same mass but opposite charge, i. e., antiparticle. Therefore, it is more appropriate to call "charge conjugation" the "particle-antiparticle conjugation". QED.

8.6.4 Time Reversal

Time reversal ($\mathcal{T}$) is a symmetry transformation of Dirac equation. The time reversal operator is defined as $\mathcal{T} = \gamma^1\gamma^3 K$, where K is a complex conjugation operator.

Theorem XXXXXV: Time reversal: The state $\mathcal{T}\Psi(\mathbf{x}, t)$ satisfies Dirac equation with $\partial/\partial t \to \partial/\partial(-t)$. Equivalently, the state $\mathcal{T}\Psi(\mathbf{x}, -t)$ satisfies Dirac equation.

Proof:

Dirac equation can be written in the form

$$i\partial_t\Psi(\mathbf{x}, t) = [-i\gamma^\circ\boldsymbol{\gamma}\cdot\nabla + \gamma^\circ m_\circ]\,\Psi(\mathbf{x}, t). \tag{8.6.4.1}$$

The $\mathcal{T}$ acts on the left side of Eq. (8.6.4.1)

$$\begin{aligned} \mathcal{T}(i\partial_t)\mathcal{T}^{-1}\mathcal{T}\Psi &= \gamma^1\gamma^3 K(i\partial_t)K(\gamma^1\gamma^3)^{-1}\Psi^* \\ &= -i\partial_t)\gamma^1\gamma^3\Psi^* \\ &= i\partial_{-t})\gamma^1\gamma^3\Psi^*((\mathbf{x}, t)), \end{aligned} \tag{8.6.4.2}$$

where only γ^2 is imaginary.

$$\begin{aligned}\mathcal{T}\left[-i\gamma^\circ\boldsymbol{\gamma}\cdot\nabla+\gamma^\circ m_\circ\right]\Psi(\mathbf{x},t) &= \mathcal{T}\left[-i\gamma^\circ\boldsymbol{\gamma}\cdot\nabla+\gamma^\circ m_\circ\right]\mathcal{T}^{-1}\mathcal{T}\Psi(\mathbf{x},t)\\ &= \gamma^1\gamma^3\left[i\gamma^\circ\boldsymbol{\gamma}^*\cdot\nabla+\gamma^\circ m_\circ\right](\gamma^1\gamma^3)^{-1}\gamma^1\gamma^3\Psi^*(\mathbf{x},t)\\ &= \left[-i\gamma^\circ\boldsymbol{\gamma}\cdot\nabla+\gamma^\circ m_\circ\right]\gamma^1\gamma^3\Psi^*(\mathbf{x},t). \qquad (8.6.4.3)\end{aligned}$$

From Eqs. (8.6.4.2-3) we obtain

$$i\partial_{-t}\mathcal{T}\Psi(\mathbf{x},t)=\left[-i\gamma^\circ\boldsymbol{\gamma}\cdot\nabla+\gamma^\circ m_\circ\right]\mathcal{T}\Psi(\mathbf{x},t). \qquad (8.6.4.4)$$

Therefor, the spinor wavefunction $\mathcal{T}\Psi(\mathbf{x},t)$ satisfies the time reversal Dirac equation with $\partial_t\to\partial_{-t}$.

Eq. (8.6.4.4) can be rewritten as

$$i\partial_t\mathcal{T}\Psi(\mathbf{x},t)=\left[-i\gamma^\circ\boldsymbol{\gamma}\cdot\nabla+\gamma^\circ m_\circ\right]\mathcal{T}\Psi(\mathbf{x},t). \qquad (8.6.4.5)$$

QED.

8.7 Composition of Dirac Spinor Wavepacket

8.7.1 General Form of Dirac Spinor Wavepacket-Function

In this textbook and Ref. [3] the wavepacket always represents an object (=particle), which can be large or small, heavy or light, with or without spin, and so on. In this textbook and Ref. [3] the wavefunction is a mathematical function, and thus does not carry the mass, the charge, and so on, i. e., wavefunction≠wavepacket. In other textbooks, such as Ref. [116], the so-called wavepacket is, actually, wavefunction (=guide-factor) other than the object itself.

This section will prove two theorems on the composition and the stability of Dirac spinor wavepacket (Refer to Ref. [3]). This section will also points out some common mistakes in other textbooks.

All the following proofs do not need to concern the particular form of Dirac spinor wavepacket; on the contrary, we just require the wavepacket-function is an unknown function either coordinate or momentum, but not both at the same time.

For convenience, let wavepacket-function Ψ satisfy normalization condition, i. e.,

$$\int_{-\infty}^{\infty}|\Psi(\mathbf{r},t)|^2d\mathbf{x}=\int_{-\infty}^{\infty}\Psi^*(\mathbf{x},t)\Psi(\mathbf{x},t)d\mathbf{r}=1. \qquad (8.7.1.1)$$

Since every spinor wavefunction in the eight spinor wavefunctions in Eqs. (8.5.2.3) and (8.5.2.7) has four component, four in the eight spinor wavefunctions are linearly independent. We could choose to work with $\{u_1,u_2,u_3,u_4\}$ or $\{\nu_1,\nu_2,\nu_3,\nu_4\}$, or $\cdots$. A natural choice is to choose only positive energy solutions: $\{u_1,u_2,\nu_1,\nu_2\}$ as four independent spinor wavefunctions.

Therefore, we can use every group in $\{u_1,u_2,u_3,u_4\}$, $\{\nu_1,\nu_2,\nu_3,\nu_4\}$, and $\{u_1,u_2,\nu_1,\nu_2\}$ by multiplying an appropriate plane wave, as four independent spinor wavefunction to expand the spinor wavepacket-function $\Psi(\mathbf{x},t)$ [116, 7]. Note that at this stage, we do not need to know quantum mechanics. We just need to remember that how to expand an unknown spinor function $\Psi(\mathbf{x},t)$ by using a group of complete set of spinor wavefunctions

(u_1, u_2, u_3, u_4).

$$\begin{aligned}\Psi(\mathbf{x},t) &= \int d\mathbf{k}\sum_{i=1}^{4} c_i(\mathbf{k},\omega(\mathbf{k}))u_i(\mathbf{k},\omega(\mathbf{k}))e^{i[\mathbf{k}\cdot\mathbf{x}-\omega(\mathbf{k})t]}\\ &= \int d\mathbf{k}\sum_{i=1}^{4} b_i(\mathbf{k},\omega(\mathbf{k}))\nu_i(\mathbf{k},\omega(\mathbf{k}))e^{-i[\mathbf{k}\cdot\mathbf{x}-\omega(\mathbf{k})t]}\\ &= \int d\mathbf{k}\sum_{i=1}^{2} c_i(\mathbf{k},\omega(\mathbf{k}))u_i(\mathbf{k},\omega(\mathbf{k}))e^{i[\mathbf{k}\cdot\mathbf{x}-\omega(\mathbf{k})t]}\\ &+ \int d\mathbf{k}\sum_{i=1}^{2} d_i(\mathbf{k},\omega(\mathbf{k}))\nu_i(\mathbf{k},\omega(\mathbf{k}))e^{-i[\mathbf{k}\cdot\mathbf{x}-\omega(\mathbf{k})t]}. \end{aligned} \qquad (8.7.1.2)$$

Actually, using any expansion in Eq. (8.7.1.2) obtains the same conclusions. For simplicity, we take the first expansion form of Ψ.

We make an expansion of $\omega(k_x)$ around point $\mathbf{k}_\circ = (k_{x\circ}, k_{y\circ}, k_{z\circ})$, which is the average wavenumber vector of the spinor wavepacket-function Ψ.

The relativistic form of $\omega(\mathbf{k})$ is

$$\omega(\mathbf{k}) = \frac{\sqrt{(\hbar\mathbf{k})\cdot(\hbar\mathbf{k})c^2 + m_\circ^2 c^4}}{\hbar}. \qquad (8.7.1.3)$$

The expansion of $\omega(\mathbf{k})$ at point $\mathbf{k}_\circ$ is

$$\begin{aligned}\omega(\mathbf{k}) &= \omega(\mathbf{k}_\circ) + \left.\frac{\partial\omega}{\partial k_x}\right|_{k_{x\circ}}(k_x-k_{x\circ}) + \left.\frac{\partial\omega}{\partial k_y}\right|_{k_{y\circ}}(k_y-k_{y\circ}) + \left.\frac{\partial\omega}{\partial k_z}\right|_{k_{z\circ}}(k_z-k_{z\circ})\\ &+ \frac{1}{2}\left.\frac{\partial^2\omega}{\partial k_x^2}\right|_{k_{x\circ}}(k_x-k_{x\circ})^2 + \frac{1}{2}\left.\frac{\partial^2\omega}{\partial k_y^2}\right|_{k_{y\circ}}(k_y-k_{y\circ})^2 + \frac{1}{2}\left.\frac{\partial^2\omega}{\partial k_z^2}\right|_{k_{z\circ}}(k_z-k_{z\circ})^2\\ &+ \left.\frac{\partial^2\omega}{\partial k_x\partial k_y}\right|_{k_{x\circ},k_{y\circ}}(k_x-k_{x\circ})(k_y-k_{y\circ}) + \left.\frac{\partial^2\omega}{\partial k_x\partial k_z}\right|_{k_{x\circ},k_{z\circ}}(k_x-k_{x\circ})(k_z-k_{z\circ})\\ &+ \left.\frac{\partial^2\omega}{\partial k_y\partial k_z}\right|_{k_{y\circ},k_{z\circ}}(k_y-k_{y\circ})(k_z-k_{z\circ})\\ &\equiv \omega(\mathbf{k}_\circ) + v_{gx}(k_x-k_{x\circ}) + +v_{gy}(k_y-k_{y\circ}) + v_{gz}(k_z-k_{z\circ})\\ &+ \frac{v_{gxx}}{2|\mathbf{k}_\circ|}(k_x-k_{x\circ})^2 + \frac{v_{gzz}}{2|\mathbf{k}_\circ|}(k_y-k_{y\circ})^2 + \frac{v_{gzz}}{2|\mathbf{k}_\circ|}(k_z-k_{z\circ})^2\\ &+ \frac{v_{gxy}}{|\mathbf{k}_\circ|}(k_x-k_{x\circ})(k_y-k_{y\circ}) + \frac{v_{gxz}}{|\mathbf{k}_\circ|}(k_x-k_{x\circ})(k_z-k_{z\circ})\\ &+ \frac{v_{gyz}}{|\mathbf{k}_\circ|}(k_y-k_{y\circ})(k_z-k_{z\circ}). \end{aligned} \qquad (8.7.1.4)$$

Note that the coefficients of $(k_x - k_{x\circ})$, $(k_x - k_{x\circ})^2$, and so on are constants at $\mathbf{k} = \mathbf{k}_\circ$. In Eq. (8.7.1.4), v_g represents group velocity of the spinor wavepacket, and is equal to

$$v_{g,x} = \left.\frac{\partial\omega}{\partial k_x}\right|_{k_{x\circ}}, \quad v_{g,y} = \left.\frac{\partial\omega}{\partial k_y}\right|_{k_{y\circ}}, \quad v_{g,} = \left.\frac{\partial\omega}{\partial k_z}\right|_{k_{z\circ}}. \qquad (8.7.1.5)$$

Let us make comparisons between the terms containing such as $(k_x - k_{xo})^1$, I, and $(k_x - k_{xo})^2$, II. For example, we calculate

$$\begin{aligned}
\frac{I}{II} &= \frac{\left.\frac{\partial \omega}{\partial k_{xo}}\right|_{k_{xo}}}{\left.\frac{1}{2}\frac{\partial^2 \omega}{\partial k_{xo}^2}\right|_{k_{xo}}} \frac{(k_x - k_{xo})}{(k_x - k_{xo})^2} \\
&= \frac{k_{xo}}{k_x - k_{xo}} \left(1 + \frac{\hbar^2 c^2}{m_o^2 c^2}\right) \\
&= \frac{k_{xo}}{k_x - k_{xo}} \left[1 + \left(\frac{m_o v}{\sqrt{1-(v/c)^2}}\right)^2 \frac{1}{m_o^2 c^2}\right] \\
&= \frac{k_{xo}}{k_x - k_{xo}} \left[1 + \left(\frac{v}{c}\right)^2 \frac{1}{1-(v/c)^2}\right] \\
&\approx \frac{k_{xo}}{k_x - k_{xo}} \times 1 \to \infty \times 1 \\
&>> 1,
\end{aligned} \tag{8.7.1.6}$$

where v is the velocity of the spinor wavepacket. For microscopic object, k_o is a very large number, and k_x is in the neighbourhood of k_o.

Thus in the following demonstrations of theorems, we just take the first order expansion terms. However, if one want to consider the second order expansion terms, one can refer to chapter 3 of Ref. [3].

8.7.2 Composition Theorem

Theorem XXXXXVI: Free spinor wavepacket compositions: The wavepacket-function of a spinor wavepacket in free motion consists of product with three factors: guide-factor, front-factor, and body-factor.

Proof:

For convenience, we take the first equality in Eq. (8.7.1.2). Thus,

$$\Psi(\mathbf{x}, t) = \int d\mathbf{k} \sum_{i=1}^{4} c_i(\mathbf{k}, \omega(\mathbf{k})) u_i(\mathbf{k}, \omega(\mathbf{k})) e^{i[\mathbf{k}\cdot\mathbf{x} - \omega(\mathbf{k})t]}, \tag{8.7.2.1}$$

where $\omega(\mathbf{k})$ is given by Eq. (8.7.1.4), and it is

$$\omega(\mathbf{k}) = \omega(\mathbf{k}_o) + v_{gx}(k_x - k_{xo}) + v_{gy}(k_y - k_{yo}) + v_{gz}(k_z - k_{zo}). \qquad ((8.7.2.2)$$

Substituting Eq. (8.7.2.2) into the plane wave in Eq. (8.7.2.1) yields

$$\begin{aligned}
\Psi(\mathbf{x}, t) &= e^{ik_{xo} v_g t} e^{ik_{yo} v_g t} e^{ik_{zo} v_{gz} t} e^{-i\omega(\mathbf{k}_o)t} \\
&\times \int_{-\infty}^{+\infty} d\mathbf{k} \sum_{i=1}^{4} c_i(\mathbf{k}, \omega(\mathbf{k})) u_i(\mathbf{k}, \omega(\mathbf{k})) e^{ik_x(x - v_{gx}t)} e^{ik_y(y - v_{gy}t)} e^{ik_z(z - v_{gz}t)} \\
&\equiv e^{i\mathbf{k}_o\cdot\mathbf{v}_g t} e^{-i\omega(\mathbf{k}_o)t} \int_{-\infty}^{+\infty} d\mathbf{k} \sum_{i=1}^{4} c_i(\mathbf{k}, \omega(\mathbf{k})) u_i(\mathbf{k}, \omega(\mathbf{k})) e^{i\mathbf{k}\cdot(\mathbf{x} - \mathbf{v}_g t)} \\
&= e^{i\mathbf{k}_o\cdot\mathbf{v}_g t - i\omega(\mathbf{k}_o)t} \times \Psi(\mathbf{x} - \mathbf{v}_g t, 0) \\
&= e^{i\mathbf{k}_o\cdot\mathbf{x} - i\omega(\mathbf{k}_o)t} \times e^{-i\mathbf{k}_o\cdot\mathbf{x}'} \times \Psi(\mathbf{x}', 0) \qquad \mathbf{x}' = \mathbf{x} - \mathbf{v}_g t \\
&\equiv I \times II \times III.
\end{aligned} \tag{8.7.2.3}$$

The factors I, II, and III are called guide-factor, front-factor, and body-factor of wavepacket, respectively. Note that this composition does not depend on the particular form of the spinor wavepacket-function. QED.

Corollary XIX: Free spinor wavepacket non-spreading property: Free spinor wavepacket does not spread at all with time.

Proof:

It is very important to note that the time-dependence of values of $\mathbf{x}'$ and $\mathbf{x}$ have not yet been determined uniquely. Let us determine it in terms of mankind practice experience, experiment facts, and mathematics. Because many observed facts tell us that free spinor wavepackets such as electrons can be stable in a quite long time interval, do not spread, and move with a velocity $\mathbf{v}_g$. Theorem XXXXXVI tells us that the factors I and II are oscillation functions, and factor III is the substance of spinor object. Theorem XXXXXVI, actually, does not know the factor III is decaying with time or not. Since many facts indicate that the free spinor object as an entity moves with group velocity, and does not decay, we have no think that it does not spread with time. From Eq. (8.7.2.3) we see that $\mathbf{v}_g$ is the group velocity of free spinor wavepacket. Therefore, we set $\mathbf{v} = \mathbf{v}_g$ naturally. This means that if $\mathbf{v}_g \neq 0$, then $\mathbf{x}' = \mathbf{x}'(t=0) + \mathbf{v}_g t$. Thus, $\mathbf{x} = \mathbf{x}' + \mathbf{v}_g t = \mathbf{x}'(t=0) + 2\mathbf{v}_g t$. QED.

8.7.3 Comments for Dirac Equation

To understand this subsection, readers have to read and understand chapter 3 in Ref. [3].

(1). Although the proof on the the corollary XIX is very simple. However, all other textbooks obtain opposite conclusion which say that, for example, the time of a free electron spreading automatically to a very big volume is only 10^{-36} second! Why so small? The mistake in all other textbooks are common: they assume that a free spinor object is a Gauss function, and, without any demonstrations, set the Gauss function satisfies free Dirac equation or Schrӧdinger equation, and obtain strongly spreading solution. The mistake is that the Gauss function is not a solution.

As an interesting exercise, readers can substitute the plane wave with spinor structure amplitude, and calculate the time evaluation. The readers will obtain non-spreading conclusion. For this proof, refer to chapters 3 and 5 of Ref. [3], in which you can find that for the time evaluation of a plane wave, free Schrӧdinger equation gives non-spreading conclusion; On the contrary, for the time evaluation of a Gauss function, free Schrӧdinger equation give fast spreading conclusion.

(2). The wavepacket law in chapter 1 or in chapter 3 of Ref. [3] still holds, i. e., the spontaneous equiprobability symmetry breaking rule still holds. The small changes are that we have to consider probability of every spinor and every component in a spinor.

(3). The appearance probability of antiparticle depends on the size of the spinor objects. For example, we assume the initial spinor wavefunction is

$$\Psi(\mathbf{x}, 0) \propto \begin{bmatrix} 1 \\ 0 \\ 0 \\ 0 \end{bmatrix}. \tag{8.7.3.1}$$

For this initial spinor wavefunction, the u_1 and u_3 in Eq. (8.5.2.3) gives the ratio of appearance probabilities of particle and antiparticle as

$$\begin{aligned}\mathcal{R} &= \frac{|u_1|^2}{|u_3|^2} \\ &= \left(\frac{cp}{m_\circ c^2 + E}\right)^2 \\ &= \frac{v/c}{\sqrt{1-(v/c)^2}} \frac{1}{1+\sqrt{1+(v/c)^2-(v/c)^4}}, \end{aligned} \tag{8.7.3.2}$$

where we assume that $\mathbf{p} = p_z = p$. Eq. (8.7.3.2) indicates that high speed of objects benefit the appearance of antiparticle is always.

One can also have a roughly estimations for the relation between the particle's size and $\mathcal{R}$. Suppose that the particle's size $L >> \lambda_c = \hbar/(m_\circ c) =$ *Compton wavelength*. In this case, $p = \hbar k \lesssim \hbar/L << m_\circ c$. Therefore,

$$\mathcal{R} << \frac{m_\circ c^2}{2m_\circ c^2} = \frac{1}{2}. \tag{8.7.3.3}$$

Eq. (8.7.3.3) indicates that when the particle's size is larger than a Compton wavelength, then the antiparticle solutions are negligible.

In the case that the particle's size $L << \lambda_c = \hbar/(m_\circ c) =$ *Compton wavelength*, $p = \hbar k \lesssim \hbar/L >> m_\circ c$. Therefore,

$$\mathcal{R} \approx \frac{cp}{m_\circ c^2 + cp} \to 1. \tag{8.7.3.4}$$

Eq. (8.7.3.4) indicates that when the particle is confined to a region less than a Compton wavelength, then the antiparticle solutions play an important role.

(4). In all other textbooks such as Ref. [116], the so-called spinor wavepackets are actually the spinor wavefunction other than the Dirac spinor object itself. Please do not make confusion.

Chapter 9

Bases of Quantum Field Theory

9.1 Introduction

Quantum field theory studies the quantization of classical field. The classical fields contain:

(1). The wavefunction of single particle in both nonrelativistic and relativistic quantum mechanics;

(2). The electromagnetic field of Maxwell equation;

(3). Any fields, which are functions of spacetime and are contained in dynamical equations of motion. Such as:

(i). One can look upon the metric tensor of Einstein gravitation equation as a field, makes quantization, and obtain quantized gravitational field equation. But, until now an appropriate quantization of Einstein field equations is unknown;

(ii). Gauge field, which is a variation of the electromagnetic theory. The gauge field theory (Yang-Mills equations) replaces the the Abelian gauge group of electromagnetism by a non-Abelian one.

The wavepacket-only theory has proved that to describe any objects (massive particle or massless particle), we have to use either canonical (position) coordinates, or equivalently, the canonical conjugate momenta. The wavepacket-only theory further proved that there are canonical commutation relations between canonical coordinate and corresponding canonical conjugate momentum. Based solely on these canonical commutation relations, chapter 1 derived all the three formulations of quantum (particle) mechanics, and chapter 8 derived the relativistic quantum (particle) mechanics. From quantum mechanics, we can obtain, for example, the energy quantization of bound state, the symmetry of multiparticle wavefunction of an identical multiparticle system, and so on. One often calls the quantization in nonrelativistic and relativistic quantum mechanics the first quantization. In principle, all objects have to be described by either coordinates or momenta, and thus have to use quantum mechanics, but for the large size and mass, the classical mechanics is a very good approximation of quantum mechanics. This aspect looks like the special theory of relativity. Although, in principle, any objects have to obey the special theory of relativity, but if theirs velocities are low enough, then the nonrelativistic mechanics (Newton mechanics) is a very good approximation of the relativistic mechanics.

This chapter will address the contemporary or newest basic concepts of quantum field theory. Due to the scale of this textbook, we can only state the basic concepts. For example, we do not analyze the requirements for the action that results in Maxwell equations as a equations of motion: Lorentz invariant, Poincaré invariant, gauge invariant, parity and time reversal invariant, and no more than second order in derivatives. This chapter only proves two new basic theorem for quantum field theory, illustrates the basic concepts and the six steps to quantize any classical field, and points out the common and existent serious misunderstanding and mistakes in the quantum field theory in AUNRTB. We belief deeply that when the present quantum field theories are improved along the line pointed out by this chapter, the quantum field theories will obtain a reliable and huge developments.

From this chapter, you will see that:

Klein-Gorden quantum field describes the field with spin zero, and is a scalar field. If the scalar field is real, then this scalar field assigns a real number to every point in spacetime. Scalar means that Alice [who uses coordinates x^μ $\underline{x} \equiv (x^\mu) \equiv (ct, x, y, z)$ and calls the field $\Phi(\underline{x})$] and Bob [who uses coordinates $\overline{x}^\mu$, related to Alice's coordinates by $\overline{x}^\mu = \Lambda^\mu{}_\nu x^\nu + a^\nu$, and calls the field $\overline{\Phi}(\overline{x})$], agree on the numerical value of the field: $\Phi(\underline{x}) = \overline{\Phi}(\overline{x})$. This then implies that the equation of motion for $\Phi(\underline{x})$ must be the same as that for $\overline{\Phi}(\overline{x})$

Dirac quantum field describes the field with spin one half, and is a spinor field.

Maxwell electromagnetic quantum field describes the field with spin one, zero rest mass, and is a vector field.

9.2 Theorems and Steps

9.2.1 Basic Theorems of Quantum Field Theory

For given classical fields such as the electromagnetic field in Maxwell equation and the guide-factor (= wavefunction) of the single particle Schrödinger, Klein-Gorden and Dirac equation, we take the four-vector potential of electromagnetic field and the single particle wavefunction as the field coordinates. Then, we choose appropriate Lagrangian densities (Refer to appendix II: classical field theory.). From these Lagrangian densities we derive Euler-Lagrange equations, which are just the classical Maxwell field equations, Schrödinger equation, and so on. Note that the field coordinate Ψ and potential vector $\mathbf{A}$ themselves are functions of $(\mathbf{r}, t) \equiv (\mathbf{x}, t)$. Therefore, the Lagrangian density is a function of a function, and is called functional instead of function.

Theorem XXXXXVII: Commutation relations of fields: Any independent field coordinate and corresponding canonical conjugate field momentum satisfy its own canonical commutation relation. This commutation relation does not require the hermiticity of the field coordinate and corresponding canonical conjugate field momentum. The types of canonical commutation relations are of, at least, two types: canonical commutation relation and canonical anticommutation relation.

Proof:

Every step in the following proof does not require the hermiticity of the field coordinate and corresponding canonical conjugate field momentum. The Lagrangian density $\mathcal{L}$ is generally a functional of $(\Psi^*(\mathbf{r}, t),\ \Psi(\mathbf{r}, t),\ \partial_\mu \Psi^*, \partial_\mu \Psi)$, $(\partial_\mu) \equiv (\partial/\partial ct, \partial/\partial x, \partial/\partial y\ , \partial/\partial z)$ (refer to appendix). The canonical conjugate field momentum corresponding to field coordinate Ψ is (On the notations refer to appendix II.)

$$\pi(\mathbf{r}, t) = \frac{\partial \mathcal{L}}{\partial \dot{\Psi}(\mathbf{r}, t)}. \tag{9.2.1.1}$$

We just need to imagine a functional $\mathcal{F}(\Psi)$, and require that the Fourier transformation can be made for $\mathcal{F}(\Psi)$ by taking Ψ as an independent variable. According to the theorem I in chapter 1, which says that the dimension of the product of canonical coordinate and corresponding canonical conjugate momentum π is certainly [energy·time], we have

$$\mathcal{F}(\Psi) = \frac{1}{\sqrt{2\pi}} \int d\Psi \mathcal{F}(\pi) e^{i\Psi\pi/\hbar}. \tag{9.2.1.2}$$

$$\mathcal{F}(\pi) = \frac{1}{\sqrt{2\pi}} \int d\pi \mathcal{F}(\Psi) e^{-i\Psi\pi/\hbar}. \qquad (9.2.1.3)$$

Note that from now we use π to express two different quantities. Let us seek the average value of field momentum π

$$\begin{aligned} <\pi> &= \int d\pi \mathcal{F}^*(\pi)\pi\mathcal{F}(\pi) \\ &= \int d\Psi \mathcal{F}^*(\Psi)\left(-i\hbar\frac{\partial}{\partial\Psi}\right)\mathcal{F}(\Psi). \end{aligned} \qquad (9.2.1.4)$$

From Eq. (9.2.1.4) we see that although in the field coordinate, Ψ, representation, the Ψ is a number, but the canonical conjugate field momentum π in the canonical field coordinate representation cannot be a number, and it is instead

$$\pi_i(\mathbf{r}) = -i\hbar\frac{\partial}{\partial\Psi_i(\mathbf{r})}, \qquad (9.2.1.4)'$$

where, for generality, we assume that i can be larger than one to include multicomponent field. For example, if $\overrightarrow{\Psi}$ is a vector, then $i = 1, 2, 3$. Using Eq. (9.2.1.4)' and equal time condition yields

$$\begin{aligned} [\Psi_i(\mathbf{r},t), \pi_j(\mathbf{r}',t)]\mathcal{F}[\Psi(\mathbf{r}')] &\equiv [\Psi_i(\mathbf{r},t)\pi_j(\mathbf{r}',t) - \pi_j(\mathbf{r}',t)\Psi(\mathbf{r})]\mathcal{F}[\Psi_i(\mathbf{r}',t)] \\ &= -i\hbar\left[\Psi_i(\mathbf{r},t)\frac{\partial}{\partial\Psi_j(\mathbf{r}')} - \frac{\partial}{\partial\Psi_j(\mathbf{r}',t)}\Psi_i(\mathbf{r},t)\right]\mathcal{F}[\Psi(\mathbf{r}')] \\ &= i\hbar\left(\frac{\partial\Psi_i(\mathbf{r},t)}{\partial\Psi_j(\mathbf{r}',t)}\right)\mathcal{F}[\psi(\mathbf{r}')] \\ &= i\hbar\delta^{\bigstar}_{ij}(\mathbf{r}-\mathbf{r}')\mathcal{F}[\Psi(\mathbf{r}')]. \end{aligned} \qquad (9.2.1.5)$$

Therefor, the commutation relations are

$$[\Psi_i(\mathbf{r},t), \pi_j(\mathbf{r}',t)] = i\hbar\delta^{\bigstar}_{ij}(\mathbf{r}-\mathbf{r}'). \qquad (9.2.1.5)'$$

$$[\Psi_i(\mathbf{r},t), \Psi_j(\mathbf{r}',t)] = [\pi_i(\mathbf{r},t), \pi_j(\mathbf{r}',t)] = 0.. \qquad (9.2.1.5)''$$

The factor $\delta^{\bigstar}(\mathbf{r}-\mathbf{r}')$ in last equality of Eq. (9.2.1.5) comes from the following general property of functional derivative. For easy to understand, we illustrate with an example. Assume that $\Psi_i(\mathbf{r},t)$, $i = 1, 2, 3$ is a transverse field vector, i. e., if we make Fourier expansion, then the wave vector $\mathbf{k} \perp \overrightarrow{\Psi}$.

$$\begin{aligned} \Psi_i^{\perp}(\mathbf{r},t) &= \int d\mathbf{r}'\delta_{ij}(\mathbf{r}-\mathbf{r}')\Psi_j(\mathbf{r}',t) \\ &= \int d\mathbf{r}'\left(\delta^{\perp}_{ij}(\mathbf{r}-\mathbf{r}') + \delta^{\parallel}_{ij}(\mathbf{r}-\mathbf{r}')\right)\Psi_j(\mathbf{r}',t) \\ &= \int d\mathbf{r}'\delta^{\perp}_{ij}(\mathbf{r}-\mathbf{r}')\Psi_j(\mathbf{r}',t). \end{aligned} \qquad (9.2.1.6)$$

Using Eq. (9.2.1.6) yields

$$\frac{\delta\Psi_i(\mathbf{r},t)}{\delta\Psi_j(\mathbf{r}',t)} = \frac{\partial\Psi_i(\mathbf{r}',t)\delta^{\perp}_{ij}(\mathbf{r}-\mathbf{r}')}{\partial\Psi_j(\mathbf{r}',t)} = \delta^{\perp}_{ij}(\mathbf{r}-\mathbf{r}'). \qquad (9.2.1.6)'$$

Substituting Eq. (9.2.1.6)′ into the third equality of Eq. (9.2.1.5) yields the last equality of Eq. (9.2.1.5).

The meaning of transverse delta function $\delta_{ij}^{\perp}(\mathbf{r}-\mathbf{r}')$ and the longitudinal delta function $\delta_{ii}^{\parallel}(\mathbf{r}-\mathbf{r}')$ are given in appendix.

$$\delta_{ij}(\mathbf{r}-\mathbf{r}') = \delta_{ij}^{\perp}(\mathbf{r}-\mathbf{r}') + \delta_{ij}^{\parallel}(\mathbf{r}-\mathbf{r}'). \qquad (9.2.1.7)$$

In section 9.6 we will explain Eq. (9.2.1.7) by more detail calculations.

It is easy to demonstrate that, generally,

$$[\Psi(\mathbf{r},t), \pi(\mathbf{r}',t) + \mathcal{G}\{\Psi'(\mathbf{r}',t)\}] = i\hbar\delta^{\star}(\mathbf{r}-\mathbf{r}'), \qquad (9.2.1.8)$$

where $\mathcal{G}\{\Psi'(\mathbf{r},t)\}$ is an arbitrary functional of $\Psi'(\mathbf{r},t)$. Eq. (9.2.1.8) is like the known formula in chapter 1

$$[x, p_x + f(x)] = i\hbar, \ [x,x] = 0, \ [p_x, p_x] = 0. \qquad (9.2.1.8)'$$

Next we show that there is, in principle, one other type of commutation relation, i. e., the so called anticommutation relation:

$$[\Psi(\mathbf{r},t), \pi(\mathbf{r}',t)]_{+} \equiv \Psi(\mathbf{r},t)\pi(\mathbf{r}') + \pi(\mathbf{r}',t)\Psi(\mathbf{r},t) = i\hbar\delta^{\star}(\mathbf{r}-\mathbf{r}'), \qquad (9.2.1.9)$$

Of course, to prove that Eq. (9.2.1.9) is a possible commutation relation, we just need to enumerate an example. Next we give this example.

We look upon the wavefunction of Schrödinger equation as a classical field coordinate, and seek the corresponding canonical conjugate field momentum. The Schrodiger equation is

$$i\hbar\frac{\partial\Psi(\mathbf{r},t)}{\partial t} = \left(-\frac{\hbar^2}{2m}\nabla^2 + V\right)\Psi(\mathbf{r},t). \qquad (9.2.1.10)$$

If we look upon the field coordinate Ψ in Eq. (9.2.1.10) as an operator, then Eq. (9.2.1.10) is just the Schrödinger equation in second quantization representation.

If we take the Lagrangian density is

$$\begin{aligned}\mathcal{L}_S(\Psi, \partial_\mu\Psi, \Psi^*, \partial_\mu\Psi^*) &= i\hbar\Psi^*(\mathbf{r},t)\dot{\Psi}(\mathbf{r},t) - \frac{\hbar^2}{2m}\nabla\Psi^*(\mathbf{r},t)\cdot\nabla\Psi(\mathbf{r},t) \\ &- V(\mathbf{r},t)\Psi^*(\mathbf{r},t)\Psi(\mathbf{r},t). \qquad (9.2.1.11)\end{aligned}$$

$\Psi(\mathbf{r},t)$ can be a complex field. Generally, $\Psi(\mathbf{r},t)$ and $\Psi^*(\mathbf{r},t)$ are independent of each other. Applying the principle of least action (Hamilton variational principle)

$$\delta\mathcal{A} = \delta\int d^4x\mathcal{L}_S(\Psi, \partial_\mu\Psi, \Psi^*, \partial_\mu\Psi^*) = 0. \qquad (9.2.1.12)$$

where

$$d^4x \equiv dx^{\circ}dx^1dx^2dx^2 = dctd\mathbf{r} = dctd\mathbf{x}$$

and making variation over Ψ^* yield Euler-Lagrange equation

$$\partial_\mu\frac{\partial\mathcal{L}_S}{\partial(\partial_\mu\Psi^*)} - \frac{\partial\mathcal{L}_S}{\partial\Psi^*} = 0. \qquad (9.2.1.13)$$

With the introduction of the fourth coordinate $x^{\circ} = ct$ in the place of time coordinate, the action integral defined by Eq. (9.2.1.12) changes only by a multiplicative factor but does not affect the formulation of the principle of least action which yields the same Euler-Lagrange equation of motion in Eq. (9.2.1.13). Please note that the only term which depends on x^μ is the first term in Eq. (9.2.1.13). This term is not affected by the change of scale of any x^μ. Substituting Eq. (9.2.1.11) into Eq. (9.2.1.13) yields Eq. (9.2.1.10). Therefore, the choice of the $\mathcal{L}_S$ is appropriate.

The corresponding canonical conjugate momentum $\pi(\mathbf{r},t)$ is

$$\pi(\mathbf{r},t) = \frac{\partial \mathcal{L}_S}{\partial \dot{\Psi}(\mathbf{r},t)} = i\hbar\Psi^*(\mathbf{r},t). \tag{9.2.1.14}$$

Due to that the Lagrangian density does not contain $\dot{\Psi}^*$, the canonical conjugate field momentum of the independent field coordinate Ψ^* is

$$\pi^*(\mathbf{r},t) = \frac{\partial \mathcal{L}_S}{\partial \dot{\Psi}^*(\mathbf{r},t)} \equiv 0. \tag{9.2.1.15}$$

π in the field coordinate representation is an operator. Thus, Eq. (9.2.1.14) becomes

$$\pi(\mathbf{r},t) = \frac{\partial \mathcal{L}_S}{\partial \dot{\Psi}(\mathbf{r},t)} = i\hbar\Psi^+(\mathbf{r},t). \tag{9.2.1.14$'$}$$

We have shown in chapter 2 that:

$$[\Psi(\mathbf{r}), \Psi^+(\mathbf{r}')]_+ = \delta(\mathbf{r}-\mathbf{r}'). \tag{2.6.1.5}$$

Therefore, from Eqs. (2.6.1.5) and (9.2.1.14)$'$ we have an another type of commutation relation:

$$[\Psi(\mathbf{r}), \pi(\mathbf{r}')]_+ = i\hbar\delta^\bigstar(\mathbf{r}-\mathbf{r}'), \tag{9.2.1.9}$$

which is called canonical anticommutation relation.

Corollary XX: Canonical quantization with constraints: The canonical commutation relations of field [Eqs. (9.2.1.8) and (9.2.1.9)] can contain some constraint conditions in spacetime for the field coordinates, and thus can be used to constrained fields.

Proof:
Many gauge fields including the electromagnetic field are constrained field. All the constraint conditions in spacetime are operations of differentiation or integration for the field coordinates and field momenta over spacetime, or may be algebraic operations for the field variables. But, all these operations are independent of field variables Ψ and π. In the representation of field coordinates Ψ, all the above mentioned operations can be looked upon as constants C, C', $\cdots$, independent of field variables Ψ and π. We can make a transformation $\Psi \to \Psi' = C\Psi + C'$, and so on. The canonical commutation relations, Eqs. (9.2.1.8) and (9.2.1.16), are still hold for Ψ' and π'. We call these canonical quantization the canonical quantization with constraints.

According to this corollary, we suggest that for the quantization of the constrained classical fields, including the gauge fields, people do not need to use the farfetched Gupta-Bleuler indefinite metric operator method [121], and the tedious and overelaborate Dirac bracket method [122]. In section 9.6 we will take the Maxwell electromagnetic field, which belongs to one kind of gauge fields, as a quantization example of constrained system to address how to use this corollary.

Theorem XXXXXVIII: Heisenberg uncertainty relation of field: If and only if both the canonical field coordinate and corresponding canonical conjugate field momentum are hermitian, and both satisfies canonical commutation relation, then there is uncertainty relation between them:

$$\sqrt{\overline{(\Delta\Psi)^2}}\sqrt{\overline{(\Delta\pi)^2}} \geqq \frac{1}{2}|\overline{[\Psi,\pi]}|, \qquad (9.2.1.17)$$

where

$$\Delta\Psi = \Psi - \overline{\Psi}, \qquad \overline{\Psi^+\Psi} = \int d\Psi \mathcal{F}^*(\Psi)\Psi^+\Psi\mathcal{F}(\Psi). \qquad (9.2.1.18)$$

Proof:

$$\begin{aligned} & (\Psi\mathcal{F}(\Psi) + \pi\mathcal{F}(\Psi), \Psi\mathcal{F}(\Psi) + \pi\mathcal{F}(\Psi)) \\ = & (\Psi\mathcal{F} + \pi\mathcal{F}, \Psi\mathcal{F} + \pi\mathcal{F}) \\ = & (\Psi\mathcal{F}, \Psi\mathcal{F}) + (\Psi\mathcal{F}, \pi\mathcal{F}) + (\pi\mathcal{F}, \Psi\mathcal{F}) + (\pi\mathcal{F}, \pi\mathcal{F}). \end{aligned} \qquad (9.2.1.19)$$

Using Cauchy-Schwarz inequality yields

$$\begin{aligned} (\Psi\mathcal{F}, \Psi\mathcal{F}) + (\pi\mathcal{F}, \pi\mathcal{F}) & \geq \frac{1}{2}|(\Psi\mathcal{F}, \pi\mathcal{F}) + (\pi\mathcal{F}, \Psi\mathcal{F})| \\ & = \frac{\hbar}{2}\left|\int d\Psi \left[\Psi\mathcal{F}^*\left(\frac{d}{d\Psi}\mathcal{F}\right) + \left(\frac{d}{d\Psi}\mathcal{F}^*\Psi\mathcal{F}\right)\right]\right| \\ & = \frac{\hbar}{2}\left|\int d\Psi\Psi\frac{d}{d\Psi}\left(\mathcal{F}^*\mathcal{F}\right)\right| \\ & = \frac{\hbar}{2}\left|\int d\Psi\mathcal{F}^*\mathcal{F}\right| \\ & = \frac{\hbar}{2}. \end{aligned} \qquad (9.2.1.20)$$

Eq. (9.2.1.20) can be rewritten as

$$\sqrt{\overline{\Psi^2} \times \overline{\pi^2}} \geqq= \frac{\hbar}{2}. \qquad (9.2.1.21)$$

Define

$$\Delta\Psi = \Psi - \overline{\Psi}, \quad \Delta\pi = \pi - \overline{\pi}. \qquad (9.2.1.22)$$

It is easy to obtain

$$\sqrt{\overline{(\Delta\Psi)^2}}\sqrt{\overline{(\Delta\pi)^2}} \geqq \frac{\hbar}{2}. \qquad (9.2.1.23)$$

Theorem XXXXXIX: Poincaré invariance of canonical commutation relations: The canonical commutation relations are consistent with Poincaré invariant.

Proof:
Refer to Ref. [123]. QED.

9.2.2 Standard Six Steps of Field Quantization

From theorem XXXXXVII and corollary XIX, the procedures to quantize any classical field in terms of canonical quantization method are the next six standard steps:

(1). First step. Write the classical field equation, which you want to quantize;

(2). Second step. Look upon the classical fields as field coordinates. Seek a appropriate and suitable Lagrangian density, from which the classical field equation can be derived; This Lagrangian density is a functional of field coordinates and first order derivative of the field coordinates at spacetime points, must be Hermitian, and must be Lorentz covariant. We require also that this Lagrangian density to be quadratic in field coordinates and theirs Hermitian conjugate field momenta, because this will lead to a linear equation of motion, with a solutions satisfying superposition theorem. Often, we want the plane wave superposition solutions because plane waves describe free particles, and thus is the starting point for a theory of interacting particles.

About this second point, we would like to say here some more.

(i). If we encounter a problem that the canonical field coordinates, such as electromagnetic four vector potential $\underline{A} \equiv [\phi(\mathbf{r}, t), \mathbf{A}(\mathbf{r}, t)]$, suffers constraint conditions, then the treatments at this second step are as follows. We illustrate with an example. Suppose that the constraint is $\nabla \cdot \mathbf{A}(\mathbf{r}, t) = 0$. This is a transverse field condition. We take a new four vector potential $\underline{A}'$, which is a transverse field coordinates. For $\underline{A}'(\mathbf{r}, t)$ the constraint condition is satisfied automatically. Use the new four vector potential $\underline{A}' = (\phi, \mathbf{A}')$ we can also write an appropriate Lagrangian density, from which we can also obtain classical Maxwell equation set. Using the canonical commutation relations given by Eq. (9.2.1.8) for bosons [Eq. (9.2.1.16) for fermions], the task of quantization of classical Maxwell field equation is complemented, i. e., the classical Maxwell field equation set with constraint condition becomes a quantum Maxwell field equation set. We well address the quantization of constrained field in detail in section 9.6.

(ii). If the classical field equation of motion comes from single particle quantummechanical equation of either nonrelativistic or relativistic, then we take the wavefunctions in Schrödinger picture as quantum field coordinates in Heisenberg picture. If the classical field equation of motion does not come from single particle quantummechanical equation of either nonrelativistic or relativistic, then we take the classical field quantities, for example, the four vector potential of electromagnetic field, as quantum field coordinates in Heisenberg picture.

By the way, here we have to emphasize that the field quantization is equivalent to quantize the guider-factor of the wavepacket-only theory, and thus the motion of body-factor of an object will be guided by the quantized field coordinate according to the spontaneous equiprobability symmetry breaking rule. Many, if not all, until now references and textbooks think that in comparison with the object (particle), the field is more basic. Some references and textbooks even confirm that there are only various fields in our universe. This textbook thinks that the object looks like a blind mother, and the guide-factor, including both the field in quantum field theory and wavefunction in quantum mechanics, looks like a son with eye. The mother and the son are always together, depend on each other for survival. Note that here the guide is only a probabilistic guide, and is not deterministic. Therefore, the object's motion does not obey various conservation laws, and general and special theory of relativity. Ref. [3] call this object's motion is a still more microscopic process. Only in average meaning, various conservation laws might be violated.

(3). Third step. According to definition of classical mechanics, write the corresponding canonical conjugate momentum;

(4). Fourth step. Based on the theorem XXXXXVII, you can say that the field coordinate and the corresponding canonical conjugate field momentum satisfy definite canonical commutation relations, then the previous classical field equation, mentioned in the first and third steps, becomes naturally a quantized field equation. At this time we accomplish the task to quantize the classical field equation, and enter into the quantum field theory. Note that at this step the quantum field equation is still in the representation of the field coordinate representation, and thus the field coordinate is still a number. We call this step the canonical quantization of field. In essence, the canonical quantization of field is same as the canonical quantization of particle. Some textbooks call the canonical quantization of particle the classical canonical quantization (refer to exercise). Some textbooks call canonical quantization of field and particle the second and first quantization, respectively. We should remember that they a the same in essence.

Let us interpret this point (4) in more detail. As we know, our Lagrangian density is appropriate. Why "appropriate"? Because if we put this Lagrangian density into Euler-Lagrange equation, we can obtain the previous classical field equation mentioned in the first step. Now we just say that the field canonical coordinate and field corresponding canonical momentum satisfy canonical commutation relation (commutation or anticommutation), but the Lagrangian density does not have a wee bit of changes formally. If we put the Lagrangian density into the Eukler-Lagrangian equation, it is natural that we obtain the same form of motion equation as the previous in the first step, but in physics the previous classical field equation becomes quantum field equation.

The above formulation of field canonical quantization is called Lagrangian formulation of field quantization. In fact, there is an another canonical quantization formulation, which is called Hamiltonian formulation of field canonical quantization. We will explain the Hamiltonian formulation, and make comparison between the two formulations of field quantization in section 9.3. The Hamiltonian formulation of field quantization requires a precondition, i. e., the canonical conjugate field momentum can be expressed as a function of canonical field coordinates and velocities. This precondition cannot be satisfied in many cases, and in these cases although the field quantization still can be done, but it becomes more trouble.

We will interpret by Klein-Gorden equation in section 9.4 that to get physically meaningful results, some times we can only choose the canonical commutation relation other then the canonical anticommutation relation, and by nonrelativistic Schrödinger equation in section 9.3 that to get physically meaningful results, some times both the commutation and anticommutation relations give physical meaningful results. In this aspect, the field canonical quantization (or the second quantization) is different from the first quantization. For the latter, just in terms of canonical quantization (use the position coordinate and momentum commutation relations) the classical nonrelativistic particle mechanics can become quantum nonrelativistic particle mechanics (i. e., quantum mechanics) by any one in the three formulations of classical nonrelativistic quantum mechanics (see chapter 1) (Schrödinger equation = nonrelativistic quantum mechanics), and the classical relativistic particle mechanics becomes relativistic quantummechanical Klein-Gorden or Dirac equation of particles (see chapter 8). We will demonstrate as well that the quantization of classical Maxwell (field) equation gives physically meaningful results only in case of using the commutation relations other then the anticommutation relation.

Our experiences on this fourth step are as follows: To get the physically meaningful results, the classical relativistic field equations with spin zero or integers and with spin half-integers can only be quantized by canonical commutation relation (e. g., Klein-Gorden equation, Maxwell equation set) and canonical anticommutation relations (e. g., Dirac equation), respectively. On the contrary, the classical nonrelativistic field equation such as Schrödinger field equation can be quantized by both the canonical commutation and anticommutation relations;

That, we set the field coordinate and corresponding canonical conjugate momentum satisfy canonical commutation relations, means the field coordinate and field momentum become operators of Heisenberg picture.

It is important to have clearly in mind the meaning of time derivatives in classical and quantum field theories [8]. In classical particle theory, both total and partial time derivatives were define in connection with a function $F(x_i, p_i, t)$ of the coordinates, momenta, and the time; these derivatives are related through equation of motion:

$$\frac{dF}{dt} = \frac{\partial F}{\partial t} + \{F, H\}_{Poisson\ bracket}. \tag{9.2.2.1}$$

Similarly, both derivatives were defined for a Heisenberg picture operator and related to each other as in equation:

$$\frac{d\Omega_H}{dt} = \frac{\partial \Omega_H}{\partial t} + \frac{1}{i\hbar}[\Omega_H, H]. \tag{9.2.2.2}$$

In classical field theory, $\Psi(\mathbf{r})$ is the analog of x_i, and the only time derivative that can be defined is $\partial\Psi/\partial t$; we refer to it as $\dot{\Psi}$ in analogy with $\dot{x}_i$ in the particle case. Thus, in the classical Hamiltonian equations of motion of the field, we interpret $\dot{\Psi}$, and also $\dot{\pi}$, as partial time derivatives. However a functional $\mathcal{F}(\Psi, \pi, t)$ can depend explicitly on the time as well as on the field, so that it is important to distinguish between $d\mathcal{F}/dt$ and $\partial\mathcal{F}/\partial t$ in the field equation of motion. The same situation appears in quantum field theory. No distinction can be made between $d\Psi/dt$ and $\partial\Psi/\partial t$, and both referred to as $\dot{\Psi}$. On the other hand, a Heisenberg picture operator can depend explicitly on the time, and the distinction between the two time derivatives must be made in the equation:

$$\frac{d\mathcal{F}_H}{dt} = \frac{\partial \mathcal{F}_H}{\partial t} + \frac{1}{i\hbar}[\mathcal{F}_H, H]. \tag{9.2.2.3}$$

(5). Fifth step. Expand the field coordinates by any complete bases, which are the solutions of the previous single particle equations (e. g., the nonrelativistic single particle Schrödinger equation or relativistic (Dirac, Klein-Gorden and so on) or the free classical field equation (e. g., free Maxwell equation). For the nonrelativistic single particle Schrödinger equation, the complete bases can be the Bloch waves, the plane wave, and so on. Here, we note that the expansions of field coordinates mean that we use the expansion coefficients to express the field coordinates. This further means that we go from the field coordinate representation to, for example, energy representation of single energy. In sections 9.3 and 9.4 we will prove two theorems, which indicate that this representation transformation leads to that the expansion coefficients become creation and annihilation operators. Thus, the field coordinates become operators as well. If the field coordinate is looked upon as an operator, then the the expansion coefficient is annihilation operator of that mode. In case of relativistic quantum mechanics, such as Klein-Gorden and Dirac single particle equations, we take the plane wave solutions of the free particle still is convenient.

In practice, one can do this step before the above fourth step. In this case the previous fourth step becomes fifth step. However, if you like to do so, then remember that the field coordinate at your fourth step is still a numerical constant, and thus you should write Ψ^* instead of Ψ^+. Just after you think that the field coordinate and corresponding canonical conjugate momentum satisfy a definite commutation relation, the Fourier expansion coefficients become annihilate and creation operators.

(6). Sixth step. If you stop at the fifth step, then the power of quantum field theory has not yet manifested. Only after introducing the interaction between fields the power of quantum field theory just manifests. For example, interaction between electron and photon fields gives rise to a variety of phenomena such as Compton effect, electron-positron pair creation, electron-positron pair annihilation, electron-positron scattering, Bremmstrahlung radiation, muon pair production in electron-positron collision, and hadron production. Due to the scale of this textbook we can only introduce the correct concepts of the elementary quantum field theory, critique the main misunderstanding and mistakes in the quantum field theory in AUNRTB, help the readers to establish the correct concepts, and cannot make more discussions and calculations on the interactions between fields. We sincerely hope that readers along our lines, pointed in section 9.7, to promote the development the quantum field theory successfully.

9.3 Quantization of Schrödinger Equation

(1). First step. We want to quantize the following classical nonrelativistic Schrödinger equation:

$$i\hbar\frac{\partial\Psi(\mathbf{r},t)}{\partial t}=\left(-\frac{\hbar^2}{2m}\nabla^2+V\right)\Psi(\mathbf{r},t). \tag{9.3.1}$$

(2). Second step. Look upon $\Psi(\mathbf{r})$ as field coordinate, and choose the following Lagrangian density:

$$\begin{aligned}\mathcal{L}_S(\Psi,\partial_\mu\Psi,\Psi^*,\partial_\mu\Psi^*) &= i\hbar\Psi^*(\mathbf{r},t)\dot{\Psi}(\mathbf{r},t)-\frac{\hbar^2}{2m}\nabla\Psi^*(\mathbf{r},t)\cdot\nabla\Psi(\mathbf{r},t)\\ &- V(\mathbf{r},t)\Psi^*(\mathbf{r},t)\Psi(\mathbf{r},t).\end{aligned} \tag{9.3.2}$$

If we substitute Eq. (9.3.2) into the following Euler-Lagrange equation

$$\partial_\mu\frac{\partial\mathcal{L}_S}{\partial(\partial_\mu\Psi)}-\frac{\partial\mathcal{L}_S}{\partial\Psi}=0, \tag{9.3.3}$$

then we obtain Eq. (9.3.1). Therefore, the choice of the $\mathcal{L}_S$ in Eq. (9.3.2) is appropriate.

(3). Third step. From the definition, the corresponding canonical conjugate momentum $\pi(\mathbf{r},t)$ is

$$\begin{aligned}\pi(\mathbf{r},t) &= \frac{\partial\mathcal{L}_S}{\partial\dot{\Psi}(\mathbf{r},t)}=i\hbar\Psi^*(\mathbf{r},t)\\ &\rightarrow i\hbar\Psi^+(\mathbf{r},t).\end{aligned} \tag{9.3.4}$$

The second equality of Eq. (9.3.4) comes from that the π is an operator, and thus the $\Psi^*(\mathbf{r},t)$ have to transformed into an operator.

Due to that the Lagrangian density does not contains $\dot{\Psi}^*$, the corresponding canonical conjugate momentum of the independent field coordinate Ψ^* is:

$$\pi^*(\mathbf{r},t)=\frac{\partial\mathcal{L}_S}{\partial\dot{\Psi}^*(\mathbf{r},t)}\equiv 0. \tag{9.3.5}$$

(4). Fourth step. Set the field coordinate Ψ and the corresponding canonical conjugate momentum π satisfy the canonical commutation relation, proved by theorem XXXXXVII.

$$[\Psi(\mathbf{r}),\pi(\mathbf{r}')]=i\hbar\delta(\mathbf{r}-\mathbf{r}')=[\Psi(\mathbf{r}),i\hbar\Psi^+(\mathbf{r}')]. \tag{9.3.6}$$

The previous classical field equation Eq. (9.3.1) becomes quantized nonrelativistic Schrödinger field equation, which can describe the bosons.

Set the field coordinate Ψ and the corresponding canonical conjugate momentum π satisfy canonical anticommutation relation

$$[\Psi(\mathbf{r}),\pi(\mathbf{r}')]_+=i\hbar\delta(\mathbf{r}-\mathbf{r}')=[\Psi(\mathbf{r}),i\hbar\Psi^+(\mathbf{r}')]_+. \tag{9.3.7}$$

The previous classical field equation Eq. (9.3.1) becomes quantized nonrelativistic Schrödinger field equation, which can describe the fermions.

From the second quantization in chapter 2 we know that in the case of nonrelativistic Schrödinger equation, both the commutation and anticommutation relations give physically meaningful results.

The above formulation of field canonical quantization is called Lagrange formulation. Next, we introduce the Hamiltonian field quantization formulation. Eq. (9.3.4) gives the field conjugate momentum π. If we can solve this equation to obtain the $\dot{\Psi}$ as a function $\dot{\Psi} = f(\Psi, \nabla\Psi, \pi, \Psi^*, \nabla\Psi^*)$, then this is equivalent to introducing new variable π to replace the old variable $\dot{\Psi}$. Therefore, the essence of introducing π is a transformation of the field variables. To make this transformation, we often use the Legendre transformation:

$$\mathcal{H}(\underline{x}) = \pi(\underline{x})\dot{\Psi}(\underline{x}) - \mathcal{L}_S(\underline{x}), \tag{9.3.8}$$

where $\mathcal{H}(\underline{x})$ is the Hamiltonian density, and its variables are $(\pi,\ \Psi)$. The total Hamiltonian of a system is

$$H(t) = \int d\mathbf{r}\mathcal{H}(\underline{x}) = \int d\mathbf{r}\left(\pi(\underline{x})\dot{\Psi}(\underline{x}) - \mathcal{L}_S(\underline{x})\right). \tag{9.3.9}$$

From Eq. (9.3.9) we enter into the Hamiltonian formulation. If the system has constraint conditions, then we call this system the singular Lagrangian system or constrained Hamiltonian system. For these singular Lagrangian systems we cannot solve the Eq. (9.3.4) to get the function of $\dot{\Psi} = f(\Psi, \nabla\Psi, \pi, \Psi^*, \nabla\Psi^*)$. In this case to use the Hamiltonian formulation to quantize the classical field is not convenient. This is the basic reason, why people like to use the field quantization formulation of the Lagrangian other than the Hamiltonian. For our present case, we can use the Hamiltonian formulation.

In Hamiltonian formulation, the classical field coordinate equation of motion is

$$\dot{\Psi} = \frac{d\Psi(\mathbf{r},t)}{dt} = \{\Psi(\mathbf{r},t), H(t)\}_{Poisson\ bracket}. \tag{9.3.10}$$

We can prove that the Hamiltonian field quantization formulation is same as the Heisenberg formulation of quantum mechanics (refer to Ref. [3].), i. e., using the canonical commutation relations to replace the classical Poison bracket. Therefore, the equation of motion of quantized field coordinate Ψ is

$$\dot{\Psi} = \frac{d\Psi(\mathbf{r},t)}{dt} = \frac{1}{i\hbar}\left[\Psi(\mathbf{r},t), H(t)\right]. \tag{9.3.11}$$

From Eq. (2.6.4.6) we have

$$H(t) = \int d\mathbf{r}'\Psi^+(\mathbf{r}',t)\left[-\frac{\hbar^2}{2m}\nabla^2 + V(\mathbf{r}',t)\right]\Psi(\mathbf{r}',t). \tag{9.3.12}$$

Here, we should use the comment on the time derivative of Ψ in section 9.3 [8]. In field theory, Ψ is an analog of position coordinate $\mathbf{r}_i$, and the only time derivative that can be define is $\partial\Psi/\partial t$; we refer to it as $\dot{\Psi}$ in analogy with $\dot{\mathbf{r}}_i$ in the particle case. Thus, in the Hamiltonian equation of motion of the field coordinate Ψ in Eq. (9.3.11), we interpret $\dot{\Psi}$ and also $\dot{\pi}$, as partial time derivatives. However a functional $\mathcal{F}(\Psi, \pi, t)$ can depend explicitly on the time as well as on the field, so that it is important to distinguish between $d\mathcal{F}/dt$ and $\partial\mathcal{F}/\partial t$.

Substituting Eq. (9.3.12) into Eq. (9.3.11), we obtain the equation of motion of field operator:

$$\begin{aligned}\dot{\Psi}(\mathbf{r},t) &= \int d\mathbf{r}'\frac{1}{i\hbar}\left[\Psi(\mathbf{r},t),\Psi^+(\mathbf{r}',t)\left(-\frac{\hbar^2}{2m}\nabla'^2+V(\mathbf{r}',t)\right)\Psi(\mathbf{r}',t)\right] \\ &= \frac{1}{i\hbar}\int d\mathbf{r}'\left\{\delta(\mathbf{r}-\mathbf{r}')\left(-\frac{\hbar^2}{2m}\nabla'^2+V(\mathbf{r}',t)\right)\Psi(\mathbf{r}',t)\right\} \\ &= \frac{1}{i\hbar}\left(-\frac{\hbar^2}{2m}\nabla^2+V(\mathbf{r},t)\right)\Psi(\mathbf{r},t). \end{aligned} \quad (9.3.13)$$

Eq. (9.3.13) is same as the Schrödinger field equation from the Lagrangian field quantization.

(5). Look upon the field coordinate as a field annihilation operator, which can annihilate a particle at any space point, and the Hermitian conjugate operator represents a field creation operator, which can create a particle at any space point. Take the eigenfunctions $\psi_k(\mathbf{r})\exp(-iE_k t\hbar)$ of the classical Schrödinger equation as orthonormalized complete basis. Expand the field annihilation and creation operators

$$\Psi(\mathbf{r},t) = \sum_k a_k\psi_k(\mathbf{r})e^{-iE_kt/\hbar}. \quad (9.3.14)$$

$$\Psi^+(\mathbf{r},t) = \sum_k a_k^+\psi_k^*(\mathbf{r})e^{iE_kt/\hbar}. \quad (9.3.15)$$

Utilizing the orthonormal property of $\{\psi_k\}$, we can obtain

$$a_k = \int d\mathbf{r}\Psi(\mathbf{r},t)\psi_k^*(\mathbf{r})e^{iE_kt/\hbar}. \quad (9.3.16)$$

$$a_k^+ = \int d\mathbf{r}\Psi^+(\mathbf{r},t)\psi_k(\mathbf{r})e^{-iE_kt/\hbar}. \quad (9.3.17)$$

If $\Psi(\mathbf{r},t)$, $\Psi^+(\mathbf{r},t)$ satisfy canonical commutation relations, then substituting Eqs. (9.3.14-15) into the commutation relations yields

$$[a_k, a_{k'}^+] = \delta_{kk'}, \quad [a_k, a_{k'}] = [a_k^+, a_{k'}^+] = 0. \quad (9.3.18)$$

If $\Psi(\mathbf{r},t)$, $\Psi^+(\mathbf{r},t)$ satisfy canonical anticommutation relations, then substituting Eqs. (9.3.14-15) into the anticommutation relations yields

$$[a_k, a_{k'}^+]_+ = \delta_{kk'}, \quad [a_k, a_{k'}]_+ = [a_k^+, a_{k'}^+]_+ = 0. \quad (9.3.19)$$

(6). The six step in case of Schrödinger equation is familiar.

9.4 Quantization of Klein-Gorden Equation

9.4.1 On Relativistic Field

In relativistic field theory, all the quantities that define the field and the associated equations should be covariant under Lorentz transformation. This means that the quantities should be either world scalar or world vector or world tensors. The action integral S, the Lagrangian density, and the Hamiltonian density are world scalar. The four dimensional volume element d^4x is invariant under Lorentz transformation. Let us now show that the second rank tensor constructed from the Lagrangian density using the Euler-Lagrange equation of motion is a world tensor of second rank, the components of which yield the momentum and energy density of the field.

Given the Lagrangian density

$$\mathcal{L} = \mathcal{L}(\Psi, \frac{\partial \Psi}{\partial x^\mu}, x^\mu), \tag{9.4.1.1}$$

the total derivative of the Lagrangian density is given by

$$\frac{d\mathcal{L}}{dx^\mu} = \frac{\partial \mathcal{L}}{\partial \Psi}\frac{\partial \Psi}{\partial x^\mu} + \frac{\partial \mathcal{L}}{\partial\left(\frac{\partial \Psi}{\partial x^\nu}\right)}\frac{\partial}{\partial x^\mu}\left(\frac{\partial \Psi}{\partial x^\nu}\right) + \frac{\partial \mathcal{L}}{\partial x^\mu}. \tag{9.4.1.1$'$}$$

From the Euler-Lagrange equation Eq. (9.2.1.13) we obtain the expression

$$\frac{\partial \mathcal{L}}{\partial \Psi} = \frac{\partial}{\partial x^\nu}\frac{\partial \mathcal{L}}{\partial\left(\frac{\partial \Psi}{\partial x^\nu}\right)}. \tag{9.4.1.2}$$

Substituting this Eq. (9.4.1.2) into Eq. (9.4.1.1), we get

$$\begin{aligned}\frac{d\mathcal{L}}{dx^\mu} &= \frac{\partial}{\partial x^\nu}\frac{\partial \mathcal{L}}{\partial\left(\frac{\partial \Psi}{\partial x^\nu}\right)}\frac{\partial \Psi}{\partial x^\mu} + \frac{\partial \mathcal{L}}{\partial\left(\frac{\partial \Psi}{\partial x^\nu}\right)}\frac{\partial}{\partial x^\mu}\left(\frac{\partial \Psi}{\partial x^\nu}\right) + \frac{\partial \mathcal{L}}{\partial x^\mu} \\ &= \frac{d}{dx^\nu}\left\{\frac{\partial \mathcal{L}}{\partial\left(\frac{\partial \Psi}{\partial x^\nu}\right)}\frac{\partial \Psi}{\partial x^\mu}\right\} + \frac{\partial \mathcal{L}}{\partial x^\mu}. \end{aligned} \tag{9.4.1.2$'$}$$

On rearrangement, we get

$$\frac{d}{dx^\nu}\left\{\frac{\partial \mathcal{L}}{\partial\left(\frac{\partial \Psi}{\partial x^\nu}\right)}\frac{\partial \Psi}{\partial x^\mu} - \mathcal{L}\delta_{\mu\nu}\right\} = -\frac{\partial \mathcal{L}}{\partial x^\mu}. \tag{9.4.1.3}$$

The quantity within the curly bracket is a second rank tensor, denoted by the symbol $T^{\mu\nu}$.

$$T^{\mu\nu} = \frac{\partial \mathcal{L}}{\partial\left(\frac{\partial \Psi}{\partial x^\nu}\right)}\frac{\partial \Psi}{\partial x^\mu} - \mathcal{L}\delta_{\mu\nu}. \tag{9.4.1.4}$$

If the Lagrangian density $\mathcal{L}$ does not depend explicitly on x^μ, then

$$\frac{\partial \mathcal{L}}{\partial x^\mu} = 0. \tag{9.4.1.5}$$

Eq. (9.4.1.5) means that this field is free from forces, in which there are no sources or sinks or any interaction with other fields or particles. It follows from Eq. (9.4.1.3) that for a free field

$$\sum_\nu \frac{dT^{\mu\nu}}{dx^\nu} = 0. \tag{9.4.1.6}$$

It can easily be verified that

$$\begin{aligned} T^{\circ\circ} &= \frac{\partial \mathcal{L}}{\partial\left(\frac{\partial \Psi}{\partial x^\circ}\right)}\frac{\partial \Psi}{\partial x^\circ} - \mathcal{L} \\ &= \frac{\partial \mathcal{L}}{\partial \dot{\Psi}}\dot{\Psi} - \mathcal{L} \\ &= \pi\dot{\Psi} - \mathcal{L} = \mathcal{H}(\underline{x}), \end{aligned} \tag{9.4.1.7}$$

where $\mathcal{H}(\underline{x})$ is Hamiltonian density (see Eq. (9.3.8)).

The other components of the tensor are

$$\begin{aligned} T^{k\circ} &= \frac{\partial \mathcal{L}}{\partial\left(\frac{\partial \Psi}{\partial x^\circ}\right)}\frac{\partial \Psi}{\partial x^k} = c\frac{\partial \mathcal{L}}{\partial \dot{\Psi}}\frac{\partial \Psi}{\partial x^k} \quad (k = 1,2,3) \\ &= c\pi\frac{\partial \Psi}{\partial x^k} \equiv c\mathcal{P}^k, \end{aligned} \tag{9.4.1.8}$$

where the $\mathcal{P}^k$ is the defined three components of momentum density.

9.4.2 One Component Real Field

(1). First step. Write the classical field equation. We want to quantize a simplest Klein-Gorden equation: one component and real field. The field equation is

$$
\begin{aligned}
(\Box+\kappa^2)\Phi(\underline{x}) &\equiv (\partial^\mu\partial_\mu+\kappa^2)\Phi(\underline{x}) \\
&\equiv \left(\frac{1}{c^2}\frac{\partial^2}{\partial t^2}-\nabla^2+\frac{m^2c^2}{\hbar^2}\right)\Phi(\underline{x})=0. \qquad (9.4.2.1)
\end{aligned}
$$

Hereafter, for convenience, we express the rest mass $m_\circ$ as m, and use natural units ($\hbar = c = 1$). In natural units, Eq. (9.4.2.1) becomes

$$(\partial^\mu\partial_\mu+m^2)\Phi(\underline{x})=0. \qquad (9.4.2.2)$$

(2). Second step. To choose an appropriate Lagrangian density from which the Eq. (9.4.2.1) can be derived. It is easy to verify that

$$
\begin{aligned}
\mathcal{L} &= \frac{1}{2}\left[\sum_\mu\left(\frac{\partial\Phi}{\partial x_\mu}\right)^2-m^2\Phi^2\right]+\Omega_0 \\
&= \frac{1}{2}[\partial^\mu\Phi(\underline{x})\partial_\mu\Phi(\underline{x})-m^2\Phi^2(\underline{x})]+\Omega_0 \qquad (9.4.2.3)
\end{aligned}
$$

is the appropriate Lagrangian density.

(3). Third step. Take $\Phi(\underline{x})$ as the field coordinate. Define the corresponding canonical conjugate field momentum:

$$\pi(\underline{x})=\frac{\partial\mathcal{L}}{\partial\dot{\Phi}}=\partial_\circ\Phi=\dot{\Phi}. \qquad (9.4.2.4)$$

(4). Fourth step. According to the theorem in section 9.2, we can try both the commutation and anticommutation relations. In section 9.4.3 we shall show that for the classical Klein-Gorden field equation the anticommutation relation cannot give physically meaningful results. Therefore, here we choose the commutation relation. After choosing the canonical commutation relation, the previous classical Klein-Gorden field equation Eqs. (9.4.2.1-2) become quantized Klein-Gorden field equation.

(5). If we stop at the fourth step, then although we accomplish the task of quantization of Klein-Gorden equation, but we still stop at the field coordinate representation. The power of quantum field theory is in the using the particle number representation or Fock space, in which the variation of particle number can be described.

We should take the eigenfunctions of the previous classical field equation in the first step, which is a single particle relativistic quantummechanical equation, as a complete basis. For convenience, we assume that the eigenfunctions are orthonormal. It is obvious that the plane waves are the solutions of Eq. (9.4.2.2):

If the particle is in a box with volume V, then the $\mathbf{k}$ take discrete values. In this case we should take the box normalization for the plane waves

$$\int_V d\mathbf{r}\frac{1}{\sqrt{V}}e^{i\mathbf{k}\cdot\mathbf{r}}\frac{1}{\sqrt{V}}e^{i\mathbf{k}'\cdot\mathbf{r}}=\delta_{\mathbf{k},\mathbf{k}'}.$$

The possible solutions are

$$\phi_k(\underline{x})=\frac{1}{\sqrt{V}}e^{-i\underline{k}\cdot\underline{x}}=\frac{1}{\sqrt{V}}e^{i\mathbf{k}\cdot\mathbf{r}-i\omega_k t}=\frac{1}{\sqrt{V}}e^{i\mathbf{k}\cdot\mathbf{r}+i\omega_k t} \ or\ ,\ \omega_k=+\sqrt{k^2+m^2} \qquad (9.4.2.5a)$$

However, if the particle is in an infinitely large space, then the $\mathbf{k}$ take continuous values. In this case we should use Dirac delta function normalization for the plane waves

$$\int_{-\infty}^{\infty} d\mathbf{r}\frac{1}{\sqrt{2\pi}}e^{i\mathbf{k}\cdot\mathbf{r}}\frac{1}{\sqrt{2\pi}}e^{i\mathbf{k}'\cdot\mathbf{r}} = \delta(\mathbf{k}-\mathbf{k}').$$

The possible solution is

$$\phi_k(\underline{x}) = \frac{1}{(2\pi)^{3/2}}e^{-i\underline{k}\cdot\underline{x}} = \frac{1}{(2\pi)^{3/2}}e^{i\mathbf{k}\cdot\mathbf{r}-i\omega_k t} \; or \; = \frac{1}{(2\pi)^{3/2}}e^{i\mathbf{k}\cdot\mathbf{r}+i\omega_k t}, \tag{9.4.2.5b}$$

where

$$\underline{k} = (k^\circ, k^1, k^2, k^3) = (\omega_k, \mathbf{k}), \quad \underline{x} = (x^\circ, x^1, x^2, x^3) = (t, x, y, z) = (t, \mathbf{x}) = (t, \mathbf{r}), \tag{9.4.2.6}$$

$$\underline{k}\cdot\underline{x} = k^\mu x_\mu = k^\circ x^\circ - \mathbf{k}\cdot\mathbf{r} = \omega_k t - \mathbf{k}\cdot\mathbf{x}. \tag{9.4.2.7}$$

$$\underline{k}^2 = k^\mu k_\mu = \omega_k^2 - \mathbf{k}\cdot\mathbf{k} = \omega_k^2 - k^2 = m^2. \tag{9.4.2.8}$$

A four-momentum $\underline{k}$ that obeys Eq. (9.4.2.8) is said to be on the mass shell, or on shell for short, if

$$\omega_k^2 = k^2 + m^2. \tag{9.4.2.9}$$

We use the convention of representing the four-vector by $\underline{k}$ and $\underline{x}$, instead of the three dimensional vector by $\mathbf{k}$ and $\mathbf{x}$ ($=\mathbf{r}$). Note that if the k and r appear in the subscripts, then they represent three dimensional vectors. Now we expand the field coordinate by $\phi_k(\underline{x})$, which means that we make a representation transformation from the field coordinate representation Φ into the particle number representation, the single particle states are the eigenstates of energy and momentum. After representation transformation, the $\Phi(\underline{x})$ becomes an operator.

Next, we want to determine the normalization constant N_k, and thus treat Φ as a function. The general solution of free Klein-Gorden equation is the superposition of the positive and negative energy solution. It is then

$$\Phi(\underline{x}) = \sum_{\mathbf{k}} N_k\left(a(\mathbf{k})e^{i\mathbf{k}\cdot\mathbf{r}-i\omega_k t} + b(\mathbf{k})e^{i\mathbf{k}\cdot\mathbf{r}+i\omega_k t}\right), \tag{9.4.2.10a}$$

where the second term comes from negative solution, $a(\mathbf{k})$ and $b(\mathbf{k})$ are arbitrary functions of the wave vector $\mathbf{k}$, and N_k will be used to do the normalization of $\Phi(\underline{x})$. We require that $\Phi(\underline{x})$ is a real function. We must impose $\Phi^*(\underline{x}) = \Phi(\underline{x})$, where

$$\begin{aligned}\Phi^*(\underline{x}) &= \sum_{\mathbf{k}} N_k\left(a^*(\mathbf{k})e^{-i\mathbf{k}\cdot\mathbf{r}+i\omega_k t} + b^*(\mathbf{k})e^{-i\mathbf{k}\cdot\mathbf{r}-i\omega_k t}\right)\\ &= \sum_{\mathbf{k}} N_k\left(a^*(\mathbf{k})e^{-i\mathbf{k}\cdot\mathbf{r}+i\omega_k t} + b^*(-\mathbf{k})e^{-i\mathbf{k}\cdot\mathbf{r}-i\omega_k t}\right),\end{aligned} \tag{9.4.2.10b}$$

In the second term of the second equality, we have changed the dummy integration variable from $\mathbf{k}$ to $-\mathbf{k}$. Comparing Eqs. (9.4.2.10a-b), one obtains that $\Phi^* = \Phi$ requires $b^*(-\mathbf{k}) = a(\mathbf{k})$. Imposing this condition, we can rewrite Φ as

$$\begin{aligned}\Phi(\underline{x}) &= \sum_{\mathbf{k}} N_k\left[a(\mathbf{k})\frac{1}{\sqrt{V}}e^{i\mathbf{k}\cdot\mathbf{r}-i\omega_k t} + a^*(-\mathbf{k})\frac{1}{\sqrt{V}}e^{i\mathbf{k}\cdot\mathbf{r}+i\omega_k t}\right]\\ &= \int d\mathbf{k}N_k\frac{1}{(2\pi)^{3/2}}\left[a(\mathbf{k})e^{i\mathbf{k}\cdot\mathbf{r}-i\omega_k t} + a^*(\mathbf{k})e^{-i\mathbf{k}\cdot\mathbf{r}+i\omega_k t}\right]\\ &= \int d\mathbf{k}N_k\frac{1}{(2\pi)^{3/2}}\left[a(\mathbf{k})e^{-i\underline{k}\underline{x}} + a^*(\mathbf{k})e^{i\underline{k}\underline{x}}\right].\end{aligned} \tag{9.4.2.10c}$$

Now let us interpret that what does the normalization mean? There are three methods to determine the normalization constant N_k [124, 7, 125]. In the next, we introduce the method in Ref. [124]. In nonrelativistic case, probability density is ρ_{n-r}, and

$$\int d\mathbf{r}\rho_{n-r} = \int d\mathbf{r}\Phi^*\Phi = 1, \tag{9.4.2.10d}$$

and the matrix element of any operator O is

$$<\Phi_1|O|\Phi_2>= \int d\mathbf{r}\Phi_1^*(\mathbf{r},t)O\Phi_2(\mathbf{r},t). \tag{9.4.2.10e}$$

However, referring to Eq. (8.3.15), in case of relativistic Klein-Gorden equation we define the matrix element of any operator O is

$$<\Phi_1|O|\Phi_2>= i\int d\mathbf{r}\left[\Phi_1^*\frac{\partial(O\Phi_2)}{\partial t} - \frac{\partial\Phi_1^*}{\partial t}O\Phi_2\right]. \tag{9.4.2.10f}$$

Now let us consider a simplest plane wave,

$$\Phi_k(\mathbf{r},t) = N_k\frac{1}{(2\pi)^{3/2}}e^{i\mathbf{k}\cdot\mathbf{r}-i\omega_k t}. \tag{9.4.2.10g}$$

Substituting Eq. (9.4.2.10g) into Eq. (9.4.2.10f), one obtains

$$\begin{aligned}
\delta(\mathbf{k}_1-\mathbf{k}_2) &= <\Phi_{k_1}|\Phi_{k_2}> \\
&= i\int d\mathbf{r}\left\{N_{k_1}\frac{1}{(2\pi)^{3/2}}e^{-i\mathbf{k}_1\cdot\mathbf{r}+i\omega_{k_1}t}\frac{\partial}{\partial t}N_{k_2}\frac{1}{(2\pi)^{3/2}}e^{i\mathbf{k}_2\cdot\mathbf{r}-i\omega_{k_2}t}\right. \\
&- \left.\frac{\partial}{\partial t}N_{k_1}\frac{1}{(2\pi)^{3/2}}e^{-i\mathbf{k}_1\cdot\mathbf{r}+i\omega_{k_1}t}N_{k_2}\frac{1}{(2\pi)^{3/2}}e^{i\mathbf{k}_2\cdot\mathbf{r}-i\omega_{k_2}t}\right\} \\
&= 2\omega(\mathbf{k}_1)N_{k_1}^2\delta(\mathbf{k}_1-\mathbf{k}_2).
\end{aligned} \tag{9.4.2.10h}$$

From Eq. (9.4.2.10h) we obtain

$$N_k = \frac{1}{\sqrt{2\omega_k}},\ \Phi_k(\mathbf{r},t) = \frac{1}{\sqrt{2\omega_k}}\frac{1}{(2\pi)^{3/2}}e^{i\mathbf{k}\cdot\mathbf{r}-i\omega_k t} = \frac{1}{\sqrt{2\omega_k}}\frac{1}{(2\pi)^{3/2}}e^{-i\underline{k}\underline{x}}. \tag{9.4.2.10i}$$

(For reference, In Ref. [125] $N_k = 1/[(2\pi)^{3/2}2\omega_k]$, and in Ref. [7] $N_k = \sqrt{m_\circ c^2/2\omega_k}$.) This expression of N_k in Eq. (9.4.2.10i) is accepted by many references and textbooks [121].

Substituting the expression of N_k in Eq. (9.4.2.10i) into Eq. (9.4.2.10c) and considering that Φ is an operator, one obtains

$$\begin{aligned}
\Phi(\underline{x}) &= \sum_{\mathbf{k}}\frac{1}{\sqrt{2\omega_k}}\left[a(\mathbf{k})\frac{1}{\sqrt{V}}e^{i\mathbf{k}\cdot\mathbf{r}-i\omega_k t} + a^+(-\mathbf{k})\frac{1}{\sqrt{V}}e^{i\mathbf{k}\cdot\mathbf{r}+i\omega_k t}\right] \\
&= \sum_{\mathbf{k}}\frac{1}{\sqrt{2\omega_k}}\left[a(\mathbf{k})\frac{1}{\sqrt{V}}e^{-i\underline{k}\ \underline{x}} + a^+(\mathbf{k})\frac{1}{\sqrt{V}}e^{i\underline{k}\ \underline{x}}\right] \\
&= \int d\mathbf{k}\frac{1}{\sqrt{2\omega_k}}\frac{1}{(2\pi)^{3/2}}\left[a(\mathbf{k})e^{i\mathbf{k}\cdot\mathbf{r}-i\omega_k t} + a^+(\mathbf{k})e^{-i\mathbf{k}\cdot\mathbf{r}+i\omega_k t}\right] \\
&= \int d\mathbf{k}\frac{1}{\sqrt{2\omega_k}}\frac{1}{(2\pi)^{3/2}}\left[a(\mathbf{k})e^{-i\underline{k}\ \underline{x}} + a^+(\mathbf{k})e^{i\underline{k}\ \underline{x}}\right] \\
&\equiv \Phi^+(\underline{x}) + \Phi^-(\underline{x}),
\end{aligned} \tag{9.4.2.11}$$

where $\Phi^+(\underline{x})$, which carries the annihilation operators, represents the positive energy part of the decomposition of the field coordinate $\Phi(\underline{x})$, and $\Phi^-(\underline{x})$, which carries the creation operator, correspond to the negative energy part of the decomposition of the field coordinate $\Phi(\underline{x})$. $\Phi(\underline{x})$ can be also expressed as

$$\begin{aligned} \Phi(\underline{x}) &= \frac{1}{(2\pi)^{3/2}} \int_{k^\circ>0} \frac{d\mathbf{k}}{\sqrt{2\omega_k}} a(\mathbf{k}) e^{-i\underline{k}\cdot\underline{x}} \\ &+ \frac{1}{(2\pi)^{3/2}} \int_{k^\circ>0} \frac{d\mathbf{k}}{\sqrt{2\omega_k}} a^+(\mathbf{k}) e^{i\underline{k}\cdot\underline{x}} \\ &\equiv \Phi^+(\underline{x}) + \Phi^-(\underline{x}). \end{aligned} \tag{9.4.2.12}$$

Using the Fourier decomposition Eq. (9.4.2.10) of the field, the conjugate momentum given by Eq. (9.4.2.4) can be written in the expanded form:

$$\pi(\underline{x}) = \frac{1}{\sqrt{V}} \sum_{\mathbf{k}} \frac{-i\omega_k}{\sqrt{2\omega_k}} \left(a(\mathbf{k}) e^{i\underline{k}\cdot\underline{x}} - a^+(\mathbf{k}) e^{-i\underline{k}\cdot\underline{x}} \right). \tag{9.4.2.13}$$

The field is quantized by invoking the equal-time commutation relations of Klein-Gorden field

$$[\Psi(\mathbf{r},t), \pi(\mathbf{r}',t)] = i\hbar\delta(\mathbf{r}-\mathbf{r}'); \tag{9.4.2.14}$$

$$[\Psi(\mathbf{r},t), \Psi(\mathbf{r}',t)] = [\pi(\mathbf{r},t), \pi(\mathbf{r}',t)] = 0. \tag{9.4.2.15}$$

They, in turn, yield the commutation relations between the operators $a(\mathbf{k})$ and $a^+(\mathbf{k})$ (See the exercise.).

$$[a(\mathbf{k}), a^+(\mathbf{k}')] = \delta(\mathbf{k}-\mathbf{k}'); \tag{9.4.2.16}$$

$$[a(\mathbf{k}), a(\mathbf{k}')] = [a^+(\mathbf{k}), a^+(\mathbf{k}')] = 0. \tag{9.4.2.17}$$

If we use the method in Ref. [125], then the Eq. (9.4.2.16) becomes $[a(\mathbf{k}), a^+(\mathbf{k}')] = (2\pi)^3 2\omega_k \delta(\mathbf{k}-\mathbf{k}')$.

There is a very close analogy between the simple harmonic oscillator. The operator $a(\mathbf{k})$ and $a^+(\mathbf{k})$ are interpreted as the annihilation and creation operators for a field quantum with mode $\mathbf{k}$. The number operator $n(\mathbf{k})$ is defined as

$$n(\mathbf{k}) = a^+(\mathbf{k}) a(\mathbf{k}), \tag{9.4.2.18}$$

which operating on the basic state in the Fock space

$$|n_1, n_2, \cdots, n_k, \cdots > \tag{9.4.2.19}$$

yields the number of quanta associated with the momentum $\mathbf{k}$.

$$n(\mathbf{k})|n_1, n_2, \cdots, n_k, \cdots >= n_k |n_1, n_2, \cdots, n_k, \cdots > . \tag{9.4.2.20}$$

From the definition, the expression of the Hamiltonian density, $\mathcal{H}(\underline{x})$, is

$$\begin{aligned} \mathcal{H}(\underline{x}) &= \pi(\underline{x})\dot{\Phi}(\underline{x}) - \mathcal{L}(\underline{x}) \\ &= \partial_\circ \Phi \partial^\circ \Phi - \frac{1}{2}\{\partial_\mu \Phi(\underline{x}) \partial^\mu \Phi(\underline{x}) - m^2 \Phi^2(\underline{x})\} - \Omega_0 \\ &= \frac{1}{2}\left\{\pi^2 + (\nabla\Phi)^2 + m^2\Phi^2\right\} - \Omega_0, \end{aligned} \tag{9.4.2.21}$$

where $\mathcal{L}$ is given by Eq. (9.4.2.3). Eq. (9.4.2.21) gives a positive-definite Hamiltonian density, which is a reasonable result. Using Eqs. (9.4.2.21) and (9.4.2.10), we can write the total Hamiltonian H in terms of the coefficients, $a(\mathbf{k})$ and $a^+(\mathbf{k})$:

$$
\begin{aligned}
H &= -\Omega_0 V + \int d\mathbf{r}\mathcal{H}(\underline{x}) = -\Omega_0 V + \frac{1}{2}\int d\mathbf{r}\left(\pi^2 + (\nabla\Phi)^2 + m^2\Phi^2\right) \\
&= -\Omega_0 V + \frac{1}{2}\int \frac{d\mathbf{k}}{(2\pi)^{3/2}\sqrt{2\omega_k}}\frac{d\mathbf{k}'}{(2\pi)^{3/2}\sqrt{2\omega_{k'}}}d\mathbf{r} \\
&\quad \Big[\left(-i\omega_k a(\mathbf{k})e^{-i\underline{k}\cdot\underline{x}} + i\omega_k a^+(\mathbf{k})e^{i\underline{k}\cdot\underline{x}}\right)\left(-i\omega_{k'}a(\mathbf{k}')e^{-i\underline{k}'\cdot\underline{x}} + i\omega_{k'}a^+(\mathbf{k}')e^{i\underline{k}'\cdot\underline{x}}\right) \\
&+ \left(i\mathbf{k}a(\mathbf{k})e^{-i\underline{k}\cdot\underline{x}} - i\mathbf{k}a^+(\mathbf{k})e^{i\underline{k}\cdot\underline{x}}\right)\left(i\mathbf{k}'a(\mathbf{k}')e^{-i\underline{k}'\cdot\underline{x}} - i\mathbf{k}'a^+(\mathbf{k}')e^{i\underline{k}'\cdot\underline{x}}\right) \\
&\quad m^2\left(a(\mathbf{k})e^{-i\underline{k}\cdot\underline{x}} + a^+(\mathbf{k})e^{i\underline{k}\cdot\underline{x}}\right)\left(a(\mathbf{k}')e^{-i\underline{k}'\cdot\underline{x}} + a^+(\mathbf{k}')e^{i\underline{k}'\cdot\underline{x}}\right)\Big] \\
&= -\Omega_0 V + \frac{1}{2}(2\pi)^3\int \frac{d\mathbf{k}}{(2\pi)^{3/2}\sqrt{2\omega}}\frac{d\mathbf{k}'}{(2\pi)^{3/2}\sqrt{2\omega'}} \\
&\quad \big[\delta(\mathbf{k}-\mathbf{k}')(+\omega_k\omega_{k'} + \mathbf{k}\cdot\mathbf{k}' + m^2) \\
&\times \left(a^+(\mathbf{k})a(\mathbf{k}')e^{-i(\omega_k-\omega_{k'})t} + a(\mathbf{k})a^+(\mathbf{k}')e^{+i(\omega_k-\omega_{k'})t}\right) \\
&+ \delta(\mathbf{k}+\mathbf{k}')(-\omega_k\omega_{k'} - \mathbf{k}\cdot\mathbf{k}' + m^2) \\
&\times \left(a(\mathbf{k})a(\mathbf{k}')e^{-i(\omega_k+\omega_{k'})t} + a^+(\mathbf{k})a^+(\mathbf{k}')e^{+i(\omega_k+\omega_{k'})t}\right)\Big] \\
&= -\Omega_0 V + \frac{1}{2}\int d\mathbf{k}\frac{1}{2\omega_k} \\
&\quad \left[(\omega_k^2 + k^2 + m^2)\left(a^+(\mathbf{k})a(\mathbf{k}) + a(\mathbf{k})a^+(\mathbf{k})\right)\right. \\
&+ \left.(-\omega_k^2 + k^2 + m^2)\left(a(\mathbf{k})a(-\mathbf{k}) + a^+(\mathbf{k})a^+(-\mathbf{k})\right)\right] \\
&= -\Omega_0 V + \frac{1}{2}\int d\mathbf{k}\omega_k\left(a^+(\mathbf{k})a(\mathbf{k}) + a(\mathbf{k})a^+(\mathbf{k})\right) \\
&= -\Omega_0 V + \int d\mathbf{k}\omega_k\left(a^+(\mathbf{k})a(\mathbf{k}) + \frac{1}{2}\right) \\
&= -\Omega_0 V + \sum_{\mathbf{k}}\omega_k\left(a^+(\mathbf{k})a(\mathbf{k}) + \frac{1}{2}\right).
\end{aligned}
\tag{9.4.2.22}
$$

To get the second equality, we used

$$\int_{-\infty}^{\infty} d\mathbf{r}e^{i\mathbf{q}\cdot\mathbf{r}} = (2\pi)^3\delta(\mathbf{q}). \tag{9.4.2.23}$$

To get the third equality, we integrated over $\mathbf{k}'$. The fourth equality then follows from $\omega = (k^2+m^2)^{1/2}$. Also, we were careful to keep the ordering of $a(\mathbf{k})$ and $a^+\mathbf{k}'$ unchanged throughout. The fifth equality used the commutation relation in Eq. (9.4.2.16). The last equality comes from that if we use different normalization for the discrete and continuous values of $\mathbf{k}$ in the plane wave expression.

The field coordinate can be expressed by $a(\mathbf{k})$ and $a^+(\mathbf{k})$ through Eq. (9.4.2.11). We can also invert this formula to get $a(\mathbf{k})$ and $a^+(\mathbf{k})$ in terms of $\Phi(\underline{x})$. We have from Eq. (9.4.2.11)

$$\frac{1}{(2\pi)^{3/2}}\int d\mathbf{r}e^{-i\mathbf{k}\cdot\mathbf{r}-i\omega_k t}\Phi(x^\circ,\mathbf{r}) = \frac{1}{\sqrt{2\omega_k}}a(\mathbf{k}) + \frac{1}{\sqrt{2\omega_k}}e^{2i\omega t}a^+(-\mathbf{k}), \tag{9.4.2.24}$$

$$\int d\mathbf{r}e^{-i\mathbf{k}\cdot\mathbf{r}}\pi(x^\circ,\mathbf{r}) = \int d\mathbf{r}e^{-i\mathbf{k}\cdot\mathbf{r}}\frac{\partial}{\partial t}\Phi(x^\circ,\mathbf{r}) = -i\frac{\sqrt{\omega_k}}{\sqrt{2}}a(\mathbf{k}) + i\frac{\sqrt{\omega_k}}{\sqrt{2}}a^+(-\mathbf{k}), \qquad (9.4.2.25)$$

We can combine these to get

$$a(\mathbf{k}) = \frac{1}{(2\pi)^{3/2}}\int d\mathbf{r}\frac{1}{\sqrt{2\omega_k}}e^{i\mathbf{k}\cdot\mathbf{r}-i\omega_k t}\left[i\pi(x^\circ,\mathbf{r}) + \omega_k\Phi(x^\circ,\mathbf{r})\right]. \qquad (9.4.2.26)$$

$$a^+(-\mathbf{k}) = \frac{1}{(2\pi)^{3/2}}\int d\mathbf{r}\frac{1}{\sqrt{2\omega_k}}e^{-i\mathbf{k}\cdot\mathbf{r}+i\omega_k t}\left[-i\pi(x^\circ,\mathbf{r}) + \omega_k\Phi(x^\circ,\mathbf{r})\right]. \qquad (9.4.2.27)$$

We can rewritten the Hamiltonian H in Eq. (9.4.2.22) as

$$\begin{aligned} H &= \sum_k \omega_k a^+(\mathbf{k})a(\mathbf{k}) + \sum_k \omega_k\frac{1}{2} - \Omega_0 V \\ &\equiv \sum_k \omega_k n(\mathbf{k}) + \mathcal{E}_0 - \Omega_0 V, \end{aligned} \qquad (9.4.2.28)$$

where the second term represents total zero point energy of all oscillators. If we make the summation in $\mathcal{E}_0$, then it is infinitely large. This is because, we have one oscillator for each value of momentum and we have an infinite number of such oscillators, each having a zero point energy. Thus the ground state energy becomes infinite. If we transform the summation into integration, and integrate only to a maximum value Λ, a number known as the ultraviolet cutoff, then we find

$$\mathcal{E}_0 = \frac{V\Lambda^4}{16\pi^2}, \qquad (9.4.2.29)$$

where we have assumed $\Lambda >> m$ (in ordinary unit $\Lambda >> mc^2$). If we want to set the energy of ground state is zero, then we can set $\mathcal{E}_0 = \Omega_0 V$.

9.4.3 On the Choice of Commutation Relation

All the relativistic field equations can be quantized only by either commutation relation or anticommutation relation. In case of Klein-Gorden field equation, we demonstrate by two methods that if we use the anticommutation relation to quantize the real Klein-Gorden field equation, then the results are not reasonable.

(1). First method. Let us use Eq. (9.4.2.11) to calculate the following anticommutation relation between field coordinates:

$$\begin{aligned} [\Phi(\underline{x}),\Phi(\underline{x}')]_+ &= \int \frac{d\mathbf{k}d\mathbf{k}'}{2(2\pi)^3\sqrt{\omega\omega'}} \\ &\quad [a(\mathbf{k})e^{-i\underline{k}\cdot\underline{x}} + a^+(\mathbf{k})e^{i\underline{k}\cdot\underline{x}}, a(\mathbf{k}')e^{-i\underline{k}'\cdot\underline{x}'} + a^+(\mathbf{k}')e^{i\underline{k}'\cdot\underline{x}'}]_+ \\ &= \int \frac{d\mathbf{k}d\mathbf{k}'}{2(2\pi)^3\sqrt{\omega\omega'}} \\ &\quad \left\{[a(\mathbf{k}),a^+(\mathbf{k}')]_+ e^{i\underline{k}'\cdot\underline{x}'-i\underline{k}\cdot\underline{x}} + [a^+(\mathbf{k}),a(\mathbf{k}')]_+ e^{-i\underline{k}'\cdot\underline{x}'+i\underline{k}\cdot\underline{x}}\right\} \\ &= \int \frac{d\mathbf{k}}{2(2\pi)^3\omega}\left\{e^{i\mathbf{k}\cdot(\mathbf{r}-\mathbf{r}')-i\omega(t-t')} + e^{-i\mathbf{k}\cdot(\mathbf{r}-\mathbf{r}')+i\omega(t-t')}\right\} \\ &= \int \frac{d\mathbf{k}}{2(2\pi)^3\omega}e^{i\mathbf{k}\cdot(\mathbf{r}-\mathbf{r}')}\left\{e^{-i\omega(t-t')} + e^{+i\omega(t-t')}\right\} \\ &= \int \frac{d\mathbf{k}}{(2\pi)^3\omega}e^{i\mathbf{k}\cdot(\mathbf{r}-\mathbf{r}')}\cos[\omega(t-t')] \equiv \Delta_1(\underline{x}_1 - \underline{x}_2). \end{aligned} \qquad (9.4.3.1)$$

For the equal time $t = t'$ and space-like interval $(\underline{x} - \underline{x}')^2 = c^2(\Delta t)^2 - (\Delta \mathbf{r})^2 < 0$, we have

$$\begin{aligned}\Delta_1(\underline{x} - \underline{x}')|_{t=t'} &= \int \frac{d\mathbf{k}}{(2\pi)^3\omega} e^{i\mathbf{k}\cdot(\mathbf{r}-\mathbf{r}')} \\ &= \int \frac{1}{(2\pi)^3} \int d\mathbf{k} \frac{1}{\sqrt{k^2 + (mc/\hbar)^2}} e^{i\mathbf{k}\cdot(\mathbf{r}-\mathbf{r}')} \\ &= \frac{mc}{2\pi^2\hbar|\mathbf{r}|} K_1\left(\frac{|\mathbf{r}|mc}{\hbar}\right) \neq 0, \quad \mathbf{r}^2 = (\mathbf{r} - \mathbf{r}')^2 > 0, \end{aligned} \tag{9.4.3.2}$$

where $\lambda_c = \hbar/(mc)$ is the Compton wavelength of a particle, m is the inertial mass and represents the Newtonian response of mass to forces, and K_ν is the modified Bessel function. $\lambda_c = 2.4263102175 \times 10^{-10}$ cm. From Ref. [29]

$$nK_1(nx) = \int_0^\infty dk \frac{k \sin kx}{\sqrt{k^2 + n^2}}, \tag{9.4.3.3}$$

$[\Phi(\underline{x}), \Phi(\underline{x}')]_+ \neq 0$ is not reasonable, because this means that there is the uncertainty relation between $\Phi(\underline{x})$ and $\Phi(\underline{x}')$. In the condition of $t = t'$, this uncertainty correlation implies any perturbation at $\mathbf{r}$ will have infinite speed to propagate to point $\mathbf{r}'$, and thus violate the special theory of relativity. Therefore, we have to use commutation relation, $[\Phi(\underline{x}), \Phi(\underline{x}')] = 0$ for the Klein-Gorden field. In nonrelativistic case, we can use both commutation and anticommutation relations, because the speed can be infinite in principle.

(2). Second method. If we use anticommutation relation $[a(\mathbf{k}), a^+(\mathbf{k})]_+ = 1$ in Eq. (9.4.2.21), then

$$H = \frac{1}{2}\sum_k \left(a^+(\mathbf{k})a(\mathbf{k}) + a(\mathbf{k})a^+(\mathbf{k})\right) - \Omega_0 V = -\Omega_0 V \tag{9.4.3.4}$$

is a simple constant. Clearly, there is something wrong with using anticommutation relation.

9.4.4 Complex Scalar Field

(1). First step. Write the classical Klein-Gorden field equation. Both the complex scalar field $\Phi(\underline{x})$ and its complex conjugate $\Phi^*(\underline{x})$ satisfy the Klein-Gorden equation

$$(\partial^\mu\partial_\mu + m^2)\Phi(\underline{x}) = 0; \quad (\partial^\mu\partial_\mu + m^2)\Phi^*(\underline{x}) = 0. \tag{9.4.4.1}$$

(2). Second step. Choose an appropriate Lagrangian density for this complex field. We take

$$\mathcal{L} = \partial^\mu\Phi(\underline{x})\partial_\mu\Phi^*(\underline{x}) - m^2\Phi(\underline{x})\Phi^*(\underline{x}). \tag{9.4.4.2}$$

It is easy to verify that if we substitute Eq. (9.4.4.2) into the Euler-Lagrange equation, then we can get Eq.(9.4.4.1).

If we express $\Phi(\underline{x})$ by another two independent real fields as

$$\Phi(\underline{x}) = \frac{1}{\sqrt{2}}(\Phi_1(\underline{x}) + i\Phi_2(\underline{x})); \; \Phi^*(\underline{x}) = \frac{1}{\sqrt{2}}(\Phi_1(\underline{x}) - i\Phi_2(\underline{x})). \tag{9.4.4.3}$$

Then, we have

$$(\partial^\mu\partial_\mu + m^2)\Phi_r(\underline{x}) = 0. \quad (r = 1,\ 2) \tag{9.4.4.4}$$

The corresponding appropriate Lagrangian density is

$$\mathcal{L} = \frac{1}{2}\sum_{r=1}^{2}\left(\partial^\mu\Phi_r(\underline{x})\partial_\mu\Phi_r(\underline{x}) - m^2\Phi_r(\underline{x})\Phi_r(\underline{x})\right). \tag{9.4.4.5}$$

(3). Third step. According to definition, write the conjugate momentum of each real field coordinate:

$$\pi_r(\underline{x}) = \frac{\partial \mathcal{L}}{\partial \dot{\Phi}_r} = \partial_\circ \Phi_r = \dot{\Phi}_r. \tag{9.4.4.6}$$

(4). Fourth step. According to the theorem in section 9.2 we always can say that the canonical field coordinates and corresponding canonical conjugate field momenta satisfy commutation relations. Therefore, the classical field equations Eq. (9.4.4.4) become the quantum field equation.

(5). Fifth step. Although in the representation of field coordinates the field coordinates are the numerical numbers, but in the representation of single particle complete basis of the single particle Klein-Gorden equation, the field coordinates become operators. The expansions are

$$\Phi_r(\underline{x}) = \frac{1}{(2\pi)^{3/2}} \int d\mathbf{k} \frac{1}{\sqrt{2\omega_k}} \left\{ a_r(\mathbf{k}) e^{-i\underline{k}\cdot\underline{x}} + a_r^+(\mathbf{k}) e^{i\underline{k}\cdot\underline{x}} \right\}, \tag{9.4.4.7}$$

where from the commutation relations between Φ and π (remember that these commutation relations are given by the theorem in 9.2), we can show that

$$[a_r(\mathbf{k}), a_s^+(\mathbf{k}')] = \delta_{rs}\delta(\mathbf{k} - \mathbf{k}'); \tag{9.4.4.8}$$

$$[a_r(\mathbf{k}), a_s(\mathbf{k}')] = [a_r^+(\mathbf{k}), a_s^+(\mathbf{k}')] = 0. \tag{9.4.4.9}$$

If we make the Fourier expansion for Φ:

$$\Phi(\underline{x}) = \frac{1}{(2\pi)^{3/2}} \int d\mathbf{k} \frac{1}{\sqrt{2\omega_k}} \left\{ a(\mathbf{k}) e^{-i\underline{k}\cdot\underline{x}} + b^+(\mathbf{k}) e^{i\underline{k}\cdot\underline{x}} \right\}, \tag{9.4.4.10}$$

then it is easy to verify that

$$a(\mathbf{k}) = \frac{1}{\sqrt{2}}\{a_1(\mathbf{k}) + ia_2(\mathbf{k})\}; \quad b(\mathbf{k}) = \frac{1}{\sqrt{2}}\{a_1(\mathbf{k}) - ia_2(\mathbf{k})\}, \tag{9.4.4.11}$$

and

$$[a(\mathbf{k}), a^+(\mathbf{k}')] = [b(\mathbf{k}), b^+(\mathbf{k}')] = \delta(\mathbf{k} - \mathbf{k}'), \tag{9.4.4.12}$$

and all other commutation relations with different combinations vanish. From Eq. (9.4.4.11) we see that a, b, and thus the complex scalar field Φ represent two types of particles.

The Hermite conjugate of Eq. (9.4,4.10) is

$$\Phi^+(\underline{x}) = \frac{1}{(2\pi)^{3/2}} \int d\mathbf{k} \frac{1}{\sqrt{2\omega_k}} \left\{ a^+(\mathbf{k}) e^{i\underline{k}\cdot\underline{x}} + b(\mathbf{k}) e^{-i\underline{k}\cdot\underline{x}} \right\}, \tag{9.4.4.13}$$

The Lagrangian density in Eq. (9.4.4.2) does not change under the transformation

$$\Phi \to \Phi e^{i\alpha}, \quad \Phi^+ \to \Phi^+ e^{-i\alpha}, \tag{9.4.4.14}$$

where α is a real number. This is called a $U(1)$ transformation, a transformation by a unitary 1×1 matrix. In terms of Φ_1 and Φ_2, this transformation reads

$$\begin{pmatrix} \Phi_1 \\ \Phi_2 \end{pmatrix} \to \begin{pmatrix} \cos\alpha & \sin\alpha \\ -\sin\alpha & \cos\alpha \end{pmatrix} \begin{pmatrix} \Phi_1 \\ \Phi_2 \end{pmatrix}. \tag{9.4.4.15}$$

If we think of Φ_1, Φ_2 as a two-component vector, then Eq. (9.4.4.15) is just a rotation of this vector in the plane by angle α. Eq. (9.4.4.15) is called an $SO(2)$ transformation, a transformation by an orthogonal 2×2 matrix with a special value of the determinant (1 other than -1)..

The invariance under the $U(1)$ transformation is called global gauge invariance if α does not depend on $\underline{x}$, is called local gauge invariance if α depend on $\underline{x}$.

For the electromagnetic field, **A** and ϕ is called gauge field. And the fact that **E** and **B** are not affected by a gauge transformation is referred to as gauge invariance.

To providing a rigorous proof of the connection between gauge invariance and conservation of current four vector is beyond the scope of concepts of quantum field theory [125]. If the readers have interest on the physics of this connection, then they can look at the Ref. [31].

9.4.5 Current Four-Vector

(6). Sixth step. Make more discussions. Define the Nother current four vector $j^\mu(\underline{x})$ for the complex scalar field by [125, 121]

$$\begin{aligned} j^\mu(\underline{x}) &= -ie\left(\frac{\partial\mathcal{L}}{\partial(\partial\Phi/\partial x^\mu)}\Phi - \frac{\partial\mathcal{L}}{\partial(\partial\Phi^*/\partial x^\mu)}\Phi^*\right) \\ &= ie\{(\partial_\mu\Phi^*)\Phi - (\partial_\mu\Phi)\Phi^*\}, \end{aligned} \tag{9.4.5.1}$$

Using Klein-Gorden Eqs. (9.4.4.1-2) yields

$$\partial_\mu j^\mu(\underline{x}) = iem^2[\Phi^*(\underline{x}), \Phi(\underline{x})] = 0. \tag{9.4.5.2}$$

The four vector $j^\mu(\underline{x})$ is the Nother current four vector and $j^\circ(\underline{x})$ is the electric charge density. Eq. (9.4.5.2) is a continuty equation, and means charge conservation.

Therefore, the total charge of the field is

$$\begin{aligned} Q &= \int d\mathbf{r} j^\circ(\underline{x}) \\ &= e\int d\mathbf{r}\{(\partial_\circ\Phi^*)\Phi - (\partial_\circ\Phi)\Phi^*\} \\ &= \frac{e}{2}\sum_k\{[a^+(\mathbf{k}), a(\mathbf{k})]_+ - [b^+(\mathbf{k}), b(\mathbf{k})]_+\} \\ &= e\sum_{\mathbf{k}}\{a^+(\mathbf{k})a(\mathbf{k}) - b^+(\mathbf{k})b(\mathbf{k})\} \\ &\equiv e\sum_k\{[n_+(\mathbf{k}) - n_-(\mathbf{k})\},\ \ \{n_\pm(\mathbf{k}) = 0, 1, 2, \cdots, \infty\}, \end{aligned} \tag{9.4.5.3}$$

where the the second equality comes from Eq. (9.4.5.1), the third equality comes from Eqs. (9.4.4.10) and (9.4.4.12-13), and the fourth equality comes from Eq. (9.4.4.11). Thus, the operators a, a^+, n_+ denote the annihilation, creation, and number operators for the field quanta of electric charge e and the operators b, b^+, n_- denote the annihilation, creation, and number operators for the field quanta of electric charge $-e$.

From Eq. (9.4.5.3) we see that the complex scalar field indicates that there are two types of quanta, one with charge $+e$ and the other with charge $-e$. Apart from the sign of charge, the two types of the quanta have identical mass and other properties. So, they have to be associated with the particle and its antiparticle.

As an example of particle and antiparticle pair, we can cite π^+ and π^- meson, which can be described by a quantized complex Klein-Gorden field. For a quantized real Klein-Gorden field, the charge operator Q is zero and such a field can describe π° meson.

9.5 Quantization of Dirac Equation

We still use the same six steps to quantize the classical Dirac field equation as that for the classical Klein-Gorden field equation quantization.

(1). First step. Write the classical Dirac field equation. In chapter 8 we have obtained the Dirac equation, which is a single particle relativistic wave equation. Now we just simply think that the single particle wavefunction is a classical field, then we get the classical Dirac field equation, which is the Eq. (8.4.4).

$$(c\vec{\alpha}\cdot\mathbf{p}+\beta m_\circ c^2)\psi = E\psi. \qquad (8.4.4)' = (9.5.1)'$$

Let us use the Feynman's notation to rewrite Eq. (8.4.4)′. Multiply the Dirac equation by β from the left

$$(c\beta\vec{\alpha}\cdot\mathbf{p}+m_\circ c^2)\Psi = \beta E\Psi. \qquad (9.5.1)$$

Using the natural units $\hbar = 1$, $c = 1$ and introducing γ matrices

$$\gamma_\circ = \beta = \begin{bmatrix} I & 0 \\ 0 & I \end{bmatrix}, \quad \vec{\gamma} = \beta\vec{\alpha}\begin{bmatrix} 0 & \vec{\sigma} \\ -\vec{\sigma} & 0 \end{bmatrix}, \qquad (9.5.2)$$

where $\vec{\sigma}$ is Pauli matrix vector. We can rewrite the Dirac equation after rearrangement in the form

$$\underbrace{p}\,\Psi(\mathbf{r},t) = m\Psi(\mathbf{r},t), \quad with \quad \underbrace{p} = \gamma_\circ E - \vec{\gamma}\cdot\mathbf{p}, \qquad (9.5.3)$$

or

$$(i\gamma^\mu\partial_\mu - m)\Psi(\underline{x}) = 0. \qquad (9.5.3)'$$

The Hermitian adjoint of Dirac equation Eq. (9.5.3)′ is

$$\overline{\Psi}(\underline{x})\left(i\gamma^\mu\overleftarrow{\partial}_\mu + m\right) = 0, \qquad (9.5.4)$$

where $\overline{\Psi}(\underline{x}) = \Psi^+(\underline{x})\gamma_\circ$. The left arrow sign indicates that the operator acts on the function located on the left.

(2). Second step. Choose an appropriate Lagrangian density. It can be easily verified that the Dirac equation Eq. (9.5.3)′ and its Hermitian adjoint Eq. (9.5.4) can be obtained from the Lagrangian density

$$\mathcal{L}(\underline{x}) = \overline{\Psi}(\underline{x})(i\gamma^\mu\partial_\mu - m)\Psi(\underline{x}), \qquad (9.5.5)$$

and Euler-Lagrange equation. From the Euler-Lagrange equation for $\overline{\Psi}$

$$\frac{\partial\mathcal{L}}{\partial\overline{\Psi}} = \partial_\mu\left(\frac{\partial\mathcal{L}}{\partial(\partial_\mu\overline{\Psi})}\right). \qquad (9.5.6)$$

Therefore,

$$\frac{\partial\mathcal{L}}{\partial\overline{\Psi}} = (i\gamma^\mu\partial_\mu - m)\Psi(\underline{x}) = 0. \qquad (9.5.7)$$

In a similar way, we can consider the Euler-Lagrange equation for Ψ and obtain Eq. (9.5.4). [Although the Lagrangian density in Eq. (9.5.5) is not Hermitian, but one can show that it is still equivalent to a Hermitian form.]

(3). Third step. Find the canonical conjugate momentum from the appropriate Lagrangian density just according to definition.

$$\pi(\underline{x}) = \frac{\partial\mathcal{L}}{\partial\dot{\Psi}} = i\overline{\Psi}(\underline{x})\gamma_\circ = i\Psi^+(\underline{x}). \qquad (9.5.8)$$

(4). Fourth step. According to the theorem in section 9.2 we can say that the field coordinates and corresponding canonical conjugate field momenta satisfy commutation relations. After trying that which type of commutation relation is correct, we find that the quantization of classical Dirac field is correct just for anticommutation relation, which will be illustrated in the following convenient place. When we write the anticommutation relation between field coordinate and corresponding canonical conjugate momentum, we should note that from Eqs. (8.5.2-4) the field coordinate is a four component spinor, each component has to be treated as independent field variable. Therefore, there is no meaning in trying to write the anticommutation relation $[\Psi(\mathbf{r}), \pi(\mathbf{r})]_+$. We can only write the anticommutation relations between their components $[\Psi_\alpha(\mathbf{r}), \pi_\beta(\mathbf{r}')]_+$.

$$[\Psi_\alpha(\mathbf{r}), \pi_\beta(\mathbf{r}')]_+ = [\Psi_\alpha(\mathbf{r}), i\Psi_\beta^+(\mathbf{r}')]_+ = iI_{\alpha,\beta}\delta(\mathbf{r}-\mathbf{r}'), \quad (\alpha,\ \beta = 1,2,3,4) \tag{9.5.9}$$

where $I_{\alpha,\beta}$ is the unit 4×4 matrix.

(5). Fifth step. Make Fourier expansion for the field coordinates. In section 8.2 of chapter 8 we look upon the Dirac equation as a single particle equation, and have found its plane wave solutions. Now we take these plane wave solutions as our basis. For the statement convenience, we change some notations. Any plane wave solution in Eqs. (8.4.4) is of four components. They are:

$$\Psi_\uparrow^+(\underline{x}) = u_1(\mathbf{p})\frac{1}{\sqrt{V}}e^{-i\underline{p}\cdot\underline{x}},\ \ \Psi_\downarrow^+(\underline{x}) = u_2(\mathbf{p})e^{-i\underline{p}\cdot\underline{x}}$$

$$\Psi_\uparrow^-(\underline{x}) = u_3(\mathbf{p})\frac{1}{\sqrt{V}}e^{-i\underline{p}\cdot\underline{x}},\ \ \Psi_\downarrow^-(\underline{x}) = u_4(\mathbf{p})\frac{1}{\sqrt{V}}e^{-i\underline{p}\cdot\underline{x}}, \tag{9.5.10}$$

where $\underline{p}\cdot\underline{x} = E_p t - \mathbf{p}\cdot\mathbf{r}$, $\underline{p}^2 = p^\mu p_\mu = E_p^2 - \mathbf{p}^2 = m^2$, $\underline{x} = (t, x^1, x^2, x^3) = (t, x, y, z)$, $\underline{p} = (E, \mathbf{p})$, and $u_r(\mathbf{p})$ $(r = 1,2,3,4)$ are a normalized vectors expressed by a 4×1 matrixes, and subscript $+$ and $-$ represent positive energy $E_k = \omega_k = +\sqrt{\mathbf{p}^2+m^2}$ and negative energy $E_k = \omega_k = -\sqrt{\mathbf{p}^2+m^2}$. Set

$$v_1(\mathbf{p}) = u_3(-\mathbf{p}),\ \ v_2(\mathbf{p}) = u_4(-\mathbf{p}). \tag{9.5.11}$$

The field coordinate $\Psi(\underline{x})$ and $\overline{\Psi}(\underline{x})$ are expanded in terms of plane waves in box as

$$\Psi(\underline{x}) = \sum_p \sqrt{\frac{1}{2E_p}}\sum_{r=1}^{2}\left\{c_r(\mathbf{p})u_r(\mathbf{p})\frac{1}{\sqrt{V}}e^{-i\underline{p}\cdot\underline{x}} + d_r^+(\mathbf{p})v_r(\mathbf{p})\frac{1}{\sqrt{V}}e^{+i\underline{p}\cdot\underline{x}}\right\}, \tag{9.5.12}$$

$$\overline{\Psi}(\underline{x}) = \sum_p \sqrt{\frac{1}{2E_p}}\sum_{r=1}^{2}\left\{c_r^+(\mathbf{p})\overline{u}_r(\mathbf{p})\frac{1}{\sqrt{V}}e^{i\underline{p}\cdot\underline{x}} + d_r(\mathbf{p})\overline{v}_r(\mathbf{p})\frac{1}{\sqrt{V}}e^{-i\underline{p}\cdot\underline{x}}\right\}, \tag{9.5.13}$$

where the expansion coefficients $c_r(\mathbf{p})$ and $c_r^+(\mathbf{p})$ are to be interpreted as annihilation and creation operators of the fermion and $d_r(\mathbf{p})$ and $d_r^+(\mathbf{p})$ are to be interpreted as annihilation and creation operators of the antifermion.

From the equal time anticommutation relation in Eq. (9.5.9) and the expansion expression Eq. (9.5.12) we deduce the following anticommutation relations:

$$\begin{gathered}[c_r(\mathbf{p}), c_s^+(\mathbf{p}')]_+ = \delta_{rs}\delta(\mathbf{p}-\mathbf{p}'),\ \ [d_r(\mathbf{p}), d_s^+(\mathbf{p}')]_+ = \delta_{rs}\delta(\mathbf{p}-\mathbf{p}') \\ [c_r(\mathbf{p}), c_s(\mathbf{p}')]_+ = 0,\ \ [c_r^+(\mathbf{p}), c_s^+(\mathbf{p}')]_+ = 0, \\ [d_r(\mathbf{p}), d_s(\mathbf{p}')]_+ = 0,\ \ [d_r^+(\mathbf{p}), d_s^+(\mathbf{p}')]_+ = 0,\end{gathered} \tag{9.5.14}$$

and all the anticommutation relations between operators with different energies or, say, between fermion and antifermion are equal to zero.

(6). Sixth step. Make more discussions. Let us seek the total charge that resides in the field. For this, let us first according to the definition seek the four vector current $j^{\mu}(\underline{x})$ of the Dirac field.

$$\begin{aligned} j^{\mu}(\underline{x}) &= -ie\left(\frac{\partial\mathcal{L}}{\partial\Psi_{,\mu}}\Psi - \frac{\partial\mathcal{L}}{\partial\Psi^{+}_{,\mu}}\Psi^{+}\right) \\ &= e\overline{\Psi}(\underline{x})\gamma^{\mu}\Psi(\underline{x}). \end{aligned} \tag{9.5.15}$$

The zeroth component of the four current j^{μ} yields the charge density.

$$\begin{aligned} Q &= \int_V d\mathbf{r} j^{\circ}(\underline{x}) \\ &= e\int_V d\mathbf{r}\overline{\Psi}(\underline{x})\gamma^{\circ}\Psi(\underline{x}) = e\int_V d\mathbf{r}\Psi^{+}(\underline{x})\Psi(\underline{x}) \\ &= e\sum_{\mathbf{p}}\sum_{r=1}^{2}\left(c_r^{+}(\mathbf{p})c_r(\mathbf{p}) + d_r(\mathbf{p})d_r^{+}(\mathbf{p})\right) \\ &= e\sum_{\mathbf{p}}\sum_{r=1}^{2}\left(c_r^{+}(\mathbf{p})c_r(\mathbf{p}) - d_r^{+}(\mathbf{p})d_r(\mathbf{p}) + 1\right). \end{aligned} \tag{9.5.16}$$

In the third equality we have not yet chosen the commutation relations for the operators c_r, d_s. In the fourth equality we choose the anticommutation relation, and the sign of the total charge in the field are either positive or negative, and we know the signs of charge of fermion and antifermion are opposite. This is physically acceptable.

On the other hand, if we choose the canonical commutation relation, then Eq. (9.5.16) becomes

$$Q = e\sum_{\mathbf{p}}\sum_{r=1}^{2}\left(c_r^{+}(\mathbf{p})c_r(\mathbf{p}) + d_r^{+}(\mathbf{p})d_r(\mathbf{p}) + 1\right). \tag{9.5.17}$$

The total Q in Eq. (9.5.17) is positive definite. This is not physically acceptable. Therefore, for fermion system we have to use the canonical anticommutation relation.

9.6 Quantization of Maxwell Field Equation

The most common (and important) spin one particle is the photon. After quantization of classical Maxwell field equation set we will obtain photon as a quantum in electromagnetic field. Maxwell equation set must be supplemented by formulae that give us the dynamics of the charges and currents. For the quantization, we will treat the charges and currents as specified sources, and focus on the dynamics of the electromagnetic fields.

(1). First step. Write the field equation which we want to be quantized. Write the classical Maxwell field equation set as [We use in unrationalized Gaussian (CGS) units]

$$\nabla\cdot\mathbf{B} = 0, \tag{9.6.1}$$

$$\nabla\times\mathbf{E} + \frac{1}{c}\frac{\partial\mathbf{B}}{\partial t} = 0, \tag{9.6.2}$$

$$\nabla\times\mathbf{B} - \frac{1}{c}\frac{\partial\mathbf{E}}{\partial t} = \frac{4\pi\mathbf{j}}{c}, \tag{9.6.3}$$

and

$$\nabla \cdot \mathbf{E} = 4\pi\rho, \tag{9.6.4}$$

The electric and magnetic field strength can also be expressed by the scalar potential ϕ and the vector potential $\mathbf{A}$ as

$$\mathbf{B} = \nabla \times \mathbf{A}, \quad \mathbf{E} = -\nabla\phi - \frac{1}{c}\frac{\partial \mathbf{A}}{\partial t}. \tag{9.6.5}$$

The potentials uniquely determine the field strengths, but the field strengths does not uniquely determine the potentials. Define four component vector potential or gauge field $\underline{A} = (A^\mu) = (A^\circ, A^1, A^2, A^3)$, $A^\circ = \phi$, and an electromagnetic field intensity, which is a second rank antisymmetric tensor, $F^{\mu\nu}$:

$$F^{\mu\nu} = \frac{\partial A^\nu}{\partial x_\mu} - \frac{\partial A^\mu}{\partial x_\nu} = \partial^\mu A^\nu - \partial^\nu A^\mu. \tag{9.6.6}$$

Since $F^{\mu\nu} = -F^{\nu\mu}$ and the diagonal elements are zero, there are only six independent components and it can be shown that they correspond to $\mathbf{B}$ and $\mathbf{E}$.

The four equations of Maxwell equation set can be written by using $F^{\mu\nu}$ $(\mu, \nu, \lambda = 0, 1, 2, 3)$ as

$$\frac{\partial F^{\mu\nu}}{\partial x_\lambda} + \frac{\partial F^{\nu\lambda}}{\partial x_\mu} + \frac{\partial F^{\lambda\mu}}{\partial x_\nu} = 0, \tag{9.6.7}$$

which represents Eqs. (9.6.1-2).

$$\sum_{\mu=0}^{3} \partial_\mu F^{\mu\nu} = \frac{4\pi j^\nu}{c}, \tag{9.6.8}$$

which represents Eqs. (9.6.3-4).

(2). Second step. Choose an appropriate Lagrangian density. The appropriate Lagrangian density is

$$\begin{aligned}\mathcal{L} &= -\frac{F_{\lambda\rho}F^{\lambda\rho}}{16\pi} - j_\lambda A^\lambda = \frac{1}{8\pi}(\mathbf{E}^2 - \mathbf{B}^2) - j_\lambda A^\lambda \\ &= -\frac{1}{8\pi}\partial^\mu A^\nu \partial_\mu A_\nu + \frac{1}{8\pi}\partial^\mu A^\nu \partial_\nu A_\mu - j_\lambda A^\lambda.\end{aligned} \tag{9.6.9}$$

Substituting Eq. (9.6.9) into the following Euler-Lagrangian equation

$$\frac{\partial \mathcal{L}}{\partial A_\mu} - \partial_\nu \frac{\partial \mathcal{L}}{\partial(\partial_\nu A_\mu)} = 0 \tag{9.6.10}$$

yields the Maxwell field equation Eq. (9.6.8).

When we use four vector potential $\underline{A} = (A^\mu) = (A^\circ, A^1, A^2, A^3) = (\phi, \mathbf{A})$ to make calculations and discuss problems on the electric and magnetic intensity, we should remember that the $\underline{A}$ has not yet be determined solely. To make $\underline{A}$ certain, we should choose a certain gauge. If we take Lorentz gauge (= Lorentz condition = Lorentz restriction = Lorentz constraint), i. e.,

$$\partial_\mu A^\mu \equiv \sum_{\mu=0}^{3} \frac{\partial A^\mu}{\partial x^\mu} = 0, \tag{9.6.11}$$

where $(\partial_\mu) \equiv (\partial/\partial x^\mu) = (\partial^\circ, \partial^1, \partial^2, \partial^3) \equiv (\partial/(c\partial t), \partial/\partial x, \partial/\partial y, \partial/\partial z) \equiv (\partial/(c\partial t), \nabla)$, and $(\partial^\mu) \equiv (g^{\mu\nu}\partial_\nu) = (\partial/(c\partial t), -\nabla)$, and $\underline{A} = (A^\mu) = (A^\circ, A^1, A^2, A^3) \equiv (\phi, \mathbf{A})$, then Eq. (9.6.8) becomes

$$\Box A^\mu = \left(\frac{1}{c^2}\frac{\partial^2}{\partial t^2} - \nabla^2\right) A^\mu = \frac{4\pi j^\mu}{c}, \quad (\mu = 0, 1, 2, 3). \tag{9.6.12}$$

where $\underline{j} = (j^\mu) = (j^\circ, j^2, j^2, j^3)$. $j^\circ = c\rho$. Since charge ρ and current density $\mathbf{j}$ are the components of a four current vector and $\partial_\mu \partial^\mu = \Box$ is a Lorentz scalar, both the Eqs. (9.6.12) and (9.6.8) are Lorentz covariant equations.

The electromagnetic field is a constrained system. The constraint condition is given by the gauge condition in Eq. (9.6.11). According to the corollary in section 9.2, we discuss the gauge condition. This is a condition that we require $\underline{A} = (A^\mu(\underline{x}))$ to satisfy. The idea is that there should be only one $\underline{A}$ that results in a given $F^{\mu\nu}$ and that also satisfies the gauge condition. In this section we work in a gauge:

$$\nabla \cdot \mathbf{A} = 0, \tag{9.6.13}$$

which is called Coulomb gauge, also known as radiation gauge or transverse gauge or physical gauge. [Beside Coulomb gauge, there are also Lorentz (Landau) gauge ($\partial_\mu A^\mu = 0$), temporal gauge $A^\circ = 0$, axial gauge $A^3 = 0$ [122].]

The physical meaning of Coulomb gauge $\nabla \cdot \mathbf{A}(\underline{x}) = 0$ is as follows. If we make Fourier expansion, then $\nabla \cdot \mathbf{A} = 0$ means wave vector $\mathbf{k} \perp \mathbf{A}$.

According to the corollary in section 9.2 we know that we can choose a new field appropriates $\mathbf{A}'$, which has already satisfied the imposing constrained condition of transverse field. For seeking the appropriate field coordinates, we make a transformation

$$A^i(\underline{x}) \to A^{i'}(\underline{x}) = \left(\delta_{ij} - \frac{\partial_i \partial_j}{\Delta}\right) A^j(\underline{x}), \tag{9.6.14}$$

where $\Delta = \sum_{i=1}^3 \partial_i^2$, $\underline{x} = (ct, x, y, z) = (ct, x^1, x^2, x^3)$, $\partial_i = \partial/\partial x^i$, and the repeated index j indicates summation over j from 1 to 3. Let us show that $\mathbf{A}'$ is a transverse field vector.

$$\begin{aligned}
\nabla \cdot \mathbf{A}'(\underline{x}) &= \sum_{i=1}^3 \frac{\partial}{\partial x^i}\left(\delta_{ij} - \frac{\partial_i \partial_j}{\Delta}\right) A^j(\underline{x}) \\
&= \frac{1}{(2\pi)^{3/2}} \sum_{i=1}^3 \partial_i \left(\delta_{ij} - \frac{\partial_i \partial_j}{\Delta}\right) \int d\mathbf{k} A^j(\mathbf{k}, t) e^{i\mathbf{k}\cdot\mathbf{r}} \\
&= \frac{1}{(2\pi)^{3/2}} \int d\mathbf{k} \sum_{i=1}^3 \partial_i \left(\delta_{ij} - \frac{\partial_i \partial_j}{\Delta}\right) A^j(\mathbf{k}, t) e^{i\mathbf{k}\cdot\mathbf{r}} \\
&= \frac{1}{(2\pi)^{3/2}} \int d\mathbf{k} \sum_{i=1}^3 ik^i \left(\delta_{ij} - \frac{k^i k^j}{k^2}\right) A^j(\mathbf{k}, t) e^{i\mathbf{k}\cdot\mathbf{r}} \\
&= \frac{1}{(2\pi)^{3/2}} \int d\mathbf{k} i \sum_{j=1}^3 A^j(\mathbf{k}, t) \sum_{i=1}^3 \left\{k^i \left(\delta_{ij} - \frac{k^i k^j}{k^2}\right)\right\} e^{i\mathbf{k}\cdot\mathbf{r}} \\
&= \frac{1}{(2\pi)^{3/2}} \int d\mathbf{k} i \sum_{j=1}^3 A^j \times 0 e^{i\mathbf{k}\cdot\mathbf{r}} = 0.
\end{aligned} \tag{9.6.15}$$

Eq. (9.6.15) shows that **A**' satisfies the Coulomb gauge constraint condition. Now we replace the **A** in Eq. (9.6.9) by **A**' and still use $A^\circ = \phi$, then we have new Lagrangian density. From now on, whenever we write A^i, we will implicitly mean the left hand side of Eq. (9.6.14).

Starting from Eq. (9.6.9), we can rewrite the Lagrangian density in terms of $\phi = A^\circ$ $(A_i) = (A_1, A_2, A_3)$, with $\mathbf{A}$ satisfying Coulomb gauge $\nabla \cdot \mathbf{A} = 0$. (For the next calculations in this section one can refer to Ref. [125, 121].)

$$
\begin{aligned}
\mathcal{L} &= \frac{1}{8\pi}\dot{A}^i\dot{A}^i - \frac{1}{8\pi}\partial_j A^i \partial_j A^i + j^i A^i \\
&+ \frac{1}{8\pi}\partial_i A^j \partial_j A^i + \frac{1}{4\pi}\dot{A}^i \partial_i \phi \\
&+ \frac{1}{8\pi}\partial_i \phi \partial_i \phi - \rho\phi. \qquad (9.6.16)
\end{aligned}
$$

Note that we have known that $\mathbf{A}$ in Eq. (9.6.16) is a transverse field. Therefore, the factor $\mathbf{j}$ in Eq. (9.6.16) should be understood as $\mathbf{j}^{\perp}$.

When we want to obtain Euler-Lagrange equation, we should write the action $\mathcal{A}$. The action is an integral, which's integrand is $\mathcal{L}$. For convenience, we can make some calculations before using the principle of least action to obtain Euler-Lagrange equation of motion. In the second equality of Eq. (9.6.16) there are two terms, the ∂_i in each term can be integrated by parts; in the first term, we will get a factor of $\partial_j(\partial_i A_i)$, and in the second term, we will get a factor of $\partial_i \dot{A}_i$. Both of these vanish by virtue of the Coulomb gauge condition, and so both of these terms can simply be dropped.

Under the condition of From Euler-Lagrange equation we have

$$-\nabla^2\phi = 4\pi\rho. \qquad (9.6.17)$$

The solution is

$$\phi(\mathbf{r}, t) = \int d\mathbf{y} \frac{\rho(\mathbf{y}, t)}{|\mathbf{r} - \mathbf{y}|}. \qquad (9.6.18)$$

This solution is unique if we impose the boundary conditions that ϕ, ρ both vanish at spatial infinity.

Eq. (9.6.18) tells us that $\phi(\mathbf{r}, t)$ is given entirely in terms of the charge density at the same time, and so has no dynamics of its own. The charge density belongs to the matter which brings charge. Thus we can take out this degree of freedom $A^{\circ} = \phi$. The remaining degrees of freedom are A^i, with $i = 1, 2, 3$, subject to the gauge condition $\nabla \cdot \mathbf{A} = 0$.

As is well known, the charge density depends only on the matter field Q and their canonical momentum P, so Eq. (9.6.18) represents an explicit solution for the auxiliary field A°.

It is therefore legitimate to substitute Eq. (9.6.18) back into the Lagrangian. After an integration by parts to turn $\partial^i\phi\partial^i\phi$ into $-\phi\nabla^2\phi = \phi 4\pi\rho$, the result is

$$\mathcal{L} = \frac{1}{8\pi}\dot{A}^i\dot{A}^i - \frac{1}{8\pi}\partial_j A^i \partial_j A^i + j^i A^i + \mathcal{L}_{coul}, \qquad (9.6.19)$$

where

$$\mathcal{L}_{coul} = -\frac{1}{2}\int d\mathbf{y} \frac{\rho(\mathbf{r}, t)\rho(\mathbf{y}, t)}{|\mathbf{r} - \mathbf{y}|}. \qquad (9.6.20)$$

(3). Third step. Seek the canonical conjugate field momenta Π^i corresponding to canonical field coordinates A^i. According to definition

$$\Pi^i = \frac{\partial \mathcal{L}}{\partial \dot{A}^i} = \frac{\dot{A}^i}{4\pi}. \qquad (9.6.21)$$

Note that $\nabla \cdot \mathbf{A} = 0$ implies $\nabla \cdot \vec{\Pi} = 0$.

(4). Fourth step. According to the corollary in section 9.2, we can have the following canonical commutation relations:

$$\begin{aligned}[A^i(\mathbf{r},t),\Pi^j(\mathbf{r}',t)] &= i\hbar\delta_{ij}(\mathbf{r}-\mathbf{r}') \\ &= i\hbar\delta^{\perp}_{ij}(\mathbf{r}-\mathbf{r}') \\ &= i\hbar\left(\delta_{ij}-\frac{\partial_i\partial_j}{\nabla^2}\right)\delta(\mathbf{r}-\mathbf{r}') \\ &= i\hbar\int\frac{d\mathbf{k}}{(2\pi)^3}e^{i\mathbf{k}\cdot(\mathbf{r}-\mathbf{r}')}\left(\delta_{ij}-\frac{k^ik^j}{k^2}\right). \qquad (9.2.1.8)'=(9.6.21)'\end{aligned}$$

Here, we make a detail explanation for the second equality. We have already proven by Eqs. (9.6.15) and (9.6.21) that both $\mathbf{A}$ and $\overrightarrow{\Pi}$ are transverse field. From the first equality and the transverse field properties, we have

$$0=\nabla_{\mathbf{r}}\cdot[A^i(\mathbf{r},t),\Pi^j(\mathbf{r}',t)]=i\hbar\delta_{ij}\nabla_{\mathbf{r}}\delta(\mathbf{r}-\mathbf{r}')\neq 0. \qquad (9.6.22)$$

$$0=\nabla_{\mathbf{r}'}\cdot[A^i(\mathbf{r},t),\Pi^j(\mathbf{r}',t)]=i\hbar\delta_{ij}\nabla_{\mathbf{r}'}\delta(\mathbf{r}-\mathbf{r}')\neq 0. \qquad (9.6.23)$$

Eqs. (9.6.22) and (9.6.23) are contract with the transverse field condition of $\mathbf{A}$ and $\overrightarrow{\Pi}$. Therefore, the delta function has to satisfy transverse condition as well. So, in $\delta_{ij}\delta=\delta^{\perp}_{ij}+\delta^{\parallel}_{ij}$ we only keep the transverse part $\delta^{\perp}_{ij}$.

When we found out the canonical commutation relations, then the previous classical Maxwell field equation set, given by the first step, becomes naturally a quantized field equation set. At this time we accomplish the task to quantize the classical Maxwell field equations, and enter into the quantum field theory. Note that at this step the quantum field equation is still in the representation of the field coordinate representation, and thus the field coordinates are still a numerical number.

(5). Fifth step. Expand the field coordinates by any complete basis. Let us seek this basis. Formally, the Lagrangian in Eqs. (9.6.19) and (9.6.9) are same. Therefore, although $\mathbf{A}$ in Eq. (9.6.19) is a transverse field, we still have

$$\Box A^i=\left(\frac{1}{c^2}\frac{\partial^2}{\partial t^2}-\nabla^2\right)A^i=\frac{4\pi j^{i\perp}}{c}, \quad (i=1,2,3). \qquad (9.6.12)'=(9.6.21)''$$

For free field $j^i=0$, the general solution of Eq. $(9.6.12)'=(9.6.21)''$ is

$$\begin{aligned}\mathbf{A}(\underline{x}) &= \sum_{\mathbf{k}}N_{\mathbf{k}}\sum_{\lambda=1}^{2}\left\{a_\lambda(\mathbf{k})\frac{1}{\sqrt{V}}e^{-i\underline{k}\cdot\underline{x}}+a^+_\lambda(\mathbf{k})\frac{1}{\sqrt{V}}e^{i\underline{k}\cdot\underline{x}}\right\}\overrightarrow{\epsilon}_\lambda(\mathbf{k}) \\ &= \int d\mathbf{k}N_{\mathbf{k}}\sum_{\lambda=1}^{2}\left\{a_\lambda(\mathbf{k})\frac{1}{(2\pi)^{3/2}}e^{-i\underline{k}\cdot\underline{x}}+a^+_\lambda(\mathbf{k})\frac{1}{(2\pi)^{3/2}}e^{i\underline{k}\cdot\underline{x}}\right\}\overrightarrow{\epsilon}_\lambda(\mathbf{k}), \qquad (9.6.24)\end{aligned}$$

where $\overrightarrow{\epsilon}_\lambda(\mathbf{k})$ is the polarization vector, is a unit vector, and $\mathbf{k}\cdot\overrightarrow{\epsilon}_\lambda(\mathbf{k})=0$, which shows that $\mathbf{A}$ is a transverse wave, and

$$N_{\mathbf{k}}=\frac{\sqrt{4\pi\hbar}}{\sqrt{2\omega_k}}$$

can guarantee that the solution in Eq. (6.6.24) satisfies the canonical commutation relation Eq. (9.6.21)′ and that one can derives Eq. (9.6.27) from Eqs. (9.6.24) and (9.6.21)′. In Eq. (9.6.24), $\underline{k}\cdot\underline{x}=k^\circ x^\circ-\mathbf{k}\cdot\mathbf{r}=\omega_k t-\mathbf{k}\cdot\mathbf{x}$ [See. Eq. (9.4.2.7)], $k^\circ=\omega_k$, $\lambda=1,2$ represents the two components of the transverse field $\mathbf{A}$.

In complete analogy with the procedures used for a scalar field in section 9.4, we can invert Eq. (9.6.24) and its time derivative to get

$$a_\lambda(\mathbf{k}) = +i\overrightarrow{\epsilon}_\lambda(\mathbf{k})\cdot\int d\mathbf{r}e^{i\underline{k}\cdot\underline{x}}\overleftrightarrow{\partial}_\circ\mathbf{A}(\underline{x}). \tag{9.6.25}$$

$$a^+_\lambda(\mathbf{k}) = -i\overrightarrow{\epsilon}^*_\lambda(\mathbf{k})\cdot\int d\mathbf{r}e^{-i\underline{k}\cdot\underline{x}}\overleftrightarrow{\partial}_\circ\mathbf{A}(\underline{x}), \tag{9.6.26}$$

where $f\overleftrightarrow{\partial}_\mu g = f(\partial_\mu g) - (\partial_\mu f)g$.

The commutation relations of the $a_\lambda(\mathbf{k})$ and $a^+_\lambda(\mathbf{k})$ operators follow from Eqs. (9.2.1.8)'=(9.6.21)', (9.6.24), (9.6.21), and $[A^i, A^j] = [\Pi^i, \Pi^j] = 0$ (at equal times). The results are

$$[a_\lambda(\mathbf{k}), a_{\lambda'}(\mathbf{k}')] = 0, \quad [a^+_\lambda(\mathbf{k}), a^+_{\lambda'}(\mathbf{k}')] = 0,$$

$$[a_\lambda(\mathbf{k}), a^+_{\lambda'}(\mathbf{k}')] = \delta_{\lambda\lambda'}\delta(\mathbf{k}-\mathbf{k}'). \tag{9.6.27}$$

The $a^+_\lambda(\mathbf{k})$ and $a_\lambda(\mathbf{k})$ represent creation and annihilation operators for photons with transverse polarization.

(6). Sixth step. Make more calculations and consider interaction. It is now straightforward to write the total Hamiltonian explicitly in terms of these operators. We find

$$\begin{aligned} H &= \int d\mathbf{r}\mathcal{H} = \int d\mathbf{r}(\Pi^i\dot{A}^i - \mathcal{L}) \\ &= \int d\mathbf{r}\left[-4\pi\Pi^i\Pi^i + \frac{1}{8\pi}\partial_j A^i\partial_j A^i - j^i A^i + \mathcal{H}_{coul}\right] \\ &= \sum_{\lambda=\pm}\sum_{\mathbf{k}}\hbar\omega_k a^+_\lambda(\mathbf{k})a_\lambda(\mathbf{k}) + 2\mathcal{E}_0 V - \int d\mathbf{r}\mathbf{j}(\mathbf{r},t)\cdot\mathbf{A}(\mathbf{r},t) + H_{coul}, \end{aligned} \tag{9.6.28}$$

where $\mathcal{E}_0$ is zero point energy per unit volume that we found for a real scalar field in section 9.4, V is the volume of space.

$$\mathcal{E}_0 = \frac{\hbar}{2}\sum_{\mathbf{k}}\omega_k. \tag{9.6.29}$$

The Coulomb Hamiltonian is

$$H_{coul} = \frac{1}{2}\int d\mathbf{r}d\mathbf{r}'\frac{\rho(\mathbf{r},t)\rho(\mathbf{r}',t)}{|\mathbf{r}-\mathbf{r}'|}. \tag{9.6.30}$$

This form of the Hamiltonian of electromagnetics is often used as the starting point for calculations of atomic transition rates, with the charges and currents treated via the non-relativistic Schrödinger equation. The Coulomb interaction appears explicitly, and the $\mathbf{j}\cdot\mathbf{A}$ term allows for the creation and annihilation of photons of definite polarization in the interaction processes between photons and source charge and current.

At last, we would like to say that the elecromagnetic field quantization method in this section is nearly same as that in Ref. [125]. But, there is a principal difference between this book and Ref. [125]. Ref. [125] wrote: "A more rigorous treatment of the quantization in Coulomb gauge can be found in Weinberg I." From Ref. [125, 122] we know that Weinberg I used the Dirac bracket method to quantize the Maxwell field. However, according to the proved corollary on the canonical commutation relations with constraints, our above quantization for the Maxwell field is rigorous and simple. We feel that the Dirac bracket method to treat the field quantization with constraints is not necessary, because Dirac method looks like to draw a snake and also to add feet to the snake. Therefore, the Dirac method is difficult to be understood, and is too much trouble. By the way, some authors like to use the Gupta-Bleuler indefinite metric operator method [116, 121, 122]. We think that the indefinite metric method draws a forced analogy, because it introduces spurious state.

Now we introduce the gauge invariance, the interaction between electromagnetic field and charged massive particles, and the gauge field theory. We make the following transformation for a system containing electromagnetic field and, for example, electron, which is called gauge transformation,

$$A^{\mu} \to A'^{\mu} = A^{\mu} - g\partial_{\mu}\kappa(\underline{x}), \quad \Psi \to \Psi' = e^{i\kappa(\underline{x})}\Psi. \tag{9.6.31}$$

The Maxwell tensor in Eq. (9.6.6) $F^{\mu\nu}$ and Maxwell equation set Eq. (9.6.8) are invariant under gauge transformation. Since $\kappa(\underline{x})$ depends on spacetime point $\underline{x} = (ct, x, y, z)$, people often call this gauge transformation local gauge transformation. The local gauge invariance belongs to one of the local symmetries.

Let us discuss the Maxwell equation set with Dirac and scalar particles by using the quantum field theory. It can readily be checked that the Lagrangian that is locally gauge invariant and has a Dirac particle, a charged scalar particle, and the zero-mass photon (Maxwell equation set) is given by [124]

$$\mathcal{L} = \overline{\Psi} i\gamma^{\mu} D_{\mu}\Psi - m\overline{\Psi}\Psi + (D_{\mu}\Phi)^{*}(D_{\mu}\Phi) - \mu^{2}\Phi^{*}\Phi - \frac{F_{\mu\nu}F^{\mu\nu}}{16\pi}, \tag{16.6.32}$$

where

$$D_{\mu} = \partial_{\mu} - ieA^{\mu}, \tag{9.6.33}$$

the first and second terms, the third and fourth terms, and the fifth term come from Eqs. (9.5.5), (9.4.4.2), and (9.6.9), respectively.

If there are Dirac field coordinate Ψ and complex (charged) field coordinate Φ, then the local gauge transformation is defined by

$$A^{\mu} \to A'^{\mu} = A^{\mu} - \frac{1}{e}\partial_{\mu}\kappa, \quad \Psi \to \Psi' = e^{i\kappa}\Psi. \tag{9.6.34}$$

$$A^{\mu} \to A'^{\mu} = A^{\mu} - \frac{1}{e}\partial_{\mu}\kappa, \quad \Phi \to \Phi' = e^{i\kappa}\Phi. \tag{9.6.34$'$}$$

From the Lagrangian density in Eq. (9.6.32), one can obtain equation of motion in quantum field and study many phenomena. This equation of motion gives dynamics called gauge interactions. The theory underlying the above formulations is called gauge theory and AUNRTB believe that all four basic interactions (strong, weak, electromagnetic, and gravitational) are described by some form of gauge theory consisting of only massless particles, included among which are the vector bosons, which are often called gauge bosons. The acquisition of mass for these gauge bosons then results from spontaneous symmetry breaking of gauge invariance. Of course, at this stage we should write a more general Lagrangian density. [In chapter 7 we have proved that the in terms of the unitary gauge and the spontaneous gauge symmetry breaking, the gauge field (i. e., the electromagnetic field) **A** gets apparent mass m_A, which called Andersaon-Higgs mechanism.]

The model in which the strong and electroweak interactions are mediated by gauge bosons is called the Standard model. Now there is a key element of the Standard model, which has not yet been discovered. There is a scalar field, called the Higgs field, that is required to break spontaneously the gauge symmetry in electroweak interaction. This Higgs field is responsible not only for the large masses of the $W^{\pm}$ and Z^0 but also for the masses of all the quarks and leptons. It should also manifest itself in the existence of, at least, one spin-0 particle called the Higgs boson. We will give some comments on the Higgs boson in section 9.7.

9.7 Future of Quantum Field Theory

On the future of quantum field theory we state the following nine basic points. All the nine points can be understood by the readers who can understand the chapters 1-8, and sections of 9.1-6 of chapter 9 in this textbook. We do not want to say more points which are beyond the scope of this textbook.

(1). The quantum field theory has to be based on the wavepacket-only theory. In principle, any objects connect with a local function called wavepacket. The wavepackets can be massive particles with rest mass and energy clusters (such as photon and phonon) without rest mass. The wavepackets can be larger (e. g., the Milky Way 10^{23} cm) or less (e. g. the Planck scale 10^{-33} cm). The mass and energy of wavepackets can be larger (e. g., the Milky Way 10^{68} GeV) or less (e. g., Binding energy of the electron in the hydrogen atom 10^{-8} GeV).

The wavepacket-only theory proves exactly by mathematics that any wavepackets are of construction: the body-factor and guide-factor. The object's Hamiltonian obeys the special relativity theory, general relativity theory, classical Maxwell equation set, and so on. The body-factor brings mass, energy, momentum, charge, spin, and so on. The guide-factor is just a mathematical function, which is only the solution of equation of motion, such as Schrödinger, Dirac, and Klein-gorden equations, the Hamiltonian of which have to obey the special relativity theory, general relativity theory, classical Maxwell equation set, and so on. The guide-factor is just the so-called wavefunction. As is well known, a pure mathematical function has no any observable meaning. Although the guide-factor does not bring mass, energy, spin, and so on, but the body-factor has to obey the guide of the guide-factor according to spontaneous equiprobability symmetry breaking rule, and every time of the actual motion of the body-factor correspond to a time of spontaneous equiprobability symmetry breaking. Just due to this spontaneous symmetry breaking our word is rich and varied. (Although Einstein said that: "I do not believe that God plays dice.", but this textbook proves that: God plays dice truly.) The quantum field theory is an extension of quantum mechanics to the systems with innumerably (continuously) infinite degrees of freedom. The quantized fields, i. e., the quantized guide-factors, act still as guider. For example, we cannot say that because we have quantum field theory, thus we have antiparticle. Actually, in appendix we have emphasized that even we do not have quantum field theory, we can still deduce the existence of the antiparticle only from the classical relativistic mechanics. We can say that quantum field theory is a good method to explain the existence of antiparticle, and, especially, to study the antiparticle's creation and annihilation and so on.

About this point (1), we find the following mistakes and misunderstandings in some references:

(i). Some references, such as Ref. [126], cannot understand correctly the field-particle duality.

The wavepacket-only theory requires the following viewpoints on the field-particle duality.

Recall the steps of field quantization: We look upon the wavefunction of single particle equation or classical field as a field, then we quantize it in terms of the proved canonical commutation relation. However, the field still indicates the probability amplitude of appearance of particle in principle. An actual particle (electron, Milky Way, photon, and so on) obeys the guidance of field, and might vanish or produce new particle or occurs at another place. In this aspect, the body-factor is main, and the field is attached. For example, if we say that this field is fulfilled by photons, then this means only that there are really many wavepackets of photons in this field. The creation and annihilation are also due to that there are the actual conditions to allow their production. The quantum field theory is only a description for the wavepacket-only world as well as quantum mechanics.

(ii). Ref. [127] thinks that: "The principle of indistinguishability of identical particle system still plays a first fundamental role in quantum field theory. The elementary particles are not individuals. They cannot be distinguished individually."

The wavepacket-only theory thinks that the above viewpoint is completely wrong. The origin of the mistake is that Ref. [127] does not know that the wavepacket has construction. The body-factors can be distinguished. For example, two electrons are in a large box. The two electrons are separated far from each other. It is obvious that we can distinguish "who is who". Although the two electrons are separated far from each other, but the two guide-factors (=single particle wavefunctions) of the two electrons are proved to have to obey the theorem of identity in principle. The basis of our proof on the identity theorem is only the invariance of the macroscopic observable quantity under the permutation of, for example, the two electron coordinates in the multiple wavefunction. Here, the permutation symmetry does not require the indistinguishability of body-factors of identical particles.

(iii). Some references feel that "Using point particle model is a defect of modern quantum field theory."

The wavepacket-only theory thinks that the above viewpoint is completely wrong. The origin of the mistake is that some references do not know that the wavepacket has construction. Actually, after field quantization the action of field is to guide the motion of the body-factor. The quantum field theory does not look upon the body-factor as a point. The quantum field theory only looks upon the spacetime as a geometric point. This should be OK.

(2). Scattering theory is a framework for studying and understanding the interaction between particles. In quantum field theory, the equations are those of quantum electrodynamics (QED), quantum chromodynamics (QCD), the Standard Model, and so on. For the scattering theory, from the experimental viewpoint the observable quantity is the cross section, and from the theoretical viewpoint the key quantity is the S matrix. However, all the theories (S matrix theory, quantum transition, decay theory, and so on) take the limit $t \to \infty$ in a wrong order.

We have listed several theorems in chapter 1, which indicates that the limit $t \to \infty$ is not physical and is illegitimate in mathematics. This obvious mistake lasts about 80 years. This wrong limit leads to the wrong spurious energy conservation, and therefore conceals the true energy nonconservation in quantum transition processes, relaxations, and scattering processes. Many models of particles depend on the analysis for scattering cross section data. Therefore, it is necessary to check if the limit $t \to \infty$ is a good approximation in every physical processes and theoretical calculations. Anyway, the Fermi golden rule as a rule has to be negatived, although the Fermi golden rule is a good approximation in some cases.

Some references think that the interaction is mediated by exchanging virtual particles. Because the energy in the processes of exchanging middle particles does not conservative, therefore people call the particle in middle processes the virtual particle, and the middle processes of exchanging the middle particle the virtual processes.

The wavepacket-only theory thinks that the above treatment is completely wrong. We think that the processes of exchanging middle particles belong also to transition processes, and thus the energies in these processes should be nonconservative. The so-called "virtual" particle is true particle actually.

(3). According to the corollary on the canonical commutation relations with constraints in section 9.2, we suggest that for the quantization of the constrained classical fields one does not need to use the farfetched Gupta-Bleuler indefinite metric operator method [121], and the tedious and overelaborate Dirac bracket method [122]. We have taken the Maxwell electromagnetic field in section 9.6 as an example of constrained system to address how to do the field quantization with constraints. The quantization of gauge field has been done successfully by using the path integral method. However, the quantization of many gauge fields has not yet been done successfully by using the method of the field canonical quantization. The author of this textbook feels that one can try to use the method of field canonical quantization in this textbook to quantize the gauge field.

(4). The quantum field theory studies interactions between particles. But, until now, the interactions are solely confined in exchanging middle bosons. Some references believe that all the four basic interactions (strong, weak, electromagnetic, and gravitational) are described by some form of gauge theory consisting of only massless particles, included among which are the vector bosons, which are often called gauge bosons.

Let us look at the superconductivity, and hope to get some instructions on the possible types of interactions. The pairing interaction in conventional superconductivity is really caused by exchanging phonon (boson) between two electrons. But, the pairing interaction in the high temperature superconductivity is caused by multiple exchange interaction between two carriers (See Ref. [6] and chapter 7). Although the exchange of middle boson (phonon) in the high temperature superconductivity still exists, but its contribution to high temperature superconductivity is exactly equal to zero.

(5). According to the theorem in section 9.2 there are uncertainty relations between the field coordinate and corresponding canonical field momentum. But these uncertainty relations have not yet been caused attention of physicists.

(6). According to wavepacket-only theory, the string is also a wavepacket, and therefore, the string has to have a certain volume. Otherwise, violate uncertainty relation theorem. Although the string theory modifies the uncertainty relation as follows:

$$\Delta x \cdot \Delta p_x \geq \frac{\hbar}{2} \rightarrow \Delta x \geq \frac{\hbar}{2\Delta p_x} + C\Delta p_x, \qquad (9.7.2)$$

where C is a constant, related to the Planck length, but this modification violates the theorem of uncertainty relation, proved in chapter 1, and is farfetched. Besides, the lengths of a string in y and z directions are zero, then these zero lengths will lead to $\Delta p_y \rightarrow \infty$ and $\Delta p_z \rightarrow \infty$, which are not reasonable obviously.

(7). Chapter 7 proves that the energy can be produced infinitely in space. Therefore, the energy is not a definite quantity in space. However, the general theory of relativity gives always that the energy is conservative.

In General theory of Relativity, the dynamical variables characterizing the gravitation field are the components of the metric tensor $g_{\mu\nu}(x^\gamma)$, and obey the Einstein gravitation field equation in which the cosmological constant term $\Lambda g_{\mu\nu}$ is contained [128].

$$R_{\mu\nu} - \frac{1}{2}g_{\mu\nu}R + \Lambda g_{\mu\nu} = \frac{8\pi G}{c^4}T_{\mu\nu}, \qquad (9.7.3)$$

where $R_{\mu\nu}$ is the Ricci curvature tensor, R the scalar curvature, $g_{\mu\nu}$ the metric tensor, Λ is the cosmological constant, G is Newton's gravitational constant, c is the speed of light in vacuum, and $T_{\mu\nu}$ is the stress-energy tensor or energy-momentum tensor. After some transformations, Einstein gravitation field equation can be interpreted as a set of equations dictating how matter/energy determines the curvature of spacetime.

In history, Einstein included the cosmological constant as a term in his field equations for general relativity because he was dissatisfied that otherwise his equations did not allow, apparently, for a static universe (i.e., one that is not expanding or contracting): gravity would cause a universe which was initially at dynamic equilibrium to contract. To counteract this possibility, Einstein added the cosmological constant term. This effort was unsuccessful for two reasons: the static universe described by this theory was unstable, and observations of distant galaxies by Hubble a decade later confirmed that our universe is, in fact, not static but expanding. So Λ was abandoned, with Einstein calling it the "biggest blunder [he] ever made". For many years the cosmological constant was almost universally considered to be 0.

Despite Einstein's misguided motivation for introducing the cosmological constant term, there is nothing inconsistent with the presence of such a term in the equations. Indeed, recent improved astronomical techniques have found that a positive value of Λ is needed to explain the accelerating universe.

Some references, including Einstein, think of the cosmological constant as an independent parameter. Now we move the cosmological constant term algebraically to the other side of Einstein field equation, written as part of the stress-energy tensor:

$$T_{\mu\nu}^{vac} = -\frac{\Lambda c^4}{8\pi G} g_{\mu\nu}. \tag{9.7.4}$$

The resulting vacuum energy density is constant and given by

$$\rho_{vac} = \frac{\Lambda c^4}{8\pi G}. \tag{9.7.5}$$

The existence of a cosmological constant is thus equivalent to the existence of a non-zero vacuum energy. The terms are now used interchangeably in general relativity. A positive vacuum energy density resulting from a cosmological constant implies a negative pressure, and vice versa. If the energy density is positive, the associated negative pressure will drive an accelerated expansion of empty space. (See dark energy and cosmic inflation for details.)

Using the upper limit of the cosmological constant, the vacuum energy in a cubic meter of free space has been estimated to be 10^{-9} Joules. However, in both quantum electrodynamics and stochastic electrodynamics, consistency with the principle of Lorentz covariance and with the magnitude of the Plank constant require it to have a much larger value of 10^{113} Joules per cubic meter. Here the definition of vacuum energy is the zero point energy of all the fields in space. Zero point energy is the lowest possible energy that a quantum mechanical system may have. Before this book some refertences propose a problem: why does not the large constant vacuum energy density cause a large cosmological constant? What cancels it out?

The supersymmetric theory gives the newest estimation of the cosmological constant Λ. It is about 1 TeV, which is about 10^{-64} in Planck units. However, this number is still about 60 orders of magnitude larger than the observational limit. Thus, we see that supersymmetry, while many authors think that it is a step in the right direction of cosmology study, does not quite solve the cosmological constant problem. Many authors think that that all theories can only give a huge cosmological constant is a physical crisis.

The viewpoint of this book is as follows. According to the result in chapter 7, due to existence of the vacuum space energy support ability, even a very small volume can provide infinitely many energies. Therefore, the above some references' problem becomes more sharp because infinitely large energy>> 10^{113} Joules. It is obvious that the way that connects the cosmological constant with the intrinsic vacuum energy density will get us absolutely nowhere.

The approach to solve this problem is obvious. We note that there is no any *a priori* reason to think that ρ_{vac} has to connect with vacuum energy density. We can think that ρ_{vac} connects with the vacuum mass fluctuation density other than the zero point energy and/or other energy forms. In this case, $T_{\mu\nu}$ has to be interpreted as stress-mass tensor or mass-momentum tensor. The latter interpretation is more better because the gravitation force directly connects with mass other than the energy. Anyway, the observation of cosmological expansion and Einstein field equation provide a method to measure the vacuum mass fluctuation.

(8). We would like doubt the uniqueness of the so the Higgs mechanism and, thus, the Standard Model of particle physics. In chapter 7 we have taken the electromagnetic field, Cooper charge pair, microscopic superconductivity, Meissner effect, Goldstone boson, and the spontaneous gauge invariance breaking as an example to introduce the gauge field. Simply speaking, in this example, by using the language of gauge field theory, the gauge field is the electromagnetic field $\{A_\mu\}$, the photon in this case can be also called gauge boson, the phase field is the phase θ of wavefunction $\Psi = \exp(i\theta)|\Psi|$ of Cooper charge pair, the phase field is called Nambu-Goldstone field or Goldstone boson, the spontaneous symmetry breaking is the appearance of microscopic superconducting state, and that the photon gets mass after the spontaneous symmetry breaking corresponds to the Meissner antimagnetic effect.

Let us tell you that how the gauge field theory to analyze this example. At first, gauge field theory thinks that after spontaneous symmetry breaking, the free energy density, or equivalently, the Lagrangian density has a term $m^2A^2/2$, in which the $m/\sqrt{2} = M_A = 1.35 \times 10^{-39}$ *gram* (See the exercise (8.B) of chapter 7.) represents the dynamical mass of gauge field (in this case it is the photon being of dynamical mass.) The cause and origin that the photon (or the gauge boson in this case) obtains (dynamical) mass are the following combination processes [124]: the massless Goldstone field (phase field) is "gauged away" or simply "eaten away". (Please recall that under unitary transformation the free energy expression of superconductor does not contain the Goldstone field. See chapter 7.) The experimental demonstration that gauge boson (photon in this case) gets mass is the Meissner effect.

This textbook thinks that the above analysis is obviously not correct. Because:

(A). In this example, that the electromagnetic field (or photon or gauge boson) obtains mass is only a false appearance, i. e., is apparent. Of course, if the photon obtains mass really in the spontaneous symmetry breaking process, and thus the massive photon cannot move to the deep interior of the superconductor, then one can explain the Meissner antimagnetic effect. However, no body really directly observed the massive photon (The last exercise (8) of chapter 7 tells us that to explain the Meissner effect, the mass acquired by the photon in the spontaneous $U(1)$ gauge symmetry breaking is about 10^{-38} g). Actually, the Meissner antimagnetic effect can be easy explained by the antimagnetic current induced by electromagnetic induction. Especially, this antimagnetic current is observed directly by many experiments. Therefore, using acquisition of mass of photon to explain the Meissner effect is only a formal formulation, and does not represents the physical reality. Due to that the massive photon in Meissner effect is a false appearance, we do not need the hypothetical massive particle such as Higgs boson to mediate.

(B). In the Standard Model, gauge bosons are defined as force carriers that mediated the strong, weak, and electromagnetic fundamental interactions. In the case of electromagnetic fundamental interaction, the gauge (field) boson is called photon. The Higgs boson is a hypothetical massive scalar elementary particle. The Higgs bosons play a unique role in the standard model by explaining why other particles, except photon and gluon, have masses. The author of this textbook thinks that we do not need to confine ourselves to consider just one boson-mediated interaction. For example, the superconductivity can be caused by both the phonon-mediated interaction and multiple exchanging-mediated interaction. Ref. [75] has been proved that the exchanging interaction can be both attractive and repulsive interaction. If we consider the latter interaction, then we do not need the Higgs boson, and at the same time the $W^{\pm}$ and Z boson can also obtain masses.

(C). Let us say something about the Weinberg-Salam electroweak theory (1960s). The discovery of $W^{\pm}$ and Z^0 boson (1983) and the neutral current weak interaction provided a great boost to the Weinberg-Salam theory. The last confirmation of the electroweak theory must however await the discovery of the Higgs scalar bosons with mass $\sim$ 10^2 GeV.

On 7 March 2012, CDF Collaborations announced that, after analyzing the full data set from the Tevatron accelerator, they found excesses in their data that might be interpreted as coming from a Higgs boson with a mass in the region of 115 to 135 GeV$/c^2$. The significance of the excesses is quantified as 2.2 standard deviations, not enough to rule out that they are due to a statistical fluctuation. The Large Hadron Collider at CERN announced results consistent with the Higgs boson on July 4, 2012 but stressed that further testing is needed to confirm the complete mechanism.

The author of this textbook feels that an another manner electroweal theory might predict the existence of $W^{\pm}$ and Z^0 boson (1983) and the neutral current as well as the W-S electroweak theory and at the same time this theory does not need the Higgs mechanism to give the mass to the $W^{\pm}$ and Z^0 boson. This new electroweak theory should use some kinds of exchanging-mediated interactions other than the sole higgs-boson-mediated interactions.

(9). On the CPT violations. Let me expose this problem in more detail.

The possible existence of CPT violation has been increasingly studied in recent years [129]-[133], because some minuscule apparent violations of CPT invariance might be observable in nature. A good and newest summary on the recent researches and history of discovery of CPT violations is given by Ref. [132]. In quantum field theory there is a famous theorem which proves that Lorentz violation will certainly lead to CPT violation. Some recent references also prove that CPT violation will certainly lead to Lorentz violation. Many references accepted *CPT violation* $\Rightarrow$ *Lorentz violation*, and, for example, Ref. [133] has modified the usual Standard Model of particle physics by adding some new terms, which are Lorentz violation, but can explain the CPT violation. The theory in Ref. [133] is called SME.

Let me introduce very simply that what is Lorentz and CPT symmetry (symmetry = invariance).

Lorentz transformations and invariances are explained in appendix A.

The CPT transformation is formed by combining three transformations: charge conjugation ($\mathcal{C}$), parity transformation ($\mathcal{P}$), and time reversal ($\mathcal{T}$). $\mathcal{C}$ converts a particle into its antiparticle. $\mathcal{P}$ transforms an object into its mirror image but turned upside down. $\mathcal{T}$ changes the direction of flow of time.

A physical system is said to have "Lorentz symmetry" if the relevant laws of physics are unaffected by Lorentz transformations (rotations and boosts). Similarly, a system is said to have "CPT symmetry" if the physics is unaffected by the combined transformation CPT. These symmetries are the basis for Einstein's relativity.

Experiments show to exceptionally high precision that all the basic laws of nature seem to have both Lorentz and CPT symmetry. Experimental results are compiled in the paper "Data Tables for Lorentz and CPT Violation" [133]. Note: CPT is the only combination of C, P, T that is presently observed to be an exact symmetry of nature.

The CPT theorem is a very general theoretical result linking Lorentz and CPT symmetry. Roughly, it states that certain theories (local quantum field theories) with Lorentz symmetry must also have CPT symmetry. These theories include all that used to describe known particle physics (for example, electrodynamics or the Standard Model) and many proposed theories (for example, Grand Unified Theories).

The CPT theorem can be used to show that a particle and its antiparticle must have certain identical properties, including mass, lifetime, and size of charge and magnetic moment. Many textbooks discuss the CPT theorem and its implications. See, for example, R.G. Sachs, The Physics of Time Reversal (University of Chicago Press, Chicago, 1987).

As is well known, all elementary particles and their nongravitational interactions are very successfully described by a theory called the Standard Model of particle physics. At the classical level, gravity is well described by Einstein's General Relativity. Both these theories have local Lorentz symmetry.

The Standard-Model Extension provides a quantitative description of Lorentz and CPT violation, controlled by a set of coefficients whose values are to be determined or constrained by experiment. Therefore SME still is only a phenomenological theory. At leat at this stage SME does not give any new physics.

This textbook keeps opposite viewpoint: The physical ideas in SME and so on might be not correct. Based on this textbook and Ref. [3], we feel that it is better to keep Lorentz invariance and general theory of relativity, and explain CPT violation by new physics. The reasons are as follows:

In chapter 3 of Ref. [3], the author points out that some references think that there is a sharp contradiction between general theory of relativity and quantum mechanics, and that this contradiction is a rigorous challenge faced by mankind in 21 century [4, 134]. However, the author of Ref. [3] concludes that the so-called "rigorous challenge" is actually a "zero challenge". For example, for a cold neutron in gravitational field, the Schrödinger equation is

$$\left(-\frac{\hbar^2}{2m}\nabla^2 + m\phi\right)\Psi = E\Psi. \tag{9.7.6}$$

It is obviously that the Hamiltonian in Eq. (9.7.6) obeys the weal equivalence principle, or generally speaking, general theory of relativity.

However, when we solve this equation, we find that it is a stationary problem, and the solution satisfies:

$$\left(-\frac{\hbar^2}{2m^2}\nabla^2 + \phi\right)\Psi_n = \frac{E_n}{m}\Psi_n \Rightarrow \frac{E_n}{m}\ is\ dependent\ on\ mass. \quad (9.7.7$$

It easy to observe that the neutron's energy in Eq. (9.7.7) does not satisfy the weak equivalence principle. According to the weak equivalence principle, we should have

$$E = \frac{mv^2}{2} + m\phi \Rightarrow \frac{E}{m}\ is\ independent\ of\ mass\ m. \tag{9.7.8}$$

The contradictions such as between Eqs. (9.7.7) and (9.7.8) can be listed by many examples (Refer to chapter 3 in Ref. [3].).

Ref. [3] thinks: When we write Schrödinger equation, we have obeyed the weak equivalence principle. However, the general theory of relativity belongs to a kind of deterministic theories; On the contrary, the solutions of Schrödinger equation represent a stochastic processes. The rule of this stochastic process is spontaneous equiprobability symmetry breaking. Thus one cannot require that a stochastic process to fit the general theory of relativity. In scientific levels, quantum mechanics is higher than the general theory of relativity.

Returning to the CPT violation, we hope to solve this problem through the following two physical approaches:

(9.1). After Eq. (A.2.29) in the appendix of this textbook, we write: "Note that not all types of forces allow a four dimensional generalization. One example of this kind is the

Coulomb force. As a static long range force, it requires an infinitely high propagational velocity, thus being in contradiction to the axioms of special relativity. Another example is the two-local-spin-mediated interaction, mentioned in chapter 7." It is obvious that if a system contains this kind of interactions, then the dynamical equation will do not keep Lorentz covariance, and thus will allow CPT violation.

(9.2). The contradiction between quantum mechanics and general theory of relativity inspired us that it is might be that the dynamical equation and the Hamiltonian obey Lorentz covariance, but the so-called CPT violation experiments might be explained by the field itself.

Chapter 10

Fractional Quantum Mechanics upon Theorems

10.1 Introduction

The term fractal was introduced into scientists' lexicon by Mandelbrot [135]. Historically, the first example of the fractional physical object was the Brownian motion, whose trajectories (paths) are nondifferentiable, self-similar curves that have a fractal dimension that is different from its topological dimension [135]. In quantum physics the first successful attempt to apply the fractality concept was the Feynman path integral approach to quantum mechanics. Feynman and Hibbs reformulated the nonrelativistic quantum mechanics as a path integral over the Brownian paths. Thus the Feynman- Hibbs fractional background leads to one formulation of standard (nonfractional) quantum mechanics [136]. Fractional quantum mechanics has been discovered by Nick Laskin (1999) as a result of expanding the Feynman path integral, from the Brownian-like to the Lévy-like quantum mechanical paths. A path integral over the Lévy-like quantummechanical paths results in a generalization of quantum mechanics. If the Feynman path integral leads to the well known Schrödinger equation, then the path integral over Lévy trajectories leads to the fractional Schrödinger equation. For many details of the fractional quantum mechanics, and the basis of fractals, fractional classical mechanics, fractional relativistic quantum mechanics, fractional quantum field theory, and some related experiments, readers can refer to Refs. [137]-[147]. Refs. [138, 139] are the summae of the fractional quantum mechanics. Refs. [146, 147, 141] are the newest applications of the fractional classical mechanics. Especially, Ref. [147] mentioned that the uncertainty relation can be used in the fractional signal analysis.

Based on the wavepacket-only theory in chapter 1 and Ref. [3], this chapter will build the fractional quantum mechanics upon theorems, and will give some new physical concepts. This chapter contains the following sections:

Section 10.2 introduces the fractional Hamiltonian, quantum Riesz fractional derivative, mathematical fractional Fourier transformation, physical fractional Fourier transformation.

Section 10.3 proves a theorem of compositions of a free fractional wavepacket.

Section 10.4 proves the fractional Schrödinger equation directly. This textbook does not prove the fractional Schrödinger equation in terms of the path integral over Lévy trajectories, because the later itself is only an assumption.

Section 10.5 proves some theorems and corollaries on fractional Heisenberg uncertainty relations.

10.2 Fractional Derivative

10.2.1 Quantum Riesz Fractional Derivative

In fact, the fractional quantum mechanics is generated by the Hamiltonian function $H_\alpha(p,x)$ of the form

$$H_\alpha(p,x) = D_\alpha |p|^\alpha + V(x), \quad 1 < \alpha \le 2, \tag{10.2.1.1}$$

where D_α is the generalized coefficient, the physical dimension of which is $[D_\alpha] = erg^{1-\alpha} cm^\alpha sec^{-\alpha}$. In Eq. (10.2.1.1), p and x are operators. One can say that Eq. (10.2.1.1) is a natural generalization of the Hamiltonian in quantum mechanics. In fractional quantum mechanics, the commutation relations between x and p, and between E and t are same as given by chapter 1. When $\alpha = 2$, $D_\alpha = 1/2m$ and Eq. (10.2.1.1) is transformed into ordinary one:

$$H_\alpha(p,x) = \frac{|p|^2}{2m} + V(x), \quad \alpha = 2, \tag{10.2.1.2}$$

A naive idea is that if we make the substitution of

$$p \to -i\hbar \frac{\partial}{\partial x}, \tag{10.2.1.3}$$

into Eq.(10.2.1.1), then we will obtain fractional derivative, and fractional Schrödinger equation. Thus, the fractional Schrödinger equation is the fractional differential equation in accordance with modern terminology. This is the main point for the term fractional Schrödinger equation or for more general term fractional quantum mechanics. The α is called Lévy index. When $\alpha = 2$, Lévy motion becomes Brownian motion. Thus, fractional quantum mechanics includes (standard) quantum mechanics as a particular Gaussian case at $\alpha = 2$.

It is easy to understand that to study fractional quantum mechanics, we need an exact definition for the fractional derivative. There are many different definitions for it actually. We use the Riesz quantum fractional derivative definition. The Riesz quantum fractional derivative is defined through its Fourier transform, and in one dimensional case it is

$$(\hbar\nabla)^\alpha \psi(x,t) = -\frac{1}{\sqrt{2\pi\hbar}} \int_{-\infty}^{\infty} dp e^{ipx/\hbar} |p|^\alpha \overline{\psi}(p,t), \quad \nabla^\alpha \equiv \frac{\partial^\alpha}{\partial x^\alpha}, \tag{10.2.1.4}$$

$$(\nabla)^\alpha \psi(x,t) = -\frac{1}{\sqrt{2\pi}} \int_{-\infty}^{\infty} dk e^{ikx} |k|^\alpha \overline{\psi}(k,t). \tag{10.2.1.5}$$

$$|p|^\alpha \to -(\hbar\nabla)^\alpha, \quad |k|^\alpha \to -(\nabla)^\alpha. \tag{10.2.1.6}$$

The $\overline{\psi}$ is the Fourier transform, and equals to

$$\overline{\psi}(p,t) = \frac{1}{\sqrt{2\pi\hbar}} \int_{-\infty}^{\infty} dx e^{-ipx/\hbar} \psi(x,t), \tag{10.2.1.7}$$

and

$$\psi(x,t) = \frac{1}{\sqrt{2\pi\hbar}} \int_{-\infty}^{\infty} dp e^{ipx/\hbar} \overline{\psi}(p,t), \tag{10.2.1.8}$$

Let me interpret how to understand Riesz quantum derivative?

Supposing $\alpha = 2$ and Substituting Eq. (10.2.1.8) into Eq. (10.2.1.7) yields

$$\begin{aligned}(\hbar\bigtriangledown)^{\alpha}\psi(x,t) &\overset{\alpha=2}{=} (\hbar\bigtriangledown)^{2}\psi(x,t)\\ &= (\hbar\bigtriangledown)^{2}\frac{1}{\sqrt{2\pi\hbar}}\int_{-\infty}^{\infty} dp e^{ipx/\hbar}\overline{\psi}(p,t)\\ &= (\hbar)^{2}\frac{1}{\sqrt{2\pi\hbar}}\int_{-\infty}^{\infty} dp e^{ipx/\hbar}\overline{\psi}(p,t)\left(\frac{ip}{\hbar}\right)^{2}\\ &= (\hbar)^{2}\frac{1}{\sqrt{2\pi\hbar}}\int_{-\infty}^{\infty} dp e^{ipx/\hbar}\overline{\psi}(p,t)(-1)\left(\frac{p}{\hbar}\right)^{2}\\ &= -\frac{1}{\sqrt{2\pi\hbar}}\int_{-\infty}^{\infty} dp e^{ipx/\hbar}\overline{\psi}(p,t)|p|^{2}. \end{aligned} \tag{10.2.1.9}$$

$$\begin{aligned}(\bigtriangledown)^{\alpha}\psi(x,t) &\overset{\alpha=2}{=} (\bigtriangledown)^{2}\psi(x,t)\\ &= (\bigtriangledown)^{2}\frac{1}{\sqrt{2\pi}}\int_{-\infty}^{\infty} dk e^{ikx}\overline{\psi}(k,t)\\ &= \frac{1}{\sqrt{2\pi}}\int_{-\infty}^{\infty} dk e^{ikx/\hbar}\overline{\psi}(k,t)(ik)^{2}\\ &= -\frac{1}{\sqrt{2\pi}}\int_{-\infty}^{\infty} dk e^{ikx/\hbar}\overline{\psi}(k,t)|k|^{2}. \end{aligned} \tag{10.2.1.9$'$}$$

Eqs. (10.2.1.9) and (10.2.1.9)' show that the definition of quantum Riesz fractional derivative does not have contradiction with the ordinary derivative in essence, and is only a natural generalization to ordinary derivative. In this textbook we do not plan to introduce the systematic theory of fractional quantum mechanics based on Eq. (10.2.1.3) due to the finite capacity space of this textbook.

In 3 dimensional case, $\triangle = \bigtriangledown \cdot \bigtriangledown$ is the Laplacian, and Eqs. (10.2.1.4.) and (10..7-8) and Eq. (10.2.1.1) become

$$(-\hbar^2\triangle)^{\alpha/2}\psi(\mathbf{r},t) = \frac{1}{\sqrt{(2\pi\hbar)^3}}\int_{-\infty}^{\infty} d\mathbf{p} e^{i\mathbf{p}\cdot\mathbf{r}/\hbar}|\mathbf{p}|^{\alpha}\overline{\psi}(\mathbf{p},t), \tag{10.2.1.4$'$}$$

$$\overline{\psi}(\mathbf{p},t) = \frac{1}{\sqrt{(2\pi\hbar)^3}}\int_{-\infty}^{\infty} d\mathbf{p} e^{-i\mathbf{p}\cdot\mathbf{r}/\hbar}\psi(\mathbf{r},t), \tag{10.2.1.7$'$}$$

and

$$\psi(\mathbf{r},t) = \frac{1}{\sqrt{(2\pi\hbar)^3}}\int_{-\infty}^{\infty} d\mathbf{p} e^{i\mathbf{p}\cdot\mathbf{r}/\hbar}\overline{\psi}(\mathbf{p},t), \tag{10.2.1.8$'$}$$

$$(-\triangle)^{\alpha/2}\psi(\mathbf{r},t) = \frac{1}{\sqrt{(2\pi)^3}}\int_{-\infty}^{\infty} d\mathbf{k} e^{i\mathbf{k}\cdot\mathbf{r}}|\mathbf{k}|^{\alpha}\overline{\psi}(\mathbf{k},t), \tag{10.2.1.4$''$}$$

$$\overline{\psi}(\mathbf{k},t) = \frac{1}{\sqrt{(2\pi)^3}}\int_{-\infty}^{\infty} d\mathbf{k} e^{-i\mathbf{k}\cdot\mathbf{r}}\psi(\mathbf{r},t), \tag{10.2.1.7$''$}$$

and

$$\psi(\mathbf{r},t) = \frac{1}{\sqrt{(2\pi)^3}}\int_{-\infty}^{\infty} d\mathbf{k} e^{i\mathbf{k}\cdot\mathbf{r}}\overline{\psi}(\mathbf{k},t), \tag{10.2.1.8$''$}$$

$$H_{\alpha}(\mathbf{p},\mathbf{r}) = D_{\alpha}|\mathbf{p}|^{\alpha} + V(\mathbf{r}) = D_{\alpha}|\mathbf{k}|^{\alpha}\hbar^{\alpha} + V(\mathbf{r}), \quad 1 < \alpha \le 2, \tag{10.2.1.1$'$}$$

We call $H_{\alpha}(\mathbf{p},\mathbf{r})$ the fractional Hamilton operator.

We would like to remind readers that if you use Hamiltonian in Eq. (10.2.1.1) and Riesz derivative, then you can extend all conclusions in chapter 1, and if you set $\alpha = 2$, then all fractional quantum mechanics will definitely be transformed into ordinary standard quantum mechanics. Alone this lone, Laskin has proved in Refs. [138, 139] many theorems and conclusions such as:

Hermiticity of fractional Hamilton
Continuity equation
Parity conservation
Schrödinger equation
Feynman path integral formulation of fractional quantum mechanics
Heisenberg uncertainty relations
Fractional Bohr atom
Quarkonium and fractional oscillator

In principle, we can extend every theorems in standard quantum mechanics into fractional quantum mechanics. This textbook does not want to do so many, and wants just to extend some important conclusions and theorems in chapter 1 and Ref. [3]. The only aim of this chapter is to establish the fractional quantum mechanics on theorems and correct concepts, and to correct some serious mistakes in both concepts and calculations in the present fractional quantum mechanics. We hope sincerely that the fractional quantum mechanics goes along the correct road from now.

Comments:

The quantum Riesz fractional derivative as a operator is just designed just for $|p|^\alpha$ in the fractional Hamiltonian of Eq. (10.2.1.1). Therefore, it has more larger limitation. In section 10.5, you will see the serious problems of this Riesz's definition.

10.2.2 Mathematical Fractional Fourier Transformation

The generalized Fourier transform in fractal space is written in the form [143]:

$$\psi(x,t) = \frac{1}{\Gamma(1+\alpha)} \int_{-\infty}^{\infty} (dk)^\alpha E_\alpha(+i^\alpha h_0 x^\alpha k^\alpha)\overline{\psi}(k,t), \tag{10.2.2.1}$$

where $h_0 = (2\pi)^\alpha/\Gamma(1+\alpha)$, $0 < \alpha \leq 1$, $E_\alpha(i^\alpha h_0 x^\alpha p^\alpha)$ is the ordinary exponential factor, the x in x^α is positive, and the Fractional Fourier transform is:

$$\overline{\psi}(k,t) = \frac{1}{\Gamma(1+\alpha)} \int_{-\infty}^{\infty} (dx)^\alpha E_\alpha(-i^\alpha h_0 x^\alpha k^\alpha)\psi(x,t). \tag{10.2.2.2}$$

Comments:

In comparison of mathematical fractional fourier transformation with the Riesz's fractional derivative, the former is more general and convenient.

10.2.3 Physical Fractional Fourier Transformation

Absorbing the advantages of the above two definitions, this subsection proposes a new(?) definition. We use the new(?), because we really do not know if this is really a new one. If after the publication of this textbook we know any reference proposed this definition, then we will put that reference in the second publication of this textbook.

Theorem XXXXXX: Operator representations: In fractional quantum mechanics, in coordinate space x, the fractional coordinate operator X^α is defined as

$$X^\alpha = |x|^\alpha, \tag{10.2.3.1}$$

where there is no the limitation of $1 < \alpha \leq 2$.

In momentum space p, the fractional momentum operator P^α is defined as

$$P^\alpha = |p|^\alpha. \qquad (10.2.3.2)$$

If we define that in coordinate space

$$\begin{aligned} P^\alpha \psi(x,t) &\equiv -i\hbar \frac{1}{\sqrt{2\pi\hbar}} \int_{-\infty}^{\infty} dp e^{ipx/\hbar} \frac{i}{\hbar} |p|^\alpha \overline{\psi}(p,t) \\ &= \frac{1}{\sqrt{2\pi\hbar}} \int_{-\infty}^{\infty} dp e^{ipx/\hbar} |p|^\alpha \overline{\psi}(p,t), \end{aligned} \qquad (10.2.3.3)$$

then

$$< P^\alpha >= \int_{-\infty}^{\infty} dp \overline{\psi}^*(p,t) |p|^\alpha \overline{\psi}(p,t), \qquad (10.2.3.4)$$

which means that Eq. (10.2.3) is consistent with Eq. (10.2.3.2), i. e., the operator representation of Eq. (10.2.3.3) is correct.

If we use wavenumber instead of momentum, then

$$\begin{aligned} P^\alpha \psi(x,t) &\equiv -i \frac{1}{\sqrt{2\pi}} \int_{-\infty}^{\infty} dk e^{ikx} \frac{i}{\hbar} |k|^\alpha \overline{\psi}(k,t) \\ &= \frac{1}{\sqrt{2\pi}} \int_{-\infty}^{\infty} dk e^{ikx} |k|^\alpha \overline{\psi}(k,t), \end{aligned} \qquad (10.2.3.3)'$$

then

$$< P^\alpha >= \int_{-\infty}^{\infty} dk \overline{\psi}^*(k,t) |k|^\alpha \overline{\psi}(k,t), \qquad (10.2.3.4)'$$

If we define that in momentum space

$$\begin{aligned} X^\alpha \overline{\psi}(p,t) &\equiv i\hbar \frac{\partial^\alpha}{\partial p^\alpha} \overline{\psi}(p,t) \\ &\equiv i\hbar \frac{1}{\sqrt{2\pi\hbar}} \int_{-\infty}^{\infty} dx e^{-ipx/\hbar} \frac{-i}{\hbar} |x|^\alpha \psi(x,t) \\ &= \frac{1}{\sqrt{2\pi\hbar}} \int_{-\infty}^{\infty} dx e^{-ipx/\hbar} |x|^\alpha \psi(x,t), \end{aligned} \qquad (10.2.3.5)$$

then

$$< X^\alpha >= \int_{-\infty}^{\infty} dx \psi^*(x,t) |x|^\alpha \psi(x,t), \qquad (10.2.3.6)$$

which means that Eq. (10.2.5) is consistent with Eq. (10.2.3.1), i. e., the operator representation of Eq. (10.2.3.5) is correct.

Prove:
As an example, let us prove Eq. (10.2.3.3) from Eq. (10.2.3.4).

$$\begin{aligned}
< P^\alpha > &\equiv \int_{-\infty}^{\infty} dx \psi^*(x,t) P^\alpha \psi(x,t) \\
&= \int_{-\infty}^{\infty} dx \frac{1}{\sqrt{2\pi\hbar}} \int_{-\infty}^{\infty} dp' e^{-ip'x/\hbar} \overline{\psi}^*(p',t) \\
& \frac{1}{\sqrt{2\pi\hbar}} \int_{-\infty}^{\infty} dp e^{ipx/\hbar} |p|^\alpha \overline{\psi}(p,t) \\
&= \frac{1}{\sqrt{2\pi\hbar}} \int_{-\infty}^{\infty} dp' \overline{\psi}^*(p',t) \\
& \frac{1}{\sqrt{2\pi\hbar}} \int_{-\infty}^{\infty} dp |p|^\alpha \overline{\psi}(p,t)(2\pi\hbar)\delta(p-p') \\
&= \int_{-\infty}^{\infty} dp \overline{\psi}^*(p,t)|p|^\alpha \overline{\psi}(p,t). \qquad (10.2.3.4)(10.3.2.7)
\end{aligned}$$

QED.
The Hermiticity of the operators X^α and P^α are easy to be proved.

10.3 Composition Theorem of Fractional Wavepacket

It is important to note that the concepts on canonical coordinate and canonical conjugate moment are the same as that of the fractional quantum mechanics. We call an object in fractional quantum mechanics the wavepacket, which is local and can have arbitrary structure. We do not need to know the mathematical form of the wavepacket-function. (As is well known, wavepacket-function≠wavefunction!). According to the corollary II, any function $\in \mathcal{M}(\mathbb{R})$ cannot depend on both coordinte and momentum at the same time. Now, we begin our discussion from that the wavepacket-function is a function of coordinate x and t.

Theorem XXXXXXI: Composition of fractional wavepacket: A free fractional wavepacket consists of three parts: guide-factor, front-factor, and body-factor.

Proof:
From Eq. (10.2.1.1) we know that the dispersion of a fractional free particle, and the fractional plane wave should be

$$W_p = \frac{1}{\sqrt{2\pi\hbar}} e^{(i\mathbf{p}\cdot\mathbf{r}-iEt)/\hbar} = \frac{1}{\sqrt{2\pi\hbar}} e^{(i\mathbf{p}\cdot\mathbf{r}-iD_\alpha|p|^\alpha t)/\hbar}, \qquad (10.3.1)$$

or,

$$W_p = \frac{1}{\sqrt{2\pi}} e^{i\mathbf{k}\cdot\mathbf{r}-i\omega(k)t} = \frac{1}{\sqrt{2\pi}} e^{i\mathbf{k}\cdot\mathbf{r}-iD_\alpha|k|^\alpha t\hbar^{\alpha-1}}, \qquad (10.3.1)'$$

In the following we just consider one dimensional case. We expands the the frequency $\omega(k) \equiv D_\alpha|p|^\alpha/\hbar = D_\alpha|k|^\alpha\hbar^{\alpha-1}$ at the average of absolute value of wavenumber of the wavepacket, $k_\circ$. For this expansion, refer to Eq. (3.4.1.12) of Ref. [3].

$$kx - \omega(k)t = kx - [\omega(k_\circ) + v_g(k-k_\circ)]t, \qquad (10.3.2)$$

where v_g is the group velocity of wavepacket.

$$v_g = \left.\frac{\partial\omega}{\partial k}\right|_{k=k_\circ}. \qquad (10.3.3)$$

Therefore, any fractional wavepacket can be expanded by the complete plane wave set as follows:

$$\begin{aligned}
\psi(x,t) &= \frac{1}{\sqrt{2\pi}}\int_{-\infty}^{\infty} dk\overline{\psi}(k)e^{ikx-i\omega(k)t} \\
&= \frac{1}{\sqrt{2\pi}}\int_{-\infty}^{\infty} dk\overline{\psi}(k,t)e^{ikx-i[\omega(k_\circ)+v_g(k-k_\circ)]t} \\
&= e^{ik_\circ v_g t-i\omega(k_\circ)t}\frac{1}{\sqrt{2\pi}}\int_{-\infty}^{\infty} dk\overline{\psi}(k)e^{ikx-iv_g kt} \\
&= e^{ik_\circ v_g t-i\omega(k_\circ)t}\psi(x-v_g t,0) \\
&\overset{x-v_g,t=x'}{=} e^{ik_\circ x-i\omega(k_\circ)t}\times e^{-ik_\circ x'}\times\psi(x',0) \\
&= e^{ip_\circ x/\hbar-i\omega(p_\circ)t}\times e^{-ip_\circ x'/\hbar}\times\psi(x',0) \\
&= e^{ip_\circ x/\hbar-iE(p_\circ)t/\hbar}\times e^{-ip_\circ x'/\hbar}\times\psi(x',0) \\
&\equiv I\times II\times III \\
&\equiv W_p(x,t,p=p_\circ)\times II\times III \\
&\equiv guide\ factor\times front\ factor\times body\ factor.
\end{aligned} \tag{10.3.4}$$

QED.

10.4 Fractional Schrödinger Equation

Until now all references derive Schrödinger equation from fractional Feynman path integral. However, Feynman path integral itself is a postulate, and thus Schrödinger equation is a postulate as well. This section will prove Schrödinger equation without any postulate.

Theorem XXXXXXII-1: Free Schrödinger equation: The guide-factor of a free fractional wavepacket satisfies Schrödinger equation:

$$i\hbar\frac{\partial}{\partial t}W_p(x,t,p=p_\circ) = -D_\alpha(\hbar\bigtriangledown)^\alpha W_p(x,t,p=p_\circ). \tag{10.4.1}$$

Proof:
Substituting the expression of $W_p(x,p,p=p_\circ)$ in Eq. (10.3.4) into Eq. (10.4.1) yields

$$\begin{aligned}
i\hbar\frac{\partial}{\partial t}W_p(x,t,p=p_\circ) &= i\hbar\frac{\partial}{\partial t}e^{ip_\circ x/\hbar-iE(p_\circ)t/\hbar} \\
&= E(p_\circ)e^{ip_\circ x-iE(p_\circ)t/\hbar} \\
&\overset{(10.2.1.1)}{=} D_\alpha|p_\circ|^\alpha W_p(x,t,p=p_\circ) \\
&\overset{(10.2.1.6)}{=} -D_\alpha(\hbar\bigtriangledown)^\alpha W_p(x,t,p=p_\circ).
\end{aligned} \tag{10.4.2}$$

QED.

Theorem XXXXXXII-2: Schrödinger equation in potential field: The guide-factor of any fractional wavepacket satisfies Schrödinger equation:

$$i\hbar\frac{\partial}{\partial t}\psi(x,t) = H_\alpha(x,t)\psi(x,t). \tag{10.4.3}$$

Proof:

(A). A simple method.

For the proof please refer to chapter 1.

From Eq. (10.4.2) we obtain

$$i\frac{\partial}{\partial t} \to E = \hbar\omega \Rightarrow i\frac{\partial}{\partial t} = \textit{total energy operator}, \tag{10.4.3}$$

which is the same as that of the standard quantum mechanics. Therefore, we obtain

$$i\hbar\frac{\partial}{\partial t}\psi(x,t) = [-D_\alpha(\hbar\nabla)^\alpha + V(x,t)]\psi(x,t) = H_\alpha(x,t)\psi(x,t). \tag{10.4.4}$$

(B). A complex method.

If a fractional object is in a general potential field $V(x,t)$, then we can expand the fractional wavepacket-function exactly as follows:

$$\begin{aligned}
\psi(x,t) &= \frac{1}{\sqrt{2\pi}}\int_{-\infty}^{\infty} dk\overline{\psi}(k)e^{ikx-i[\omega(k)+V(x,t)/\hbar]t} \\
&= e^{-iV(x,t)t/\hbar}\frac{1}{\sqrt{2\pi}}\int_{-\infty}^{\infty} dk\overline{\psi}(k,t)e^{ikx-i[\omega(k_\circ)+v_g(k-k_\circ)]t} \\
&= e^{-iV(x,t)t/\hbar}e^{ip_\circ x/\hbar-iE(p_\circ)t/\hbar} \times e^{-ip_\circ x'/\hbar} \times \psi(x',0) \\
&= e^{ip_\circ x/\hbar-i[E(p_\circ)+V(x,t)]t/\hbar} \times e^{-ip_\circ x'/\hbar} \times \psi(x',0) \\
&\equiv \mathcal{W}_p(x,t,p=p_\circ) \times II \times III \\
&\equiv \textit{guide factor} \times \textit{front factor} \times \textit{body factor}.
\end{aligned} \tag{10.3.4$'$}$$

The guide-factor satisfies the following equation:

$$i\hbar\frac{\partial}{\partial t}\mathcal{W}_p(x,t,p=p_\circ) = [-D_\alpha(\hbar\nabla)^\alpha + V(x,t)]\mathcal{W}_p(x,t,p=p_\circ) = H_\alpha\mathcal{W}_p(x,t,p=p_\circ). \tag{10.4.5}$$

We call the $\mathcal{W}_p(x,t,p=p_\circ)$ in Eq. (10.4.5) the wavefunction, and is expressed as

$$i\hbar\frac{\partial}{\partial t}\psi(x,t) = H_\alpha(x,t)\psi(x,t). \tag{10.4.6}$$

QED.

Corollary XXI: General compositions of fractional wavepacket: Any fractional wavepacket consists of three parts: guide-factor, front-factor, and body-factor.

Proof:

In potential field, the variation of guide-factor is determined by Schrödinger equation, and the other two factors, consisting of a fractional wavepacket, do not depend on x and t. Due to that the guid-factor satisfies Schödinger equation, we call the guide-factor the wavefunction. QED.

10.5 Fractional Uncertainty Relation

10.5.1 Preparation

Theorem XXXXXXIII: If $W^{\alpha/2}$ is areal number, and $\Delta\gamma$, $\Delta\beta$ are two quantities, which satisfy

$$|\Delta\gamma|^\alpha - \lambda W^{\alpha/2} + \lambda^2|\Delta\beta|^\alpha \geq 0 \tag{10.5.1.1}$$

for all real number λ, then the product of these two quantities satisfies

$$|\Delta\gamma|^{\alpha}|\Delta\beta|^{\alpha} \geq \frac{1}{4}W^{\alpha}. \tag{10.5.1.2}$$

In Eq. (10.5.1.1)

$$|\Delta\gamma|^{\alpha} \equiv \sqrt{[\langle(\gamma - \langle\gamma\rangle)\rangle^{\alpha}\rangle][\langle(\gamma - \langle\gamma\rangle)^{\alpha})]^{*}}$$
$$|\Delta\beta|^{\alpha} \equiv \sqrt{[\langle(\beta - \langle\beta\rangle)^{\alpha}\rangle][\langle(\beta - \langle\beta\rangle)^{\alpha}]^{*}}$$

Proof:

This theorem is the extension of theorem IV.

It is easy to confirm that Eq. (10.5.1.1) is correct for $\lambda \to \infty$. To determine the sign of left hand side of Eq. (10.5.1.1) for finite values of λ, let us express Eq. (10.5.1.1) in terms of its roots in λ.

$$\lambda^2|\Delta\beta|^{\alpha} - \lambda W^{\alpha/2} + |\Delta\gamma|^{\alpha} = |\Delta\beta|^{\alpha}(\lambda - \lambda_1)(\lambda - \lambda_2), \tag{10.5.1.3}$$

where λ_1 and λ_2 are the roots, which are expressed as

$$\lambda_1 = a_{\circ} - \Delta, \tag{10.5.1.4}$$

and

$$\lambda_2 = a_{\circ} + \Delta, \tag{10.5.1.5}$$

where

$$a_{\circ} = \frac{W^{\alpha/2}}{2|\Delta\beta|^{\alpha}}, \tag{10.5.1.6}$$

$$\Delta = \frac{\sqrt{W^{\alpha} - 4|\Delta\beta|^{\alpha}|\Delta\gamma|^{\alpha}}}{2|\Delta\beta|^{\alpha}}. \tag{10.5.1.7}$$

Since $|\Delta\beta|^{\alpha}$ is positive definite, Eq. (10.5.1.3) implies that

$$(\lambda - \lambda_1)(\lambda - \lambda_2) \geq 0. \tag{10.5.1.8}$$

However, left hand side of Eq. (10.5.1.8) is negative for

$$\lambda_1 < \lambda < \lambda_2. \tag{10.5.1.9}$$

To solve this apparent contradiction, we substitute Eqs. (10.5.1.4) and (10.5.1.5) for λ_1 and λ_2 in the product

$$(\lambda - \lambda_1)(\lambda - \lambda_2) = (\lambda - a_{\circ} - |\Delta)(\lambda - a_{\circ} + |\Delta) = (\lambda - a_{\circ})^2 - \Delta^2. \tag{10.5.1.10}$$

In order to satisfy Eq. (10.5.1.8), the right hand side must be ≥ 0, which implies that $\Delta^2 < 0$, i. e. Δ is pure imaginary,

$$\Delta = i|\Delta|. \tag{10.5.1.11}$$

Hence,

$$(\lambda - \lambda_1)(\lambda - \lambda_2) = (\lambda - a_{\circ})^2 + |\Delta|^2. \tag{10.5.1.12}$$

This is a positive definite quantity. We note that since W is real, $a_{\circ}$ [defined by Eq. (10.5.1.6)] is real. For Δ to be a pure imaginary we must have, from Eqs. (10.5.1.11) and (10.5.1.7),

$$4|\Delta\beta|^{\alpha}|\Delta\gamma|^{\alpha} \geq W^{\alpha}. \tag{10.5.1.13}$$

Eq. (10.5.1.13) concludes the proof of the Lemma I. QED.

10.5.2 Uncertainty Relations

Theorem XXXXXXIV: Uncertainty relations: Considering an one dimensional system. Take the origin point as the zero point of coordinate x, and the average momentum as the zero point of momentum p. We have the following uncertainty relation:

$$<\overline{\psi}||k|^{\alpha}|\overline{\psi}><\psi||x|^{\alpha}|\psi>\geq\frac{W}{4}>0 \qquad (10.5.2.1)$$

where the W is a positive number.

Proof:

$$\begin{aligned}
2<\overline{\psi}||k|^{\alpha}|\overline{\psi}><\psi||x|^{\alpha}|\psi> &= <|k|^{\alpha/2}\overline{\psi}||k|^{\alpha/2}\overline{\psi}><\psi||x|^{\alpha}|\psi> \\
&= 2<P^{\alpha/2}\psi|P^{\alpha/2}\psi><|x|^{\alpha/2}\psi||x|^{\alpha/2}\psi> \\
& \quad using\ Cauchy-Schwarz\ inequality \\
&\geq \left|<P^{\alpha/2}\psi||x|^{\alpha/2}\psi>+<|x|^{\alpha/2}\psi|P^{\alpha/2}\psi>\right| \\
&= \left|<P^{\alpha/2}\psi||x|^{\alpha/2}\psi>+(<P^{\alpha/2}\psi||x|^{\alpha/2}\psi>)^{*}\right| \\
&= \left|2Re(<P^{\alpha/2}\psi||x|^{\alpha/2}\psi>)\right| \\
&= \left|2Re(<\psi|P^{\alpha/2}||x|^{\alpha/2}\psi>)\right| \\
&\equiv \frac{W}{2}>0. \qquad (10.5.2.2)
\end{aligned}$$

The exact calculations give ($p=\hbar k$): [138, 143]

$$<\overline{\psi}||p|^{\alpha}|\overline{\psi}><\psi||x|^{\alpha}|\psi>\geq\frac{\hbar^{\alpha}}{2\alpha},\quad 1<\alpha\leq 2. \qquad (10.5.2.3)$$

$$<\overline{\psi}||p|^{\alpha}|\overline{\psi}><\psi||x|^{\alpha}|\psi>\geq\frac{\hbar^{\alpha}\Gamma^{2}(1+\alpha/2)}{4},\quad 0<\alpha\leq 2. \qquad (10.5.2.4)$$

QED.

Corollary XXII: Variables of function: For fractional wavepacket-function $\in \mathcal{M}(\mathbb{R})$ cannot depend on the two variables of Fourier pair variables at the same time, or, equivalently, any fractional wavepacket-function $\in \mathcal{M}(\mathbb{R})$ cannot depend on the two variables of canonical conjugate pair variables at the same time.

Proof:

This corollary is equivalently to say that any fractional wavepacket-function $\in \mathcal{M}(\mathbb{R})$ cannot depend on the two non-commutation variables at the same time.

The theorem XXXXXXIV of uncertainty relations [Eq. (10.5.2.3)] indicates the existence of lowest bound. If A takes certain value, then $\Delta x=0$, and thus $\Delta p=\infty$, which is not reasonable. QED.

Although this Corollary XXII can be proved very easy, but its significance is significant. As is well known, the Cartesian coordinate $\mathbf{r}$ and the canonical conjugate momentum $\mathbf{p}$ are Fourier pair variables. The fractional wavepacket in physics describes a fractional object. The wavepacket in mathematics is a function $\in \mathcal{M}(\mathbb{R})$. Corollary XXI tells us clearly that the fractional wavepacket-function can only be a function either $\mathbf{r}$ or $\mathbf{p}$, but cannot be a function of both **r** and **p** at the same time.

Corollary XXIV-1: Fractional object's volume: If any position space function $\in \mathcal{M}(\mathbb{R})$ represents the distributions of mass or energy or charge or momentum of any fractional object, including both massy and massless fractional object, then this fractional object has to occupy three dimensions in position space.

Proof:
If the fractional object occupies only two dimensions (x, y), then according to the fractional Heisenberg uncertainty relation theorem, the z direction momentum and thus the energy of fractional object are infinite, which is not reasonable.

Corollary XXIV-2: Zero point energy: Any fractional object has to have zero point energy, which is caused by the fractional Heisenberg uncertainty relation. This zero point energy is independent of temperature, and thus is not equal to zero even at 0 K.

Proof:
From uncertainty relation $\Delta p^\alpha \cdot \Delta x^\alpha \geq \hbar^\alpha/(2\alpha)$ we deduce that if $\Delta x^\alpha < \infty$, then $\Delta p^\alpha > 0$, which means that the absolute value of momentum and, thus, kinetic energy of any object are larger than zero. This energy is called zero point energy.

Let us give an example. The potential energy function is $V(x)$. Next we show that even in potential field, the zero point energy exists still.
Suppose that the x-dependence of the potential energy is

$$H = \frac{p^2}{2m} + V(x) = D_\alpha |p|^\alpha + A|x|^n, \quad (n \text{ is real}) \tag{10.5.2.5}$$

The average value on any states is

$$\begin{aligned} < H > &= D_\alpha < |p|^\alpha > + < V(x) >= D_\alpha < |p|^\alpha > + A < |x|^n > \\ &\geq D_\alpha \frac{\hbar^\alpha}{2\alpha < |x|^\alpha >} + A < |x|^n > . \end{aligned} \tag{10.5.2.6}$$

Eq. (10.5.2.6) clearly shows that the average value of H has minimum value. This minimum value is independent of temperature, and is called zero point energy.

Appendix A

Special Theory of Relativity

A.1 Physics

If the velocity of an object approaches to the light velocity, then the Newton's mechanics has to be replaced by relativistic mechanics. The special theory of relativity is based on the following axioms [148]:

(1). Constancy of the velocity of light: The vacuum velocity c of light is the same in all uniformly moving system. No information propagates faster than light.

(2). Relativity principle: Physical laws are equally valid in all inertial systems, i. e., there is no preferred reference frame.

Next we list some formulas often used. According to special theory of relativity Newton's second law still holds formally

$$\frac{d\mathbf{p}}{dt} = \mathbf{F}, \quad \mathbf{p} = m\mathbf{v}, \tag{A.1.1}$$

where $\mathbf{p}$ and $\mathbf{v}$ are physical momentum and velocity, respectively, and the mass should be

$$m = \frac{m_\circ}{\sqrt{1-\beta^2}}, \quad \beta \equiv \frac{v}{c} \tag{A.1.2}$$

where $m_\circ$ is the rest mass of a structureless point particle. As we know, the velocity of photon is c. According to Eq. (A.1.2), the rest mass of the photon has to be zero. The present observed value is that the upper limit of photon rest mass is $m_\circ < 1.2 \times 10^{-51}$ g (Luo Jun et al., 2003).

We can also derive the above formula from an appropriate Lagrangian function. The Lagrangian function is

$$L = T^* - V, \ \ T^* = m_\circ c^2(1 - \sqrt{1-\beta^2}). \tag{A.1.3}$$

Note that the T^* in Eq. (A.1.3) is not the relativistic kinetic energy T. The latter is

$$T = mc^2 - m_\circ c^2 = m_\circ c^2 \left(\frac{1}{\sqrt{1-\beta^2}} - 1 \right) = (total\ energy) - (rest\ energy). \tag{A.1.4}$$

The definition of canonical momentum (See Ref. [3]) is

$$p_x = \frac{\partial L}{\partial v_x} = \frac{\partial T^*}{\partial \beta}\frac{\partial \beta}{\partial v_x} = m_\circ c^2 \frac{\beta}{\sqrt{1-\beta^2}}\frac{\partial \beta}{\partial v_x}. \tag{A.1.5}$$

Considering

$$\beta^2 = \frac{v^2}{c^2} = \frac{v_x^2 + v_y^2 + v_z^2}{c^2}, \tag{A.1.6}$$

one obtains

$$\beta \frac{\partial \beta}{\partial v_x} = \frac{v_x}{c^2}, \quad \frac{\partial \beta}{\partial v_x} = \frac{v_x}{\beta c^2}. \tag{A.1.7}$$

Substituting Eq. (A.1.7) into Eq. (A.1.5) yields

$$p_x = \frac{m_\circ c^2}{\sqrt{1-\beta^2}} \frac{v_x}{c^2} = \frac{m_\circ v_x}{\sqrt{1-\beta^2}} = mc^2 \frac{v_x}{c^2}. \tag{A.1.8}$$

From the same method one obtains

$$p_y = \frac{m_\circ v_y}{\sqrt{1-\beta^2}} = mc^2 \frac{v_y}{c^2}, \quad p_z = \frac{m_\circ v_z}{\sqrt{1-\beta^2}} = mc^2 \frac{v_z}{c^2}. \tag{A.1.9}$$

From Eqs.(A.1.8-9) one obtains

$$p_x^2 + p_y^2 + p_z^2 = (mc^2)^2 \frac{v_x^2 + v_y^2 + v_z^2}{c^4}, \quad or \quad p^2 = (mc^2)^2 \frac{v^2}{c^4}. \tag{A.1.10}$$

From Eq. (A.1.10) we obtain

$$p = \pm (mc^2) \frac{v}{c^2}. \tag{A.1.11}$$

If we consider the minus sign in Eq. (A.1.11), the sole possibility is $m \times v < 0$. The $m < 0$ is difficult to be understood. The minus sign in Eq. (A.1.11) can be interpreted as a solution of antiparticle being simply particle "moving backwards" [149]. Relativistic quantum mechanics also explains the minus solution as existence of antiparticle with positive energy and motion along the direction opposite to the direction of motion of particle [149]. Eq. (A.1.10) can be rewritten as

$$p^2 = (mc^2)^2 \frac{\beta^2}{c^2}, \tag{A.1.12}$$

or

$$c^2 p^2 = m_\circ^2 c^4 \frac{\beta^2}{1-\beta^2} = m_\circ^2 c^4 \left(\frac{1}{1-\beta^2} - 1 \right) = (mc^2)^2 - m_\circ^2 c^4. \tag{A.1.13}$$

From Eq. (A.1.13) one obtains

$$(mc^2)^2 \equiv E^2 = m_\circ^2 c^4 + c^2 p^2. \tag{A.1.14}$$

The definition of Hamiltonian function H (See chapter 2 of Ref. [3])

$$\begin{aligned} H &= p_x \dot{x} + p_y \dot{y} + p_z \dot{z} - L \\ &= \frac{m_\circ v^2}{\sqrt{1-\beta^2}} - m_\circ c^2 (1 - \sqrt{1-\beta^2}) + V \\ &= m_\circ c^2 \left(\frac{1}{\sqrt{1-\beta^2}} - 1 \right) + V = T + V \end{aligned} \tag{A.1.15}$$

i. e., the expressions of Hamiltonian function are same in both the nonrelativistic and relativistic classical mechanics.

A.2 Notations of Special Theory of Relativity

Due to the axioms of special theory of relativity in section A1, relativistic physical events are described mathematically within a four dimensional space, where $x^\circ \equiv ct$ appears as an additional dimension on an equal footing with the three position space dimensions $(x^1,\ x^2,\ x^3) \equiv (x,\ y,\ z)$. The notations used in this textbook are nearly the same as that in Refs. [148, 121]. We use $\underline{x} \equiv (x^\circ, x^1, x^2, x^3) = (x^\circ, x, y, z) = (x^\circ, \mathbf{r}) = (x^\circ, \mathbf{x})$. This four dimensional space is called Minkowski spacetime. A point in Minkowski spacetime is called world point. A trajectory in Minkowski space is called world line. In Minkowski space-time the contravariant four-vector is defined as

$$\underline{x} = [x^\mu(t)] \equiv [x^\circ(t), x^1(t), x^2(t), x^3(t)] \equiv [ct, x(t), y(t), z(t)] = [ct, \mathbf{r}(t)] = [ct, \mathbf{x}(t)], \tag{A2.1a}$$

and the covariant four-vector is defined as

$$[x_\mu(t)] \equiv [x_\circ(t), x_1(t), x_2(t), x_3(t)] \equiv [ct, -x(t), -y(t), -z(t)] = [ct, -\mathbf{r}(t)] = [ct, -\mathbf{x}(t)]. \tag{A2.1b}$$

If it is necessary, $[x^\mu(t)]$ and $[x_\mu(t)]$ can also represent a row vector.

To define the scalar product of two four-vector, we define the non-Euclidian metric tensor $(g_{\mu\nu})$:

$$(g_{\mu\nu}) = (g^{\mu\nu}) \equiv \begin{pmatrix} 1 & 0 & 0 & 0 \\ 0 & -1 & 0 & 0 \\ 0 & 0 & -1, & 0 \\ 0 & 0 & 0 & -1 \end{pmatrix}, \quad g^{\mu\alpha} g_{\alpha\nu} = g^\mu_\nu = \delta^\mu_\nu, \tag{A2.2}$$

where the first index = row index, second index = column index, and two same indices, one of which is an upper and the other a lower index, are summed over so the summation sign is dropped (Einstein sum convention).

The scalar product of two four-vectors is defined as

$$\begin{aligned}(x^\mu)\cdot(y^\mu) &= \underline{x}\cdot\underline{y} = x^\circ y^\circ - x^1 y^1 - x^2 y^2 - x^3 y^3 \equiv x^\mu g_{\mu\nu} y^\nu \\ &= x^\mu y_\mu = x_\mu g^{\mu\nu} y_\nu = (x_\mu)\cdot(y_\mu), \end{aligned} \tag{A2.3}$$

where vectors with an upper index are called contravariant, those with a lower index are called covariant, and $(x^\circ = x_\circ,\ x^1 = -x_1,\ x^2 = -x_2,\ x^3 = -x_2)$. This can be transferred to tensors of higher rank. For example, $T^{\mu\nu}_{\rho}$ is a twofold contravariant and onefold covariant tensor of rank 3. We use $\underline{x}$ to express four-vector. We also use

$$x \ \ and \ \ \mathbf{x}$$

to express four- and three-vector, respectively. Note that $x_\mu x^\mu \neq x_\mu x_\mu$.

The metric tensor can be used to pull upper indices down and vice versa. For example:

$$x_\mu = g_{\mu\nu} x^\nu, \quad T^{\mu\nu\rho} = g^{\mu\alpha} T_\alpha^{\ \nu\rho} = g^{\mu\alpha} g^{\nu\beta} T^\gamma_{\alpha\beta} = g^{\mu\alpha} g^{\nu\beta} g^{\rho\gamma} T_{\alpha\beta\gamma} = etc.$$

Contrary to the three dimensional Euclidean case, the norm of a four-vector is not positive definite. The following cases can occur:

$$x_\mu x_\mu = x_\circ^2 - x^2 - y^2 - z^2 \begin{cases} > 0 & (time-like) \\ = 0 & (light-like) \\ < 0 & (space-like) \end{cases}$$

where the classification corresponds to the vector's position relative to the light cone $x_\mu x^\mu = 0$.

The axioms of special theory of relativity, introduced in section A1, imply that the four dimensional "distance" between two four-vectors x^μ and y^μ is preserved in any inertial system:

$$(x-y)_\mu(x-y)^\mu = (x'-y')_\mu(x'-y')^\mu. \tag{A2.4}$$

Lorentz transformations describe the relativistic transformation from one inertial system to another. They are defined by the linear transformational equation:

$$x^\mu \to x'^\mu = \Lambda^\mu{}_\nu x^\nu + a^\mu, \quad (a^\mu) = spacetime \;\; translation \tag{A2.5}$$

of contravariant vectors. Along with the conservation of distance, one obtains from this the conditional equation

$$\Lambda^\mu{}_\alpha g_{\mu\nu}\Lambda^\nu{}_\beta = g_{\alpha\beta}, \tag{A2.6}$$

as well as the corresponding transformational behavior of covariant vectors

$$x'_\mu = g_{\mu\nu}(\Lambda^\nu{}_\alpha x^\alpha + a^\nu) = x_\beta\left[\Lambda^{-1}\right]^\beta{}_\mu + a_\mu, \tag{A2.7}$$

with the inverse transformation

$$\left[\Lambda^{-1}\right]^\beta{}_\mu = g_{\mu\nu}\Lambda^\nu{}_\alpha g^{\alpha\beta} = \Lambda_\mu{}^\beta. \tag{A2.8}$$

$$\left[\Lambda^{-1}\right]^\beta{}_\mu \Lambda^\mu{}_\gamma = g_{\mu\nu}\Lambda^\nu{}_\alpha g^{\alpha\beta}\Lambda^\mu{}_\gamma = \delta^\beta_\gamma. \tag{A2.9}$$

Now let us give the transformational behavior of differential operators. Taking into account $x^\nu = [\Lambda^{-1}]^\nu{}_\mu x'^\mu$ in terms of Eq. (A2.5), we have

$$\frac{\partial x^\nu}{\partial x'^\mu} = [\Lambda^{-1}]^\nu{}_\mu \Rightarrow \frac{\partial}{\partial x'^\mu} = \frac{\partial}{\partial x^\nu}\frac{\partial x^\nu}{\partial x'^\mu} = \frac{\partial}{\partial x^\nu}[\Lambda^{-1}]^\nu{}_\mu. \tag{A2.10}$$

Therefore,

$$\frac{\partial}{\partial x^\mu} = \partial_\mu = \left(\frac{\partial}{\partial ct}, \frac{\partial}{\partial x}, \frac{\partial}{\partial y}, \frac{\partial}{\partial z}\right) = \left(\frac{\partial}{\partial ct}, \nabla\right) \tag{A2.11}$$

transforms as a covariant vector and, accordingly,

$$\frac{\partial}{\partial x_\mu} = \partial^\mu = \left(\frac{\partial}{\partial ct}, -\frac{\partial}{\partial x}, -\frac{\partial}{\partial y}, -\frac{\partial}{\partial z}\right) = \left(\frac{\partial}{\partial ct}, -\nabla\right) \tag{A2.12}$$

as a contravariant vector. Furthermore, it follows that the d'Alembert operator

$$\Box = \partial_\mu\partial^\mu = \frac{1}{c^2}\frac{\partial^2}{\partial t^2} - \nabla^2 \tag{A2.13}$$

is Lorentz-invariant, i. e., a Lorentz scalar.

Using the above notations, we write the equations of motions in Lorentz-covariant (=invariant) form. At first we reformulate the Newtonian quantities of velocity, momentum, and force in such a way that they have a defined transformational behavior under Lorentz transformations. Contrary to dt the eigentime (or, say, proper time) differential

$$d\tau = dt\sqrt{1 - \frac{1}{c^2}\left(\frac{d\mathbf{r}}{dt}\right)^2} \tag{A2.14}$$

is a Lorentz scalar since, due to the conservation of the scalar product, we have

$$d\tau^2 = dt^2 - \frac{1}{c^2}d\mathbf{r}^2 = dt'^2 - \frac{1}{c^2}d\mathbf{r}'^2. \tag{A2.15}$$

With this, the following quantities can be defined, of which the four-versions transform as x^μ:

Four-velocity u^μ:

$$u^\mu = \frac{dx^\mu}{d\tau} = \frac{dt}{d\tau}\frac{dx^\mu}{dt}, \quad \underline{u} = (u^\mu) = \frac{1}{\sqrt{1-\beta^2}}\begin{pmatrix} c \\ \mathbf{v} \end{pmatrix}, \tag{A2.16}$$

where the physical velocity is

$$\mathbf{v} = \frac{d\mathbf{r}}{dt}. \tag{A2.17}$$

Four-momentum p^μ:

$$p^\mu = m_\circ u^\mu, \quad \underline{p} = (p^\mu) = \begin{pmatrix} cm \\ \mathbf{p} \end{pmatrix} = \begin{pmatrix} E/c \\ \mathbf{p} \end{pmatrix}, \quad m = \frac{m_\circ}{\sqrt{1-\beta^2}}, \tag{A2.18}$$

where the physical momentum is

$$\mathbf{p} = m\dot{\mathbf{r}}. \tag{A2.19}$$

Four-force F^μ:

$$F^\mu = \frac{dp^\mu}{d\tau} = \frac{1}{\sqrt{1-\beta^2}}\frac{dp^\mu}{dt}, \quad \underline{F} = (F^\mu) = \frac{1}{\sqrt{1-\beta^2}}\begin{pmatrix} c\frac{dm}{dt} \\ \mathbf{F} \end{pmatrix}, \tag{A2.20}$$

where the physical force is

$$\mathbf{F} = \frac{d\mathbf{p}}{dt}. \tag{A2.21}$$

Four-wavevector of monochromatic light is

$$\underline{k} = (k^\mu) = \left(\frac{\omega}{c}, \mathbf{k}\right). \tag{A2.22}$$

where the physical wavevector

$$\mathbf{k} = \frac{2\pi}{\lambda}(n_1, n_2, n_3). \tag{A2.23}$$

Four-momentum of single photon

$$\underline{p} = (p^\mu_\gamma) = \hbar(k^\mu). \tag{A2.24}$$

Four-current vector

$$\underline{j} = (j^\mu) = (\rho c, \mathbf{j}), \tag{A2.25}$$

where for electromagnetic field ρ and $\mathbf{j}$ represent the charge density and physical current density, and for quantum mechanics ρ and $\mathbf{j}$ represent the probability density and physical probability current density, respectively. The continuity equation

$$\frac{\partial\rho(\mathbf{r},t)}{\partial t} + \nabla\cdot\mathbf{j}(\mathbf{r},t) = 0, \tag{A2.26}$$

can be written in Lorentz-covariant form

$$\partial_\mu j^\mu(\underline{x}) = 0, \quad [\underline{x} \equiv (ct, \mathbf{r})], \quad (j^\mu) = (c\rho, \mathbf{j}). \tag{A2.27}$$

In the case of a conservative force field $\mathbf{F} = -\nabla V(\mathbf{r})$, we have the energy conservation:

$$E = mc^2 + V(\mathbf{r}) = constant \Rightarrow E = mc^2 \quad for \quad V(\mathbf{r}) = 0, \tag{A2.28}$$

and, together with the definition of p^μ, the energy-momentum relation for free particles

$$p_\circ^2 - \mathbf{p}^2 = m_\circ^2 c^2 \Leftrightarrow E^2 = \mathbf{p}^2 c^2 + m_\circ^2 c^4. \qquad (A2.29)$$

Note that not all types of forces allow a four dimensional generalization. One example of this kind is the Coulomb force. As a static long range force, it requires an infinitely high propagational velocity, thus being in contradiction to the axioms of special relativity. Another example is the two-local-spin-mediated interaction, mentioned in chapter 7.

In the statement of Maxwell field equation, the symbol ϵ_{ijk} stands for

$$\begin{aligned} \epsilon_{ijk} &= +1, \ \textit{if } (i,j,k) \textit{ is an even permutation of the numbers } 1,2,3 \\ &= -1, \ \textit{if } (i,j,k) \textit{ is an odd permutation of the numbers } 1,2,3 \\ &= 0, \ \textit{if any two indices are equal.} \qquad (A2.30) \end{aligned}$$

A.3 Forms of Gamma Matrices

$$\boldsymbol{\gamma} = \begin{bmatrix} 0 & \boldsymbol{\sigma} \\ -\boldsymbol{\sigma} & 0 \end{bmatrix}, \qquad (A3.1)$$

where $\boldsymbol{\sigma} = (\sigma_x, \sigma_y, \sigma_z)$ is Pauli matrix vector.

$$\sigma_x = \begin{bmatrix} 0 & 1 \\ 1 & 0 \end{bmatrix}, \sigma_y = \begin{bmatrix} 0 & -i \\ i & 0 \end{bmatrix} \sigma_z = \begin{bmatrix} 1 & 0 \\ 0 & -1 \end{bmatrix}. \qquad (A3.1)'$$

$$\gamma^\circ = \begin{bmatrix} \mathbf{I} & 0 \\ 0 & -\mathbf{I} \end{bmatrix}, \qquad (A3.2)$$

$$\gamma^5 = \begin{bmatrix} 0 & \mathbf{I} \\ \mathbf{I} & 0 \end{bmatrix} = i\gamma^\circ\gamma^x\gamma^y\gamma^z, \qquad (A3.3)$$

$$\gamma^\mu\gamma^5 = -\gamma^5\gamma^\mu, \qquad (A3.3)'$$

$$\frac{1}{2}\left(\gamma^\mu\gamma^\nu + \gamma^\nu\gamma^\mu\right) \equiv \frac{1}{2}\left\{\gamma^\mu, \gamma^\nu\right\} = g^{\mu\nu}. \qquad (A3.4)$$

Thus the four matrixes γ^μ, $\mu = 0, 1, 2, 3$, are not simply complex number but rather entities that they obey something called a Clifford algebra. Clearly, this algebra implies that

$$(\gamma^\circ)^2 = 1 \qquad (A3.5)$$

$$\left(\gamma^i\right)^2 = -1, \quad i = 1, 2, 3 \qquad (A3.5)'$$

$$\gamma^\mu\gamma^\nu = -\gamma^\nu\gamma^\mu \ \ \textit{if } \mu \neq \nu. \qquad (A3.6)''$$

$\gamma_\circ$ is Hermitian.

$$\gamma^{1+} = (\beta\alpha_x)^+ = \alpha_x^+\beta^+ = -\beta\alpha_x = -\gamma^1. \qquad (A3.7)$$

Hence γ^1, γ^2, γ^3 are anti-hermitian.

Appendix B

Classical Field Theory

B.1 Recall on Multiparticles

This chapter is intended to serve as a necessary mathematical preparation to quantum field theory. The classical field equation is an extension of classical multiparticle theory. Let us recapitulate the derivation of the Euler-Lagrange equation of motion in case of multiparticles [3]. Then we extend it to classical field. Let $q_i(t)$, $i = 1, 2, 3, \cdots, n$ be the generalized coordinates, and $\dot{q}_i(t) = dq_i(t)/dt$ is the generalized velocities, and L is its Lagrangian.

$$L = L(q_i(t), \dot{q}_i(t), t). \qquad (B.1.1)$$

The action is

$$\mathcal{A} = \int_{t_1}^{t_2} dt L(q_i(t), \dot{q}_i(t), t). \qquad (B.1.2)$$

Make an infinitesimal change

$$q_i(t) \rightarrow q_i(t) + \delta q_i(t), \qquad (B.1.3a)$$

and set at the end-points t_1 and t_2

$$\delta q_i(t_1) = \delta q_i(t_2) = 0. \qquad (B.1.3b)$$

This variation results in

$$\dot{q}_i(t) \rightarrow \dot{q}_i(t) + \frac{d}{dt}\delta q_i(t). \qquad (B.1.4)$$

The principle of leat action asserts that

$$\begin{aligned} \delta\mathcal{A} &= \delta \int_{t_1}^{t_2} dt L(q_i(t), \dot{q}_i(t), t) \\ &= \delta \int_{t_1}^{t_2} dt \sum_i \left\{ \frac{\partial L}{\partial q_i}\delta q_i + \frac{\partial L}{\partial \dot{q}_i}\delta \dot{q}_i \right\} \\ &= \int_{t_1}^{t_2} dt \sum_i \left\{ \frac{\partial L}{\partial q_i} - \frac{d}{dt}\frac{\partial L}{\partial \dot{q}_i} \right\} \delta q_i = 0. \qquad (B.1.5) \end{aligned}$$

From Eq. (B.1.5) we obtain Euler-Lagrange equation

$$\frac{\partial L}{\partial q_i} - \frac{d}{dt}\frac{\partial L}{\partial \dot{q}_i} = 0, \quad (i = 1, 2, 3, \cdots, n). \qquad (B.1.6)$$

The definition of canonical conjugate momentum of q_i, p_i, is

$$p_i = \frac{\partial L}{\partial \dot{q}_i}, \quad \dot{p}_i = \frac{\partial L}{\partial q_i}. \tag{B.1.7}$$

If T and V represent kinetic and potential energies, respectively, then the Lagrangian and Hamiltonian are

$$L = T - V, \quad H(q_i, p_i, t) = \sum_i p_i \dot{q}_i - L. \tag{B.1.8}$$

Using Euler-Lagrange equation yields Hamilton equation of motion:

$$\dot{q}_i = \frac{\partial H}{\partial p_i}, \quad \dot{p}_i = -\frac{\partial H}{\partial q_i}. \tag{B.1.9}$$

A general dynamical variable $F = F(q_i, p_i, t)$ satisfies the following equation of motion:

$$\begin{aligned} \frac{dF}{dt} &= \frac{\partial F}{\partial t} + \sum_i \left[\frac{\partial F}{\partial q_i}\frac{\partial q_i}{\partial t} + \frac{\partial F}{\partial p_i}\frac{\partial p_i}{\partial t} \right] \\ &= \frac{\partial F}{\partial t} + \sum_i \left[\frac{\partial F}{\partial q_i}\frac{\partial H}{\partial p_i} - \frac{\partial F}{\partial p_i}\frac{\partial H}{\partial q_i} \right] \\ &\equiv \frac{\partial F}{\partial t} + \{F, H\}_{Poison\ \ bracket}. \end{aligned} \tag{B.1.10}$$

B.2 Lagrangian and Hamiltonian Formulations of Classical Field

In the above derivations, i represents degree of freedom. The classical field evidently has a continuously infinite number of degrees of freedom and is analogous to a system that consists of an infinite number of particles. It is natural, then, to use the field $\Psi(\mathbf{r}, t)$ at all points $\mathbf{r}$ as coordinates, in analogy with the particle coordinates $q_i(t)$. Pictorially, every point has a particle, and thus infinite number of particle can be expressed conveniently by a function of field. (Field=Space) Therefore, the Lagrangian formulation can be extended to the field with a little modifications. Since the field $\Psi(\mathbf{r}, t)$ depends of continuous variable $\mathbf{r}$, we have to invoke the Lagrangian density $\mathcal{L}$. We hope that the classical field theory can be extended directly to relativistic field theory, thus we write $\mathbf{r}$ and t on the same footing by using $\underline{x} = (x^\mu) = (x^\circ, x^1, x^2, x^3) = (ct, x^1, x^2, x^3) = (ct, x, y, z)$ and $(x_\mu) = (x_\circ, x_1, x_2, x_3) = (x^\circ, -x^1, -x^2, -x^3) = (ct, -x, -y, -z)$.

(1). Lagrangian formulation. By analogy with the particle case, we expect the total field Lagrangian to be a functional of the coordinate $\Psi_\alpha(\mathbf{x}, t) = \Psi_\alpha(\mathbf{r}, t)$, $\dot{\Psi}_\alpha(\mathbf{x}, t) = \dot{\Psi}_\alpha(\mathbf{r}, t)$ and $\nabla\Psi_\alpha(\mathbf{x}, t) = \nabla\Psi_\alpha(\mathbf{r}, t)$.

$$\begin{aligned} L &= \int d\mathbf{x} \mathcal{L}(\Psi_\alpha, \nabla\Psi_\alpha, \dot{\Psi}_\alpha, \underline{x}) \\ &\equiv \int d\mathbf{x} \mathcal{L}(\Psi_\alpha, \partial_\mu\Psi_\alpha, \underline{x}), \end{aligned} \tag{B.2.1}$$

where we consider the existence of many different field coordinates, Ψ_α represent

$$\Psi_\alpha = \Psi_1, \Psi_2, \Psi_3, \cdots, \tag{B.2.2}$$

$\partial_\mu\Psi$ represent

$$\partial_\mu\Psi = \frac{\partial\Psi}{\partial x^\circ}, \frac{\partial\Psi}{\partial x^1}, \frac{\partial\Psi}{\partial x^2}, \frac{\partial\Psi}{\partial x^3}, \tag{B.2.3}$$

and the double subscripts of α mean the summation over α from 1 to n. The appearance of, for example, $\nabla\Psi$ as an argument of $\mathcal{L}$ is a consequence of the continuous dependence of Ψ on $\mathbf{r}$ (=$\mathbf{x}$), which is a continuously infinite number of degrees of freedom; higher derivatives of Ψ could also be present but do not seem to arise in problems of physical interest [8, 121]. The action is

$$\mathcal{A} = \int_{t_1}^{t_2}\int_V d^4x\mathcal{L}(\Psi_\alpha, \partial_\mu\Psi_\alpha, \underline{x}), \qquad (B.2.4)$$

where $d^4 = dx^\circ dxdydz$, V is a certain domain in spacetime, bounded by a hypersurface $\mathcal{S}$.

According to the principle of least action, under conditional variation, the solution Ψ_α of motion equation makes the action to take extreme value. The conditional variation $\delta\mathcal{A}$ of action $\mathcal{A}$ can be periodic boundary conditions by using box normalization, or that the variation of the fields Ψ_α at the boundary of the spacetime is zero, i. e.,

$$\delta\Psi_\alpha(\underline{x}) = 0, \; on \;\; surface \;\; of \;\; volume \;\; V. \qquad (B.2.5)$$

The variations of field and field derivative are:

$$\Psi_\alpha(\underline{x}) \to \Psi_\alpha(\underline{x}) + \delta\Psi_\alpha(\underline{x}), \quad \frac{\partial\Psi_\alpha(\underline{x})}{\partial x_\mu} \to \frac{\partial\Psi_\alpha(\underline{x})}{\partial x_\mu} + \frac{\partial}{\partial x_\mu}\delta\Psi_\alpha(\underline{x}). \qquad (B.2.6)$$

Thus

$$\begin{aligned}
\delta\mathcal{A} &= \delta\int_{t_1}^{t_2}\int_V d^4x\mathcal{L}(\Psi_\alpha, \partial_\mu\Psi_\alpha, \underline{x}) \\
&= \int_V\int_{t_1}^{t_2} d^4x\sum_\alpha\left\{\frac{\partial\mathcal{L}}{\partial\Psi_\alpha}\delta\Psi_\alpha + \sum_\mu\frac{\partial\mathcal{L}}{\partial(\partial\Psi_\alpha/\partial x_\mu)}\frac{\partial}{\partial x_\mu}\delta\Psi_\alpha\right\} \\
&= \int_V\int_{t_1}^{t_2} d^4x\sum_\alpha\frac{\partial\mathcal{L}}{\partial\Psi_\alpha}\delta\Psi_\alpha + \sum_\alpha\sum_\mu\frac{\partial\mathcal{L}}{\partial(\partial\Psi_\alpha/\partial x_\mu)}\delta\Psi_\alpha\Bigg]_{\mathcal{S}} \\
&\quad -\int_V\int_{t_1}^{t_2} d^4x\sum_\alpha\sum_\mu\frac{\partial}{\partial x_\mu}\frac{\partial\mathcal{L}}{\partial(\partial\Psi_\alpha/\partial x_\mu)}\delta\Psi_\alpha \\
&= \int_V\int_{t_1}^{t_2} d^4x\sum_\alpha\left\{\frac{\partial\mathcal{L}}{\partial\Psi_\alpha} - \sum_\mu\frac{\partial}{\partial x_\mu}\frac{\partial\mathcal{L}}{\partial(\partial\Psi_\alpha/\partial x_\mu)}\right\}\delta\Psi_\alpha = 0, \quad (B.2.7)
\end{aligned}$$

where the third equality comes from the integrating the second term in the curly bracket of the second equality. Since this equation holds for arbitrary $\delta\Psi_\alpha$ and also valid for arbitrary domain V, we obtain from Eq. (B.2.7) the equation of motion of the field coordinates Ψ_α in spacetime

$$\frac{\partial\mathcal{L}}{\partial\Psi_\alpha} - \sum_\mu\frac{\partial}{\partial x_\mu}\frac{\partial\mathcal{L}}{\partial(\partial\Psi_\alpha/\partial x_\mu)} = 0 \;\; (\alpha = 1, 2, \cdots, n). \qquad (B.2.8)$$

(2). Hamiltonian formulation, or say canonical formulation. In this formulation, at first write the canonical conjugate momenta according to definition

$$\pi_\alpha(\underline{x}) \equiv \frac{\partial\mathcal{L}(\Psi_\alpha, \nabla\Psi_\alpha, \dot{\Psi}_\alpha, \underline{x})}{\partial\dot{\Psi}_\alpha}. \qquad (B.2.9)$$

It is easy to observe that Eq. (B.2.9) is an extension of Eq. (B.1.7).

If we can obtain from equation set Eq. (B.2.9) $\dot{\Psi}_\alpha = f(\Psi_\alpha, \nabla\Psi_\alpha, \pi_\alpha)$, then this is equivalent to introduce new variables to replace the old variables. Therefore, the operation of introducing new variables π_α is an operation of transformation of field variables. This operation is called Legendre transformation. After this transformation, one obtain Hamiltonian density

$$\mathcal{H} = \sum_\alpha \pi_\alpha \dot{\Psi}_\alpha - \mathcal{L}. \tag{B.2.10}$$

Note that in Eq. (B.2.10) the independent field variables are π_α and Ψ_α, other than $\dot{\Psi}_\alpha$ and Ψ_α. The total Hamiltonian is

$$H(t) = \int d\mathbf{r}\mathcal{H} = \int d\mathbf{r} \left[\sum_\alpha \pi_\alpha \dot{\Psi}_\alpha - \mathcal{L} \right]. \tag{B.2.11}$$

After Eq. (B.2.11) we enter into the Hamiltonian formulation. The equations of motion of the fields are called Hamilton equation or canonical equation. An arbitrary field functional A has equation of motion

$$\frac{dA}{dt} = \frac{\partial A}{\partial t} + \{A, H(t)\}_{Poison \quad brackey}. \tag{B.2.12}$$

The definition of Poison bracket for two classical field functionals is

$$\{A, B\}_{Poison \quad brackey} = \int_V d\mathbf{r} \left(\frac{\delta A}{\delta \Psi_\alpha(\underline{x})} \frac{\delta B}{\delta \pi_\alpha(\underline{x})} - \frac{\delta A}{\delta \pi_\alpha(\underline{x})} \frac{\delta B}{\delta \Psi_\alpha(\underline{x})} \right). \tag{B.2.13}$$

It is obvious that to do the Legendre transforation we have to get the explicit function $\dot{\Psi}_\alpha = f_\alpha(\Psi_\alpha, \nabla\Psi_\alpha, \pi_\alpha)$. This is a presupposition from Lagrangian formulation to Hamiltonian formulation. The condition to have the explicit function $\dot{\Psi}_\alpha = f(\Psi_\alpha, \nabla\Psi_\alpha, \pi)$ is the functional Jacobi-Hess determinant [150]

$$M_{\alpha\beta(\mathbf{r},\mathbf{r}')} = \frac{\partial^2 \mathcal{L}}{\partial \dot{\Psi}_\alpha(\mathbf{r}, t) \partial \dot{\Psi}_\beta(\mathbf{r}', t)} \delta(\mathbf{r} - \mathbf{r}') \neq 0. \tag{B.2.14}$$

If $M_{\alpha\beta} = 0$, then this system is called strange system or Hamilton constrained system. The classical Maxwell field equation set belongs to this system, and we will discuss in detail in section 9.6 of chapter 9.

Appendix C

δ Function

C.1 Definition of δ Function

The $\delta(x)$ function is defined by the following two equations:

$$\delta(x-a)=0, \quad if\ x\neq a, \tag{C.1.1}$$

and

$$\int_{-\infty}^{\infty} dx\delta(x-a)=1. \tag{C.1.2}$$

From this definition we note two points:

(1). δ function is a density distribution function, i. e., its dimension is $1/x$;

(2). δ function is not an ordinary function, for which mathematicians have many discussions on its analyticity. However, physicists use δ function just for convenience, do not take care of its analyticity. Roughly speaking, physicists treat the δ function as ordinary function.

In this appendix, we just list the properties and the representations of δ functions, and do not give demonstrations. For demonstrations, refer to Refs. [151, 10, 20, 5].

C.2 Properties of δ Function

$$\delta(-x)=\delta(x). \tag{C.2.1}$$

$$\delta(ax)=\frac{1}{|a|}\delta(x). \tag{C.2.2}$$

$$\int_{-\infty}^{\infty} dxf(x)\delta(x-a)=f(a). \tag{C.2.3}$$

$$\int_{-\infty}^{\infty} dx\delta(x-a)\delta(x-a)=\delta(x-a). \tag{C.2.4}$$

$$(x-a)\delta(x-a)=0. \tag{C.2.5}$$

If from $\phi(x)=0$ we have the roots are x_i $i=1,2,3,\cdots$ and at the same time $\phi'(x_i)\neq 0$, then

$$\delta[\phi(x)]=\sum_i\frac{\delta(x-x_i)}{|\phi'(x_i)|}=\sum_i\frac{\delta(x-x_i)}{|\phi'(x)|}. \tag{C.2.6}$$

For example,

$$\delta[(x-a)(x-b)] = \frac{1}{|a-b|}[\delta(x-a) + \delta x - b], \quad a \neq b. \qquad (C.2.6)'$$

$$\delta[x^2 - a^2] = \frac{1}{2|x|}[\delta(x-a) + \delta x + a] = \frac{1}{2|a|}[\delta(x-a) + \delta x + a]. \qquad (C.2.6)''$$

$$\int_{-\infty}^{\infty} dx \delta^{(n)}(x-a) f(x) \equiv \int_{-\infty}^{\infty} dx \frac{\partial^n \delta(x-a)}{\partial x^n} f(x) = (-1)^n f^{(n)}(a). \qquad (C.2.7)$$

$$\delta^{(n)}(-x) = (-1)^n \delta^{(n)}(x). \qquad (C.2.7)'$$

$$x\delta^{(n)}(x) = -n\delta^{(n-1)}(x). \qquad (C.2.7)''$$

$$\delta(x-a) = \frac{d\theta(x-a)}{dx}, \quad \theta(x-a) = \begin{cases} 1, & x > a \\ 0, & x < a \end{cases}. \qquad (C.2.8)$$

C.3 One-Dimensional δ Functions

$$\delta(x-a) = \lim_{\alpha\to\infty}\sqrt{\frac{\alpha}{\pi}}e^{-\alpha(x-a)^2}, \quad (C.3.1)$$

$$= \lim_{\alpha\to\infty}\frac{\alpha}{2}e^{-\alpha|x-a|^2}, \quad (C.3.2)$$

$$= \lim_{\alpha\to\infty}(1\mp i)\sqrt{\frac{\alpha}{2\pi}}e^{\pm i\alpha(x-a)^2}, \quad (C.3.3)$$

$$= \lim_{\alpha\to\infty}e^{\mp i\pi/4}\sqrt{\frac{\alpha}{\pi}}e^{\pm i\alpha(x-a)^2}, \quad (C.3.5)$$

$$= \lim_{\alpha\to 0}\frac{\alpha}{\pi[\alpha^2+(x-a)^2]}, \quad (C.3.6)$$

$$= \lim_{\alpha\to 0}Im\frac{1}{\pi[-i\alpha+(x-a)]}, \quad (C.3.7)$$

$$= \lim_{\alpha\to\infty}\frac{sin\alpha(x-a)}{\pi(x-a)}, \quad (C.3.8)$$

$$= \lim_{\alpha\to\infty}\frac{sin^2\alpha(x-a)}{\pi\alpha(x-a)^2}, \quad (C.3.9)$$

$$= \lim_{\alpha\to\infty}\frac{1-cos\alpha(x-a)}{\pi\alpha(x-a)^2}, \quad (C.3.12)$$

$$= \frac{1}{2\pi}\sum_{n=-\infty}^{\infty}e^{in(x-a)}, \quad a-\pi\le x\le a+\pi, \quad (C.3.13)$$

$$= \frac{1}{2\pi}\left(1+\sum_{n=1}^{\infty}2cosn(x-a)\right), \quad a-\pi\le x\le a+\pi, \quad (C.3.14)$$

$$= \frac{1}{2\pi}\int_{-\infty}^{\infty}dke^{ik(x-a)}, \quad (C.3.15)$$

$$= \frac{1}{2\pi}\int_{0}^{\infty}dkcos[k(x-a)], \quad (C.3.16)$$

$$= \sum_n\Psi_n^*(a)\Psi_n(x), \quad \Psi_n(x)\ are\ any$$

orthogonal complete set and discrete spectrum, (C.3.17)

$$= \sum_{n=0}^{\infty}\frac{2n+1}{2}P_n^*(a)P_n(x),$$

$P_n(x)$ *are Legendre polinominals,* $x=cos\theta$, (C.3.18)

$$= \sum_{n=0}^{\infty}\frac{1}{2^n\sqrt{\pi}n!}e^{-(x^2+a^2)/2}H_n^*(a)H_n(x),$$

$H_n(x)$ *are Hermite polinominals,* (C.3.19)

$$= \int df\Psi_f^*(x)\Psi_f(a),$$

$\Psi_f(x)$ *are any orthogonal complete set and continuous spectrum,* (C.3.20)

Some times one encounters under the integral sign the singular function

$$\zeta(x-a) = \lim_{t\to\infty} \frac{1-e^{-i(x-a)t}}{i(x-a)} = \pi\delta(x-a) - i\mathcal{P}\frac{1}{x-a}, \qquad (C.3.21)$$

where the $\mathcal{P}$-sign indicates that one needs to take the principal value of the integral.

The singular function

$$\lim_{\varepsilon\to 0} \frac{1}{(x-a)-i\varepsilon} = i\pi\delta(x-a) + \mathcal{P}\frac{1}{x-a}. \qquad (C.3.22)$$

One often uses other important functions, such as, for instance,

$$\delta^{+}(x) = \delta^{-*}(x) = \frac{1}{2\pi i} \lim_{\varepsilon\to 0} \frac{1}{x-i\varepsilon}. \qquad (C.3.23)$$

Using Eq. (C.3.23) and Eq. (C.3.6) yield

$$\delta(x) = \lim_{\alpha\to 0} \frac{\alpha}{\pi[\alpha^2+x^2]} = \delta^{+}(x) + \delta^{-}(x). \qquad (3.24)$$

$$\lim_{\alpha\to 0} \frac{\alpha}{i\pi[\alpha^2+x^2]} = \delta^{+}(x) - \delta^{-}(x). \qquad (C.3.25)$$

The $\delta(z)$ function as a function of complex variable has two simple poles at the points $\pm ix$ with residues equal to $\pm 1/2\pi i$, respectively. When integrating expressions containing $\delta(z)$, the integration path must be go between these poles. Eqs. (C.3.23-25) remain valid also for complex values of x. In that case, we have

$$\delta^{+}(-z) = \delta^{-}(z) = \delta^{+*}(z) = [\delta^{+}(z^*)]^*, \qquad (C.3.26)$$

$$\delta^{+}(z) = \frac{1}{2\pi iz}, \quad \delta^{-}(z) = \frac{-1}{2\pi iz}, \qquad (C.3.27)$$

and remember to take the path of integration above and below the points $z=0$, respectively.
The δ function has a summation formula:

$$\sum_{n=-\infty}^{\infty} \delta(x-nx_\circ) = \frac{1}{x_\circ} \sum_{m=-\infty}^{\infty} e^{\pm 2\pi im/x_\circ}. \qquad (C.3.28)$$

C.4 Three-Dimensional δ Functions

In Cartesian coordinates,

$$\delta(\mathbf{r}-\mathbf{r}_\circ) = \delta(x-x_\circ)\delta(y-y_\circ)\delta(z-z_\circ). \qquad (C.4.1)$$

In spherical coordinates,

$$\delta(\mathbf{r}-\mathbf{r}_\circ) = \frac{1}{r^2 sin\theta}\delta(r-r_\circ)\delta(\theta-\theta_\circ)\delta(\phi-\phi_\circ). \qquad (C.4.2)$$

We note also the differential equations:

$$(\nabla^2+k^2)\frac{e^{ikr}}{r} = (\nabla^2+k^2)\frac{coskr}{r} = -4\pi\delta(\mathbf{r}). \qquad (C.4.3)$$

Is case of $k=0$, Eq. (C.4.3) becomes Poisson equation,

$$(\nabla^2)\frac{1}{r} = -4\pi\delta(\mathbf{r}). \qquad (C.4.4)$$

In cylindrical coordinates,

$$\delta(\rho-\rho_\circ) = \frac{1}{\rho}\delta(\rho-\rho_\circ)\delta(\phi-\phi_\circ)\delta(z-z_\circ), \qquad (C.4.5)$$

C.5 Longitudinal and Transverse δ Functions

C.5.1 Definitions

Longitudinal δ function:

$$\begin{aligned}\delta^{\parallel}_{ij}(\mathbf{r}-\mathbf{r}') &= \frac{\partial_i\partial_j}{\nabla^2}\delta(\mathbf{r}-\mathbf{r}') \quad i,j=1,2,3\\ &= \frac{\partial_i\partial_j}{\nabla^2}\frac{1}{(2\pi)^3}\int d^3\mathbf{k}e^{i\mathbf{k}\cdot(\mathbf{r}-\mathbf{r}')}\\ &= \frac{1}{(2\pi)^3}\int d^3\mathbf{k}\frac{k_ik_j}{k^2}e^{i\mathbf{k}\cdot(\mathbf{r}-\mathbf{r}')}. \qquad (C.5.1.1)\end{aligned}$$

Let us illustrate that why we call the $\delta^{\parallel}(\mathbf{r}-\mathbf{r}')$ longitudinal δ function?

Eq. (C.5.1.1) indicates:

(i). $\delta^{\parallel}_{ij}(\mathbf{r}-\mathbf{r}')$ is no longer a scalar quantity, and is a vector. $\delta^{\parallel}_{ij}(\mathbf{r}-\mathbf{r}')$ with $i=1,2,3$ represent three components of vector, which corresponds to a fixed value of j in $j=1,2,3$;

(ii). In Eq. (C.5.1.1) $exp[i\mathbf{k}\cdot(\mathbf{r}-\mathbf{r}')]$ is a plane wave, and k_ik_j/k^2 is an amplitude. Actually, these amplitudes are vector: every value of $j=1,2,3$ corresponds to a vector with $i=1,2,3$. Next, we prove that $(k_1k_j/k^2, k_2k_j/k^2, k_3k_j/k^2)=\mathbf{u}_{k_j}k_j$, where $\mathbf{u}_{k_j}$ represents a unit vector in k_j direction.

Proof:

$$(k_1,k_2,k_3)\cdot\left(\frac{k_1k_j}{k^2},\frac{k_2k_j}{k^2},\frac{k_3k_j}{k^2}\right)=k_j\left(\frac{k_1^2}{k^2}+\frac{k_2^2}{k^2}+\frac{k_3^2}{k^2}\right)=k_j.$$

Thus for the wave amplitude vectors of $j=1,2,3$, the wave amplitude vectors $(k_1k_j/k^2, k_2k_j/k^2, k_3k_j/k^2)=$ $\mathbf{u}_{k_j}k_j$ are the three components of wave vector $\mathbf{k}$. We can conclude that $\delta^{\parallel}(\mathbf{r}-\mathbf{r}')$ in wavevector space is a longitudinal wave.

Transverse δ function:

$$\begin{aligned}\delta^{\perp}_{ij}(\mathbf{r}-\mathbf{r}') &= \left(\delta_{ij}-\frac{\partial_i\partial_j}{\nabla^2}\right)\delta(\mathbf{r}-\mathbf{r}') \quad i,j=1,2,3\\ &= \left(\delta_{ij}-\frac{\partial_i\partial_j}{\nabla^2}\right)\frac{1}{(2\pi)^3}\int d^3\mathbf{k}e^{i\mathbf{k}\cdot(\mathbf{r}-\mathbf{r}')}\\ &= \frac{1}{(2\pi)^3}\int d^3\mathbf{k}\left(\delta_{ij}-\frac{k_ik_j}{k^2}\right)e^{i\mathbf{k}\cdot(\mathbf{r}-\mathbf{r}')}. \qquad (C.5.1.2)\end{aligned}$$

Eq. (C.5.1.2) indicates:

(i). $\delta^{\perp}_{ij}(\mathbf{r}-\mathbf{r}')$ is no longer a scalar quantity, and is a vector, $i=1,2,3$ represent the three components for fixed j, which can take 1,2,3;

(ii). In Eq. (C.5.1.2) $exp(i\mathbf{k}\cdot(\mathbf{r}-\mathbf{r}'))$ is a plane wave, and $\delta_{ij}-k_ik_j/k^2$ represents the amplitude vector at a fixed j. Next, we prove that the vectors $(\delta_{1j}-k_1k_j/k^2, \delta_{2j}-k_2k_j/k^2, \delta_{3j}-k_3k_j/k^2)$ at $j=1,2,3$ are always perpendicular to $\mathbf{k}$, and thus is a transverse wave. Actually, this proof is simple. For the convenience of statement, take $j=1$.

$$(k_1,k_2,k_3)\cdot\left(1-\frac{k_1k_1}{k^2},-\frac{k_2k_1}{k^2},-\frac{k_3k_1}{k^2}\right)=k_1\left(1-\frac{k_1^2}{k^2}-\frac{k_2^2}{k^2}-\frac{k_3^2}{k^2}\right)=0.$$

For $j=2,3$ we have the same results. Thus for the wave amplitude vectors of $j=1,2,3$, the wave amplitude vectors are always perpendicular to $\mathbf{k}=(k_1,k_2,k_3)$. We conclude that $\delta^{\perp}(\mathbf{r}-\mathbf{r}')$ in wavevector space is a transverse wave.

C.5.2 Properties

(1). From the definitions, we obtain immediately

$$\delta_{ij}^{\perp}(\mathbf{r}-\mathbf{r}') + \delta_{ij}^{\parallel}(\mathbf{r}-\mathbf{r}') = \delta_{ij}\delta(\mathbf{r}-\mathbf{r}'). \qquad (C.5.2.1)$$

(2).

$$\nabla \cdot \delta_{ij}^{\perp}(\mathbf{r}-\mathbf{r}') \equiv \sum_{i=1}^{3} \frac{\partial}{\partial x_i}\delta_{ij}^{\perp}(\mathbf{r}-\mathbf{r}') = 0. \qquad (5.2.2)$$

Proof:
Substituting Eq. (C.5.1.2) into Eq. (C.5.2.2) yields

$$\begin{aligned}
\nabla \cdot \delta_{ij}^{\perp}(\mathbf{r}-\mathbf{r}') &= \sum_{i=1}^{3} \frac{\partial}{\partial x_i}\frac{1}{(2\pi)^3}\int d^3\mathbf{k}\left(\delta_{ij} - \frac{k_i k_j}{k^2}\right) e^{i\mathbf{k}\cdot(\mathbf{r}-\mathbf{r}')} \\
&= \sum_{i=1}^{3} \frac{1}{(2\pi)^3}\int d^3\mathbf{k}\left(\delta_{ij} - \frac{k_i k_j}{k^2}\right) ik_i e^{i\mathbf{k}\cdot(\mathbf{r}-\mathbf{r}')} \\
&= \frac{1}{(2\pi)^3}\int d^3\mathbf{k}\left(k_j - \frac{k^2 k_j}{k^2}\right) i e^{i\mathbf{k}\cdot(\mathbf{r}-\mathbf{r}')} \\
&= 0.
\end{aligned}$$

Eq. (C.5.2.2) indicates that the vector $\delta_{ij}^{\perp}(\mathbf{r}-\mathbf{r}')$ is a transverse vector.

(3).

$$\nabla \cdot \delta_{ij}^{\parallel}(\mathbf{r}-\mathbf{r}') \equiv \sum_{i=1}^{3} \frac{\partial \delta_{ij}^{\parallel}(\mathbf{r}-\mathbf{r}')}{\partial x_i} = \frac{\partial \delta(\mathbf{r}-\mathbf{r}')}{\partial x_j}. \qquad (C.5.2.3)$$

Proof:
Substituting Eq. (C.5.1.1) into Eq. (C.5.2.3) yields

$$\begin{aligned}
\nabla \cdot \delta_{ij}^{\parallel}(\mathbf{r}-\mathbf{r}') &= \sum_{i=1}^{3} \frac{\partial}{\partial x_i} \\
&\quad \frac{1}{(2\pi)^3}\int d^3\mathbf{k}\frac{k_i k_j}{k^2} e^{i\mathbf{k}\cdot(\mathbf{r}-\mathbf{r}')} \\
&= \sum_{i=1}^{3} \frac{1}{(2\pi)^3}\int d^3\mathbf{k}\left(\frac{k_i k_j}{k^2}\right) ik_i e^{i\mathbf{k}\cdot(\mathbf{r}-\mathbf{r}')} \\
&= \frac{1}{(2\pi)^3}\int d^3\mathbf{k}\left(\frac{k^2 k_j}{k^2}\right) i e^{i\mathbf{k}\cdot(\mathbf{r}-\mathbf{r}')} \\
&= \frac{\partial \delta(\mathbf{r}-\mathbf{r}')}{\partial x_j}.
\end{aligned}$$

(4).

$$\nabla \times \delta_{ij}^{\parallel}(\mathbf{r}-\mathbf{r}') = 0. \qquad (C.5.2.4)$$

Eq. (C.5.2.4) indicates that the vector $\delta_{ij}^{\parallel}(\mathbf{r}-\mathbf{r}')$ is a longitudinal vector for any fixed value of j.

Proof:
Refer to Eq. (C.5.2.2).

(5).

$$\int d^3\mathbf{r}\delta^{\perp}(\mathbf{r}-\mathbf{r}')\delta^{\|}(\mathbf{r}-\mathbf{r}'')=0. \qquad (C.5.2.5)$$

Proof:
Substituting the definitions into Eq. (C.5.2.4) can see that Eq. (C.5.2.4) is correct.

C.5.3 Application

Any vector field $\mathbf{A}(\mathbf{r})$ can be expressed as

$$\mathbf{A}(\mathbf{r})=\mathbf{A}^{\perp}(\mathbf{r})+\mathbf{A}^{\|}(\mathbf{r}). \qquad (C.5.3.1)$$

$$\nabla\cdot\mathbf{A}^{\perp}(\mathbf{r})=0.\ \ \nabla\times\mathbf{A}^{\|}(\mathbf{r})=0. \qquad (C.5.3.2)$$

From Eq. (C.5.3.2) one can prove that the Fourier transforms satisfy

$$\mathbf{k}\cdot\mathbf{A}^{\perp}(\mathbf{k})=0.\ \ \mathbf{k}\times\mathbf{A}^{\|}(\mathbf{k})=0. \qquad (C.5.3.3)$$

We can prove:

$$A_i^{\|}(\mathbf{r})=\sum_{j=1}^{3}\int d\mathbf{r}'\delta_{ij}^{\|}(\mathbf{r}-\mathbf{r}')A_j(\mathbf{r}'). \qquad (C.5.3.4)$$

$$A_i^{\perp}(\mathbf{r})=\sum_{j=1}^{3}\int d\mathbf{r}'\delta_{ij}^{\perp}(\mathbf{r}-\mathbf{r}')A_j(\mathbf{r}'). \qquad (C.5.3.5)$$

Proof:
For example. Substituting Eq. (C.5.1.2) into Eq. (C.5.3.5) yields

$$\begin{aligned}
\nabla\cdot\mathbf{A}^{\perp}(\mathbf{r}) &= \sum_{i=1}^{3}\frac{\partial}{\partial x_i}\sum_{j=1}^{3}\int d\mathbf{r}'\frac{1}{(2\pi)^3}\int d^3\mathbf{k}\left(\delta_{ij}-\frac{k_ik_j}{k^2}\right)e^{i\mathbf{k}\cdot(\mathbf{r}-\mathbf{r}')}A_j(\mathbf{r}') \\
&= \sum_{i=1}^{3}\frac{\partial}{\partial x_i}\sum_{j=1}^{3}\frac{1}{(2\pi)^3}\int d^3\mathbf{k}\left(\delta_{ij}-\frac{k_ik_j}{k^2}\right)e^{i\mathbf{k}\cdot(\mathbf{r})}A_j(\mathbf{k}) \\
&= \sum_{i=1}^{3}\sum_{j=1}^{3}\frac{1}{(2\pi)^3}\int d^3\mathbf{k}\left(\delta_{ij}-\frac{k_ik_j}{k^2}\right)ik_ie^{i\mathbf{k}\cdot(\mathbf{r})}A_j(\mathbf{k}) \\
&= \sum_{j=1}^{3}\frac{1}{(2\pi)^3}\int d^3\mathbf{k}\left(k_j-\frac{k^2k_j}{k^2}\right)ik_ie^{i\mathbf{k}\cdot(\mathbf{r})}A_j(\mathbf{k}) \\
&= 0. \qquad (C.5.3.6)
\end{aligned}$$

Eq. (C.5.3.4) tells us that

$$\begin{aligned}
A_i^{\|}(\mathbf{r}) &= \sum_{j=1}^{3}\int d\mathbf{r}'\delta_{ij}^{\|}(\mathbf{r}-\mathbf{r}')A_j(\mathbf{r}') \\
&= \sum_{j=1}^{3}\int d\mathbf{r}'\left(\frac{\partial_i\partial_j}{\nabla^2}\right)\delta(\mathbf{r}-\mathbf{r}')A_j(\mathbf{r}') \\
&= \sum_{j=1}^{3}\frac{\partial_i\partial_j}{\nabla^2}A_j(\mathbf{r}) \\
&\equiv \frac{\partial_i\partial_j}{\nabla^2}A_j(\mathbf{r}). \qquad (C.5.3.7)
\end{aligned}$$

The operator $\partial_i \partial_j / \nabla^2$ can be called longitudinal operator of vector.

Eq. (C.5.3.5) tells us that

$$
\begin{aligned}
A_i^{\perp}(\mathbf{r}) &= \sum_{j=1}^{3} \int d\mathbf{r}' \left(\delta_{ij} - \frac{\partial_i \partial_j}{\nabla^2} \right) \delta(\mathbf{r} - \mathbf{r}') A_j(\mathbf{r}') \\
&= \sum_{j=1}^{3} \left(\delta_{ij} - \frac{\partial_i \partial_j}{\nabla^2} \right) A_j(\mathbf{r}) \\
&\equiv \left(\delta_{ij} - \frac{\partial_i \partial_j}{\nabla^2} \right) A_j(\mathbf{r}). \qquad (C.5.3.8)
\end{aligned}
$$

The operator $\delta_{ij} - \partial_i \partial_j / \nabla^2$ can be called transverse operator of vector.

C.5.4 N-Dimensional δ Function

In cases of N-dimensional space, $N \neq 3$ and coordinates $\mathbf{q} = (q_1, q_2, \cdots, q_N)$, the above formulas in subsection 5.3 are still correct, if we make transformation $N = 3 \to N \neq 3$. Let us illustrate by theirs definitions.

N-dimensional longitudinal δ function:

$$
\begin{aligned}
\delta_{ij}^{\parallel}(\mathbf{q} - \mathbf{q}') &= \frac{\partial_i \partial_j}{\nabla^2} \delta(\mathbf{q} - \mathbf{q}') \quad i, j = 1, 2, 3, \cdots, N \\
&= \frac{\partial_i \partial_j}{\nabla^2} \frac{1}{(2\pi)^N} \int d^N \mathbf{k} e^{i\mathbf{k} \cdot (\mathbf{q} - \mathbf{q}')} \\
&= \frac{1}{(2\pi)^N} \int d^N \mathbf{k} \frac{k_i k_j}{k^2} e^{i\mathbf{k} \cdot (\mathbf{q} - \mathbf{q}')} \\
&= \frac{1}{(2\pi)^N} \int d^3 \mathbf{k} \frac{k_i}{k} \frac{k_j}{k} e^{i\mathbf{k} \cdot (\mathbf{q} - \mathbf{q}')}, \qquad (C.5.1.1)'
\end{aligned}
$$

where $\mathbf{k}/k = (k_1, k_2, \cdots, k_N)/k$, $\nabla^2 = \partial_1^2 + \partial_2^2 + \cdots + \partial_N^2$, and $\partial_i^2 \equiv \partial^2 / \partial q_i^2$.

Let us illustrate that why we call the $\delta_{ij}^{\parallel}(\mathbf{q} - \mathbf{q}')$ longitudinal δ function?

Eq. (C.5.1.1)$'$ indicates:

i). $\delta_{ij}^{\parallel}(\mathbf{q} - \mathbf{q}')$ is no longer a scalar quantity, and is a vector in N-dimensional space, $\mathbf{q}$, for any fixed value of j. $\delta_{ij}^{\parallel}(\mathbf{q} - \mathbf{q}')$ with $i = 1, 2, 3, \cdots, N$ represent N components of vector, which corresponds to a fixed value of j in $j = 1, 2, 3, \cdots, N$;

(ii). In Eq. (5.1.1)$'$ $exp[i\mathbf{k} \cdot (\mathbf{q} - \mathbf{q}')]$ is a plane wave in N-dimensional space, and $k_i k_j / k^2$ is an amplitude. Actually, these amplitudes are vector: every value of $j = 1, 2, 3, \cdots, N$ corresponds to a vector with components $i = 1, 2, 3, \quad, N$. Next, we prove that if $(k_1/k, k_2/k;, \quad , k_N/k) \equiv (u_{k_1}, u_{k_2}, \cdots, u_{k_N}) \equiv \mathbf{u_k}$, then $\mathbf{u_k}$ represents a unit vector along $\mathbf{k}$ direction.

Proof:

$$
\left(\frac{k_1}{k}, \frac{k_2}{k}, \cdots, \frac{k_N}{k} \right) \cdot \left(\frac{k_1}{k}, \frac{k_2}{k}, \cdots, \frac{k_N}{k} \right) = \left(\frac{k_1^2}{k^2} + \frac{k_2^2}{k^2} + \cdots + \frac{k_N^2}{k^2} \right) = 1.
$$

Therefore, Eq. (C.5.1.1)$'$ can also be expressed as

$$
\delta_{ij}^{\parallel}(\mathbf{q} - \mathbf{q}') = \frac{1}{(2\pi)^N} \int d^N \mathbf{k} u_{k_i} u_{k_j} e^{i\mathbf{k} \cdot (\mathbf{q} - \mathbf{q}')}. \qquad (C.5.1.1)''
$$

Any wave amplitude vectors of $j = 1, 2, 3, \cdots, N$

$$
(u_{k_1}, u_{k_2}, \cdots, u_{k_N}) u_{k_j} = \mathbf{u_k} u_{k_j}
$$

are always in the **k** direction. Therefore, $\delta_{ij}^{\|}$ represents a longitudinal wave.

N-dimensional transverse δ function:

$$\begin{aligned}
\delta_{ij}^{\perp}(\mathbf{q}-\mathbf{q}') &= \left(\delta_{ij} - \frac{\partial_i \partial_j}{\nabla^2}\right)\delta(\mathbf{q}-\mathbf{q}') \quad i,j = 1,2,3,\cdots,N \\
&= \left(\delta_{ij} - \frac{\partial_i \partial_j}{\nabla^2}\right)\frac{1}{(2\pi)^N}\int d^N \mathbf{q} e^{i\mathbf{k}\cdot(\mathbf{q}-\mathbf{q}')} \\
&= \frac{1}{(2\pi)^N}\int d^N \mathbf{k}\left(\delta_{ij} - \frac{k_i k_j}{k^2}\right) e^{i\mathbf{k}\cdot(\mathbf{q}-\mathbf{q}')} \\
&= \frac{1}{(2\pi)^N}\int d^N \mathbf{k}\left(\delta_{ij} - \frac{k_i k_j}{k^2}\right) e^{i\mathbf{k}\cdot(\mathbf{q}-\mathbf{q}')}, \qquad (C.5.1.2)'
\end{aligned}$$

where $\mathbf{k}/k = (k_1, k_2, \cdots, k_N)/k$, $\nabla^2 = \partial_1^2 + \partial_2^2 + \cdots + \partial_N^2$, and $\partial_i^2 \equiv \partial^2/\partial q_i^2$.

Let us illustrate that why we call the $\delta_{ij}^{\perp}(\mathbf{q}-\mathbf{q}')$ transverse δ function?

Eq. (C.5.1.2)′ indicates:

(i). $\delta_{ij}^{\perp}(\mathbf{q}-\mathbf{q}')$ is no longer a scalar quantity, and is a vector in N-dimensional space, $\mathbf{q}$, for any fixed value of j. $\delta_{ij}^{\perp}(\mathbf{q}-\mathbf{q}')$ with $i = 1,2,3,\cdots,N$ represent N components $i = 1,2,\cdots,N$ of vector, which corresponds to a fixed value of j in $j = 1,2,3,\cdots,N$;

(ii). In Eq. (C.5.1.2)′ $exp[i\mathbf{k}\cdot(\mathbf{q}-\mathbf{q}')]$ is a plane wave in N-dimensional space, and $\delta_{ij} - k_i k_j/k^2$ is an amplitude. Actually, these amplitudes are vector: every value of $j = 1,2,3,\cdots,N$ corresponds to a vector with components $i = 1,2,3,\cdots,N$. Next, we prove that if $(k_1/k, k_2/k, k_3/k, \cdots, k_N/k) \equiv (u_{k_1}, u_{k_2}, \cdots, u_{k_N}) \equiv \mathbf{u_k}$, then $\mathbf{u_k}$ represents a unit vector along **k** direction.

Proof:

$$\left(\frac{k_1}{k}, \frac{k_2}{k}, \cdots, \frac{k_N}{k}\right)\cdot\left(\frac{k_1}{k}, \frac{k_2}{k}, \cdots, \frac{k_N}{k}\right) = \left(\frac{k_1^2}{k^2} + \frac{k_2^2}{k^2} + \cdots + \frac{k_N^2}{k^2}\right) = 1.$$

Therefore, Eq. (C.5.1.2)′ can also be expressed as

$$\delta_{ij}^{\perp}(\mathbf{q}-\mathbf{q}') = \frac{1}{(2\pi)^N}\int d^N \mathbf{k}(\delta_{ij} - u_{k_i} u_{k_j}) e^{i\mathbf{k}\cdot(\mathbf{q}-\mathbf{q}')}. \qquad (C.5.1.2)''$$

Any wave amplitude vectors of $j = 1,2,3,\cdots,N$ is

$$\begin{aligned}
\{(\delta_{ij} - u_{k_i} u_{k_j})\} &= (0_1, 0_2, \cdots, 1_j, 0, \cdots, 0_N) - (u_{k_1}, u_{k_2}, \cdots, u_{k_N}) u_{k_j} \\
&= (-u_{k_1} u_{k_j} - u_{k_2} u_{k_j} + \cdots, 1 - u_{k_j} u_{k_j}, - \cdots, -u_{k_N} u_{k_j}) . (C.5.4.1)
\end{aligned}$$

It is easy to see that

$$\{(\delta_{ij} - u_{k_i} u_{k_j})\} \cdot (u_{k_1}, u_{k_2}, \cdot, u_{k_n}) = u_{k_j} - u_{k_j} \mathbf{u_k} \cdot \mathbf{u_k} = 0.$$

Therefore, $\delta_{ij}^{\perp}$ represents a transverse wave.

Bibliography

[1] Kastner, R. E. *The Transactional Interpretation of Quantum Mechanic*, Cambridge University Press: Cambridge, 2013.

[2] Whitker, A. *The New Quantum Age*, Oxford University Press: Oxford, 2012, p. 268.

[3] Liu, Fu-sui. *Quantum Mechanic upon Theorems*, Nova Publishers: New York, 2013.

[4] Zeng, Jin-yan. *Quantum Mechnics*, I, Scientific Press: Beijing, 2010, p 47, 56, 100, 96, 97, 155, 98, 109, 19, 34, 35, 411, 51, 392, 508, 393, 136, 182, 145.

[5] Davidov, A. S. *Quantum Mechanics*, Pergamon Press: Oxford, 1976, p 53, 58-64, 426, 413, 392.

[6] Greiner, W. *Quantum Mechanics*, Springer: Berlin, 1994, p 368, 367, 290, 284.

[7] Zhang, Y. D. *Advanced Quantum Mechanics*, I, Scientific Press: Beijing, 2009, p 79, 222, 235, 236, 252.

[8] Schiff, L. I. *Quantum Mechanics*, McGraw-Hill: New York, 1968, p 288, 492.

[9] Zelevinsky, V. *Quantum Physics*, II, WILEY-VCH: Weinheim, 2011, p 24, 209, 347.

[10] Zhang, Y. D. *Quantum Mechanics*, Beijing: Scientific Press, 2008, p 29.

[11] van Driel, A. F., Vanmakelbergh, D. *Phys. Rev. Lett.* 2005, 95, 236804.

[12] Tylor, J. R. *Scattering Theory*, Wiley: New York, 1971.

[13] Newton, R. G. *Scattering Theory of Wave and Particles*, Springer: New York, 1982.

[14] Sakurai, J. J., Napolitano, J. *Modern Quantum Mechanics*, Addison-Weley: Boston, 2011, pp 116 - 128, 271, 446, 461, 487, 519.

[15] Zeng, Jin-yan. *Quantum Mechnics*, II, Scientific Press: Beijing, 2010, p 136, 451, 466.

[16] Fetter, A. L., Walecka, J. D. *Quantum Theory of Many-Particle System*, McGraw-Hill: New York, 1971, p 15, 28.

[17] Callaway, J. *Quantum Theory of the Solid State*, Academic Press: New York, 1976.

[18] Dick, R. *Advanced Quantum Mechganics*, Springer: New York, 2012, p 219, 233, 221, 222, 226, 229, 234.

[19] Shankar, R. *Principles of Quantum Mechanics*, Plenum, New York, 1994, p 199, 587, 160.

[20] Zeng, Jin-yan. *Quantum Mechnics*, I, Scientific Press: Beijing, 2010, p 408, 394, 417.

[21] Srednicki, M. *Quantum Field Theory*, Cambridge University Press: Cambridge, 2007, p 73.

[22] Devanathan, V. *Relativistic Quantum Mechanics and Quantum Field Theory*, Alpha, Oxford, 2011, p 209, 202.

[23] Fermi, M. *Nuclear Physics*, Chicago University Press: Chicago, 1950, p 142.

[24] Squires. G. L. *Thermal Neutron Scattering*, Cambrige University Press, Cambridge, 2012, p 8.

[25] Wang, Huai-yu. *Greens Function in Condensed Matter Physics*, Scientific Press: Beijing, 2012, p 43.

[26] Kakani, S. L., Kakanu, Shubhra. *Nuclear and Particle Physics*, Anshan: London, 2008, 33, 11.

[27] Stöcker, H. *Taschenbuch der Physik*, Verlag Harri Deutsch: Berlin, 2003, p 797.

[28] Stein, E. M., Shakarchi, R. S. *Fourier Analysis*, Princeton University Press: Prenceton, 2003, p 131, 135, 158, 15, 160, 72, 80.

[29] Abramowitz, M., Stegun, I. E. *Handbook of Mathematics Functions*, National Bureau of Standard, New York, 1965, p 358, 379, 446, 450, 478, 504, 510, 231, 244, 376.

[30] Zelevinsky, V. *Quantum Physics*, I, Wiley-VCH: Weinheim, 2011, p 327, 107-110.

[31] Townsend, J. S. *Quantum Physics*, University Science Books: Sausalito, 2010, p 169, 229, 291, 230, 247, 359.

[32] Bush, P. *Time in Quantum Mechanics*, edtied by G. Muga, Springer-Verlag: Berlin, 2007, Chapter 3.

[33] Briggs, J. S. *J. Phys.: Conference Series*. 2008, 99, 012002.

[34] Hilgevoord, *J. Am. J. Phys.* 1996, 64, 12; 1998, 66, 396.

[35] Staudenmann, J.T. *Phys. Rev.* 1980, A21, 1419.

[36] Nesvizhevsky, V. V. *Nature*, (London), 2002, 415, 297.

[37] Mandelsta, L., Tamm, I. *J. Phys.* (U. S. S. R), 1945, 9, 249.

[38] Masiach, A. *Quantum Mechaics*, I, North–Holland Publishin Company: Amsterdam, 1965, p 100.

[39] Galindo, A., Pascual, P. *Quantum Mechanics*, I, Springer: Berlin, 1990, p 90.

[40] Wigner, E. P. in *Aspectsts of Quantum Theory*, edited by Salam, A. Wigner, E. P. Cambridge University Press: Cambridge, 1972, p 237.

[41] Fleming, G. N. *Nuovo Cimento*, 1973, A16, 232.

[42] Bhattacharyya, K. *J. Phys.* 1983, A16, 2993.

[43] Gislason, E. A., Sabelli, N. H., Wood, J. W. *Phys. Rev.* 1985, A31, 2078.

[44] Ngai, K. L., Plazek, D. J. *J. Polym. Sci. Polym. Phys.* 1985, 28, 2159.

[45] Ngai, K. L. *Relaxation and Diffusion in Complex Systems*, Springer: Berlin, 2011.

[46] Menon, M., O'Brien, K. P., Dixon, P. K., Wu, Lei., Nagel, C. M. *J. Non-Cryst. Solids*, 1992, 141, 61.

[47] Stickel, F., Fischer, E. M., Richert, R. *J. Chem. Phys.* 1996, 104, 2043.

[48] Weise, U. *Quantum Dissipative Systems*, Fourth Edition, World Scientific: Singapore, 2012, p 233.

[49] Mode, C. J., Sleeman, C. K. *Stochatic Processes in Genetics and Evoluation*, World Scientific: Singapore, 2012.

[50] Xiu, L. Z., Zhang, X. F. *Quantum Theory*, Scientiic Press: Beijing, 2012, p 96.

[51] Wu, Jing-Nuo., Hsieh, Wen-Feng. *Phys. Rev.* 2010, A81, 023827.

[52] Breuer, H. -P., Petruccione, F. *The Theory of Open Quantum Systems*, Oxford University pressOxford2010, p 14.

[53] Hecksher, T., Albena, I. N., Nielsen, B. O., Jeppe, C. D. *Nature Phys.* 2008, 4, 737.

[54] Gradshteyn, I. S., Ryzhik, L. M. *Table of Integrals, Series, and Products*, Academic Press: London, 1980, p 38, 1067, 1033, 318, 447, 337, 338, 317, 188, 191, 183.

[55] Aschcroft, N. W., Mermin, N. D. *Solid State Physics*, Holt, Rinehart and Winston: New York, 1976, p 465, 496, 361.

[56] Ngai, K. L. *Comments Solis State Physics*, 1979, 9, 127.

[57] Vogel, H. *Z. Phys.* 1921, 22, 645.

[58] Ngai, K. L., Rendell, R. W., Jain, H. *Phys. Rev.* 1964, B30, 2133.

[59] Pink noise, (2013 June 6), Wikipedia, The Free Encyclopedia. http://en.wikipedia.org/wiki/pink-noise.

[60] Yu, Ming-kang., Liu, Fu-sui. *Acta Physica Sinica*, 1983, 32, 594.

[61] Fonda, L., Ghirardi, G. C., Rimini, A., *Rep. Prog. Phys.* 1978, 41, 587.

[62] Bagchi, B. *Moleqular Relaxation in Liquid*, Oxford University Press: Oxford, 2012.

[63] Fleishmann, M., Pons, S. *Electroanal. Chem.* 1989, 261, 301.

[64] Jones, S. E., Palmer, E. P., Czirr, J. B., Decker, D. L. *Nature*, (London), 1989, 338, 737.

[65] Legett, A. J., Baym, G. *Phys. Rev. Lett.* 1989, 63, 191.

[66] Ichimaru, S. —em Reviews of Modern Physics, 1993, 65, 285.

[67] Taleyarkhan, R. P., Cho, J. S., West, C. D., Block, R. C. *Phys. Rev.* 2004, E69, 036109.

[68] Liu, Fu-sui., Chen, Wan-fang. *J. Phys.: Condens. Matter*, 2003, 15, 4995.

[69] Liu, Fu-sui., Chen, Wan-fang Chen. *Commun. Theor. Phys.* 2003,, 39, 209.

[70] Liu, Fu-sui., Hou, Yu-min., Chen, Wan-fang. *J. Condensed Matter Nucl. Sci.* 2007, 1, 142. http://www.iscmns.org/CMNS/publications.htm.

[71] Liu, Fu-sui. *Modern Physics Letters*, 1996, B10, 1129.

[72] Taleyarkhan, R. P., Nigmatulin, R. I., Jr. Lahey, R. T. *Science.* 2002, 295, 1868.

[73] Kruus, P. *Liquids and Solutions*, Marcel Dekker: New Yok, 1977, p 164, 399, 467, 497, 460, 57, 400

[74] Larsson, K. E. *Faraday Symp. Chem. Soc.* 1972, 6, 122.

[75] Liu, Fu-sui., Hou, Yu-min. *General Theory of Superconductivity*, Nova Punlishers: New York, 2008.

[76] Liu, Fu-sui., Chen, Wan-fang Chen. *Commun. Theor. Phys.* 2008, 50, 1237.

[77] Liu, Fu-sui., Hou, Yu-min Hou., Chen, Wan-fang Chen. *Commun. Theor. Phys.* 2007, 47, 343.

[78] Liu, Fu-sui. *Journal of Superconductivity and Novel Magnetism*, 2012, 25, 299.

[79] Tan., F., Liu, Fu-sui. *Journal of Superconductivity and Novel Magnetism*, 2012, 25, 215.

[80] Tan, F., Liu, Fu-sui. —em ISRN Condensed Matter Physics, submitted. 215.

[81] Bardeen, J., Cooper, L. N., Schrieffer, J. R. *Phys. Rev.* 1957, 106, 162; 1957, 108, 1175.

[82] Landau, L. P. *J. Phys. Moscow*, 1941, 5, 71.

[83] Tilley, D. R., Tilley, J. *Superfluidity and Superconductivity*, Adam Hilgen Ltd.: Bristol, 1986, p 48, 49, 40, 125.

[84] Peter, P. J. *Engineerring Superconductivity*, John Wiley & Sons: New Yorc, 2001.

[85] Chahara, Ken-ichi., Ohno, T., Kassai, Kozono, Y. *Appl. Phys. Lett.* 1993, 63, 1990.

[86] Helmolt, V. R., Wecker, J., Holzapfel, B., Schultz, L., Samwer, K. *Phys. Rev. Lett.* 1993, 2331.

[87] Dessau, D. S., Saitoh, T., Park, C. -H., Shen, Z. -X., Moritomo, Y., Tokura, Y., *Inter. J. Modern Phys.* 1998, 12, 3389.

[88] Dessau, D. S., Saitoh, T., C. -H. Park, C. -H., Shen, Z. -X., Villella, P., Hamada, N., Moritomo, Y., Tokura, Y., Phys. *Rev. Lett.* 1999, 81, 192.

[89] Mannella, N., Yang, W. L., Zhou, X. J., Zheng, H., Mitchell, J. F., Zaanen, J., Devereaux, Nagaosa, N., Hussain, Z., Shen, Z. -X., *Nature*, (London) 2005, 438, 474.

[90] Yu, R., Dagotto, E. *Phys. Rev.* 2008, B77 214434.

[91] Hughes, C. H., Shi, J. *Phys. Rev.* 2010, B82, 134441.

[92] Ahn, C. H., Triscone, J. -M., Mannhart, J. *Nature*, (London), 2003, 424, 1015.

[93] Moreo, A., Yunoki, S., Dagotto, E. arXiv:cond-mat/9901057v2 [cond-matt.str-el] 18 Feb. 1999.

[94] Takabataka, T., Iga, F., Yoshino, T., Echizen, Y., Katoh, K., Kobayashi, K., Higa, M., Shimizu, N., Bando, Y., Nakamoto, G., Fujii, H., Izawa, K., Suzuki, T., Fujita, T., Sera, M., Hiroi, M., Maezawa, K., Mock, S., Lohneysen, H. V., Bruckl, A., Neumaiex, K., Andres, K. J. *Magn. Magn. Mater.* 1998, 177-181, 277.

[95] Hou, Yumin., Yan, Shou-sheng. *Physica*, 2000, C341-348, 883.

[96] Kagan, Y., Kikoin, K. A., Mishchenko, A. S. *Phys. Rev.* 1997, B55, 12348.

[97] Josephson, B. D. *Phys. Lett.* 1962, 1, 251.

[98] Josephson, B. D. *Adv. Phys.* 1965, 14, 419.

[99] Hikino, S. I. *J. Phys. Soc. Jap.* 2009, 78, 014708.

[100] Ji, A. C. *Phys. Rev. Lett.* 2009, 102, 023602.

[101] Hua, M. X. arXiv.org¿quant-ph¿arXiv:08030059-08030065.

[102] Marchenkov, A. *Nature Nanotechnology*, 2007, 2, 482.

[103] Mahan, G. D. *Many-Particle Physics*, Plenum Publishers: New York, 2000, p 655, 561.

[104] Schrieffer, J. R., Brooks, J. S. *Handbook of High-Temperature Superconductivity*, Springer: Berlin, 2007, p 20, 37, ,50, 60, 260.

[105] Chun, B. S., *at al. J. Appl. Phys.* 2000, 99, 08A902.

[106] Girgis, E., *et al. J. Magn. Magm Matter*, 2000, 222, 133.

[107] Ranner, C., Ravaz, B., Genoud, J. Y., Kadowaki, K., Fisher, O. *Phys. Rev. Lett.* 1998, 80, 149.

[108] Suzuki, M., Watanabe, T. *Phys. Rev. Lett.* 1999, 82, 5361.

[109] Anderson, P. W. *Phys. Rev.* 1958, 110, 827.

[110] Xiang, T. *d-Wave Superconductivity*, Scientific Press: Beijing, 2007, p 14.

[111] Emery, V. J., Kivelson, S. A. *Nature*, (London), 1995, 374, 434.

[112] Sergei, S., Nagao, H., Aono, S. *Modern Aspect of Superconductivity*, World Scientific: Singapore, 2011.

[113] Melhem, Z. *High Temperature Superconductors for Energy Applications*, Woodhead Publishing: Oxford, 2012.

[114] Pauli, W., Weiskopf, V. *Helf. Acta*, 1934, 7, 709.

[115] Thomson, M. *Particle Physics*, (Handout 2, 2011), p 5, 70. http://www.hep.phy.cam.ac.uk/ thomson/partIIIparticles/handouts/Handout-2-2011.pdf.x

[116] Schwabl, F. *Advanced Quantum Mechanics*, 4-th Edition, Springer: New York, 2008, p 141, 196-199, 389.

[117] Bratkovskaya, E. *Lecture 10: Weak Interaction, Parity, Helicity*, in *SS2011: Introduction to Nuclear and Particle Physics*, Part 2. http://th.physik.uni-frankfurt.de/ brat/LecturesSS11/Lec-SS10.pdf.

[118] Thomson, M. *Particle Physics*, (Handout 9, 2011), p 293. http://www.hep.phy.cam.ac.uk/ thomson/partIIIparticles/handouts/Handout-9-2011.pdf.

[119] Thomson, M. *Particle Physics,* (Handout 4, 2011), p 65. http://www.hep.phy.cam.ac.uk/ thomson/partIIIparticles/handouts/Handout-4-2011.pdf.

[120] Paar, H. P. *An Introducyion to Advanced Quantum Physics*, Wiley: New York, 2010, p 115.

[121] Devanathan, V. *Relativistic Quantum Mechanics and Quantum Field Theory*, Alpha: Oxford, 2011, p 190, 143, 151.

[122] Weinberg, S. *Quantum Theory of Fields*, I, Cambridge University Press: Cambridge, 2002, p 325.

[123] Duan, P. C. Quantum Field Theory, High Education Press: Beijing, 2007, p 57.

[124] Desai, B. R. *Quantum Mechanics with Basic Field Theory*, Cambridge University Press: Cambridge, 2010, p 97, 130, 754, 768, 578.

[125] Srednicki, M. *Quantum Field Theory*, Cambridge University Press: Cambridge, 2007, p 340, 26, 28, 24, 133.

[126] Smolin, L. *Three Roads to Guantion Gravity*, John Brockman: New York, 1995, p 85, 132.

[127] Zeidler, E. *Quantum Field Theory*, Springer: Berlin, 2006, p 147, 192.

[128] Mukhanov, V. *Physical Foundations of Cosmology*, Cambridge University Press: Cambidge, 2005, p 20, 203.

[129] Koranga, B. S., Kumar, S. *Inter. J. Theor. Phys.*, DOI 10.1007/s1077-013-1518-2, on line Feb. 2013.

[130] Kundu, A., Nandi, S., Patra, S. K., Sodi, A. arXiv:1209.6063v2 [hep-ph] 2 Jan 2013. Will be published in Phys. Rev. D.

[131] Potting, R. *J. Phys.: Conf. Serries*, 2013, 447, 012009.

[132] Kostrelecky, V. A. Mewes, M. arXiv:1308.4973v1 [hep-ph] 22 Aug. 2013.

[133] Kostrelecky, V. A. *Background information on Lorentz and CPT violation*, www.physics.indiana.edu/ kostelec/faq.html?, Web page last updated August 25, 2013.

[134] Amerino-Camelia, G. *Nature*, 2000, 408, 661.

[135] Mandelbrot, B. B. *The Fractal Geometry of Nature*, Freeman: New York, 1982.

[136] Feynman, R. P., Hibbs, A. R. *Quantum Mechanics and Path Integrals*, McGraw-Hill: New York, 1965.

[137] Laskin, N. *Phys. Lett.* 2000, A268, 298.

[138] Laskin, N. *Phys. Rev.* 2000, E62, 3135. arXiv:0811.1769v1 [math-ph] 11 Nov 2008, p 15.

[139] Laskin, N. arXiv.org/pdf/1009.5533.

[140] Laskin, N. *Phys. Rev.* 2002, E66, 056108. arXiv:quant-ph/0206098v1. 14 June 2002.

[141] Laskin, N. arXiv:1302.0547v1 [math-ph] 3 February 2013.

[142] Rozmej, P., Bandowdki, B. *Computational Methods in Science and Technology*, 2010, 16, 191.

[143] Yang, Xiao-Jin., Baleanu, D., Machado, J. *Boundary Value Problems*, 2013, 6, 131.

[144] Naschie, M. S. El. *International J. of Modern Nonlinear Theory and Application*, 2013, 2, 167.

[145] Muslih, S. I., Agrawai, O. P., Baleanu, D. *Proceedings of the ASME 2009 International Design Engineering Technical Conferences & Computers and Information in Engineering Conference*, IDETC/CIE 2009, August 30 - September 2, 2009, San Diego, California, USA.

[146] Luo, Y., Chen, Y. *Fractional Order Motion Controls*, Wiley, New York, 2013.

[147] Briolle, F., Leoncini, X., Ricaud, B. arXiv:1310.06772v1 [nlin.CD] 2 Oct. 2013.

[148] Wachter, A. *Relativistic quantum Mechanics*, Springer: New York, 2011, p 347.

[149] Costella, J. p., et al. *Amer. J. of phys*, 1997, 65, 835.

[150] Hill, E. *Rev. Mod. Phys.* 1957, 23, 253.

[151] Liboff, R. L. *Introductory Quantum Mechanics*, Addison-Wesley: New York, 2003, p 857.

Index